Molecular Basis of
Cardiovascular Diseases

Molecular Basis of Cardiovascular Diseases: Implications of Natriuretic Peptides

Special Issue Editors

Speranza Rubattu
Massimo Volpe

MDPI • Basel • Beijing • Wuhan • Barcelona • Belgrade

Special Issue Editors
Speranza Rubattu
Sapienza University of Rome
Italy

Massimo Volpe
Sapienza University of Rome
Italy

Editorial Office
MDPI
St. Alban-Anlage 66
4052 Basel, Switzerland

This is a reprint of articles from the Special Issue published online in the open access journal *International Journal of Molecular Sciences* (ISSN 1422-0067) in 2019 (available at: https://www.mdpi.com/journal/ijms/special_issues/cadiovasc_peptides).

For citation purposes, cite each article independently as indicated on the article page online and as indicated below:

LastName, A.A.; LastName, B.B.; LastName, C.C. Article Title. *Journal Name* **Year**, *Article Number*, Page Range.

ISBN 978-3-03921-582-9 (Pbk)
ISBN 978-3-03921-583-6 (PDF)

© 2019 by the authors. Articles in this book are Open Access and distributed under the Creative Commons Attribution (CC BY) license, which allows users to download, copy and build upon published articles, as long as the author and publisher are properly credited, which ensures maximum dissemination and a wider impact of our publications.

The book as a whole is distributed by MDPI under the terms and conditions of the Creative Commons license CC BY-NC-ND.

Contents

About the Special Issue Editors vii

Speranza Rubattu and Massimo Volpe
Natriuretic Peptides in the Cardiovascular System: Multifaceted Roles in Physiology, Pathology and Therapeutics
Reprinted from: *Int. J. Mol. Sci.* **2019**, *20*, 3991, doi:10.3390/ijms20163991 1

Maurizio Forte, Michele Madonna, Sonia Schiavon, Valentina Valenti, Francesco Versaci, Giuseppe Biondi Zoccai, Giacomo Frati and Sebastiano Sciarretta
Cardiovascular Pleiotropic Effects of Natriuretic Peptides
Reprinted from: *Int. J. Mol. Sci.* **2019**, *20*, 3874, doi:10.3390/ijms20163874 8

Kailash N. Pandey
Genetic Ablation and Guanylyl Cyclase/Natriuretic Peptide Receptor-A: Impact on the Pathophysiology of Cardiovascular Dysfunction
Reprinted from: *Int. J. Mol. Sci.* **2019**, *20*, 3946, doi:10.3390/ijms20163946 26

Marica Bordicchia, Francesco Spannella, Gianna Ferretti, Tiziana Bacchetti, Arianna Vignini, Chiara Di Pentima, Laura Mazzanti and Riccardo Sarzani
PCSK9 is Expressed in Human Visceral Adipose Tissue and Regulated by Insulin and Cardiac Natriuretic Peptides
Reprinted from: *Int. J. Mol. Sci.* **2019**, *20*, 245, doi:10.3390/ijms20020245 47

Speranza Rubattu, Maurizio Forte, Simona Marchitti and Massimo Volpe
Molecular Implications of Natriuretic Peptides in the Protection from Hypertension and Target Organ Damage Development
Reprinted from: *Int. J. Mol. Sci.* **2019**, *20*, 798, doi:10.3390/ijms20040798 62

Daniela Maria Tanase, Smaranda Radu, Sinziana Al Shurbaji, Genoveva Livia Baroi, Claudia Florida Costea, Mihaela Dana Turliuc, Anca Ouatu and Mariana Floria
Natriuretic Peptides in Heart Failure with Preserved Left Ventricular Ejection Fraction: From Molecular Evidences to Clinical Implications
Reprinted from: *Int. J. Mol. Sci.* **2019**, *20*, 2629, doi:10.3390/ijms20112629 74

Masako Baba, Kentaro Yoshida and Masaki Ieda
Clinical Applications of Natriuretic Peptides in Heart Failure and Atrial Fibrillation
Reprinted from: *Int. J. Mol. Sci.* **2019**, *20*, 2824, doi:10.3390/ijms20112824 97

Ryuji Okamoto, Yusuf Ali, Ryotaro Hashizume, Noboru Suzuki and Masaaki Ito
BNP as a Major Player in the Heart-Kidney Connection
Reprinted from: *Int. J. Mol. Sci.* **2019**, *20*, 3581, doi:10.3390/ijms20143581 114

Zhipeng Cao, Yuqing Jia and Baoli Zhu
BNP and NT-proBNP as Diagnostic Biomarkers for Cardiac Dysfunction in Both Clinical and Forensic Medicine
Reprinted from: *Int. J. Mol. Sci.* **2019**, *20*, 1820, doi:10.3390/ijms20081820 131

Amie J. Moyes and Adrian J. Hobbs
C-Type Natriuretic Peptide: A Multifaceted Paracrine Regulator in the Heart and Vasculature
Reprinted from: *Int. J. Mol. Sci.* **2019**, *20*, 2281, doi:10.3390/ijms20092281 147

Emmanuel Eroume-A Egom
Pulmonary Arterial Hypertension Due to NPR-C Mutation: A Novel Paradigm for Normal and Pathologic Remodeling?
Reprinted from: *Int. J. Mol. Sci.* **2019**, *20*, 3063, doi:10.3390/ijms20123063 170

Valentina Cannone, Aderville Cabassi, Riccardo Volpi and John C. Burnett Jr.
Atrial Natriuretic Peptide: A Molecular Target of Novel Therapeutic Approaches to Cardio-Metabolic Disease
Reprinted from: *Int. J. Mol. Sci.* **2019**, *20*, 3265, doi:10.3390/ijms20133265 178

Massimo Volpe, Speranza Rubattu and Allegra Battistoni
ARNi: A Novel Approach to Counteract Cardiovascular Diseases
Reprinted from: *Int. J. Mol. Sci.* **2019**, *20*, 2092, doi:10.3390/ijms20092092 189

About the Special Issue Editors

Speranza Rubattu is an Associate Professor of Cardiology at the University of Rome Sapienza and a Scientific consultant at the Institute IRCCS Neuromed, Polo di Pozzilli, Sapienza University of Rome, Pozzilli (Is), Italy. She has spent several years as a research fellow at International Research Centers in the USA (Cardiovascular and Hypertension Center of Cornell University Medical College-The New York Hospital; The Laboratory of Cellular and Molecular Cardiology, Children's Hospital, Harvard Medical School, Boston; the Department of Medicine, Cardiovascular Division, Brigham and Women's Hospital, Harvard Medical School, Boston). Professor Rubattu's main research activities are: the circulation and local tissue renin-angiotensin systems and characterization of their role in hypertension; the characterization of the stroke-prone spontaneously hypertensive rat as an animal model of hypertension and stroke; the molecular genetics of target organ damage in hypertension; the analysis of the atrial natriuretic factor and of its molecular variants as determinant of cardiovascular risk; the role of mitochondrial dysfunction into the pathogenesis of target organ damage in hypertension. She has authored about 200 publications in national and international peer reviewed journals. She has served as a reviewer for many top scientific journals. She is a member of the Editorial Board of High Blood Pressure & Cardiovascular Prevention, International Journal of Molecular Sciences; International Journal of Cardiology: Hypertension. She is a member of Italian and International Scientific Societies (Italian Society of Hypertension, Italian Society of Cardiology, Italian Society of Cardiovascular Prevention, European Society of Hypertension).

Massimo Volpe is Professor of Cardiology and Chairman of Cardiology at the Specialty School of Cardiology, University of Rome "Sapienza". He is Chief of the Cardiology Division, Department of Cardiovascular and Respiratory Science, and Hypertension Unit at Sant'Andrea Hospital, Rome, Italy. He is also Dean of the Faculty of Medicine and Psychology, University of Rome Sapienza. He's primary research interests are neurohormonal control of circulation and cardiovascular structural and functional abnormalities in hypertension and heart failure. He is a member of several Italian and International Societies, a Fellow of the American Heart Association (High Blood Pressure Council), and European Society of Cardiology, and a Hypertension Specialist of the European Society of Hypertension. He has served as President of the Italian Society of Hypertension, Italian Society of Cardiovascular Prevention and Italian College of Cardiology. He has authored more than 700 publications in prestigious cardiovascular, peer-reviewed, international journals, has lectured at many international congresses and Universities and has served on the editorial boards of several journals including the American Journal of Hypertension, Journal of Hypertension, European Heart Journal and is Associate Editor of International Journal of Cardiology. He is currently Editor-in-Chief of High Blood Pressure and Cardiovascular Prevention. He has served as a reviewer for many top scientific journals and international cardiovascular research programmes. He is a member of the Italian Ministry panel for supervision of the quality of research in the Country.

Editorial

Natriuretic Peptides in the Cardiovascular System: Multifaceted Roles in Physiology, Pathology and Therapeutics

Speranza Rubattu [1,2,*] and Massimo Volpe [1,2,*]

1. Department of Clinical and Molecular Medicine, School of Medicine and Psychology, Sapienza University of Rome, 00189 Rome, Italy
2. IRCCS Neuromed, 86077 Pozzilli (Isernia), Italy
* Correspondence: rubattu.speranza@neuromed.it (S.R.); massimo.volpe@uniroma1.it (M.V.); Tel.: +39-06-33-775-979 (S.R.); Fax: +39-06-33-775-061 (S.R.)

Received: 8 August 2019; Accepted: 15 August 2019; Published: 16 August 2019

Abstract: The natriuretic peptides (NPs) family includes a class of hormones and their receptors needed for the physiological control of cardiovascular functions. The discovery of NPs provided a fundamental contribution into our understanding of the physiological regulation of blood pressure, and of heart and kidney functions. NPs have also been implicated in the pathogenesis of several cardiovascular diseases (CVDs), including hypertension, atherosclerosis, heart failure, and stroke. A fine comprehension of the molecular mechanisms dependent from NPs and underlying the promotion of cardiovascular damage has contributed to improve our understanding of the molecular basis of all major CVDs. Finally, the opportunity to target NPs in order to develop new therapeutic tools for a better treatment of CVDs has been developed over the years. The current Special Issue of the Journal covers all major aspects of the molecular implications of NPs in physiology and pathology of the cardiovascular system, including NP-based therapeutic approaches.

Keywords: natriuretic peptides; arterial hypertension; pulmonary arterial hypertension; heart failure; stroke; atrial fibrillation; ARNi; MANP

The natriuretic peptides (NPs) family includes a class of hormones [atrial (ANP), B-type (BNP) and C-type (CNP)] and their receptors [natriuretic peptide receptor-A (NPRA), receptor-B (NPRB), and receptor-C (NPRC)] needed for the physiological control of cardiovascular functions. First, the discovery of NPs provided a fundamental contribution for the understanding of the physiological regulation of blood pressure (BP) and of cardiovascular and renal functions [1]. Subsequently, abnormalities of the NPs physiological properties were implicated in the pathogenesis of major cardiovascular diseases (CVDs), such as hypertension and heart failure (HF) [2,3]. Finally, a more thorough comprehension of the molecular mechanisms linked to NPs actions through their distinct receptors has contributed to improve our understanding of key molecular mechanisms of cardiovascular homeostasis, as well as the progression of several CVDs [3,4].

As a matter of fact, the NPs system has provided over time a continuous, attractive source of new knowledge and discoveries regarding the pathogenesis, diagnosis, prognosis, and therapy of CVDs. In particular, the opportunity to target NPs in order to design new therapeutic tools for a more effective treatment of CVDs has been developed, ultimately culminating in the introduction of a new class of drugs for the management of HF, the angiotensin receptor neprilysin inhibitor (ARNi) [5,6].

The continuous interest in this field of biomedical research is documented by accumulating data produced from several expert scientific groups. This issue of the Journal collects some original and review articles on the molecular and biomedical aspects concerning NPs, with a discussion of their current clinical and therapeutic applications.

The cellular effects of NPs include the regulation of cell proliferation, angiogenesis, apoptosis, fibrosis, and inflammation [3]. Their anti-proliferative, anti-fibrotic, and anti-hypertrophic effects, including the underlying signaling pathways, were largely documented at both cardiac and vascular levels [3]. In this issue, a review of the literature presented by Forte M. et al. summarizes the current knowledge on the cardiovascular pleiotropic effects of NPs and highlights the most relevant findings that underscore the NPs system as a key player in the cardiovascular remodeling process [7]. A major strength of this aspect of NPs function was initially provided by genetically modified animal models showing that lack of either the ANP (*Nppa*) or NPRA gene (*Npr-1*) led to hypertension and marked cardiac hypertrophy, the latter being independent from high blood pressure levels [7]. In particular, as outlined in the article by Pandey K.N. [8], the gene-targeted (gene-knockout and gene-duplication) mouse models demonstrated the key roles of guanylyl cyclase/NPRA in cardiovascular disease states. Above all, we learned that lack of *Npr-1* led to salt-sensitive increases in BP whereas *Npr-1* gene duplication lowered BP and protected against high dietary salt concentrations [8]. The findings obtained in animal models were subsequently translated to the human disease [3,8]. In fact, both genetic and clinical studies could demonstrate the significant associations of variant alleles at *Nppa*, BNP gene (*Nppb*), and *Npr-1* with cardiovascular disorders in humans [3,8,9].

Interestingly, NPs control the lipid metabolism through an anti-lipolytic effect [10]. Of note, they promote mitochondria biogenesis in adipocytes and the process of "browning" of white adipocytes to increase energy expenditure [11]. Herein, a novel original mechanism underlying the anti-lipolytic effect of ANP is presented by Bordicchia M. et al. [12]. This mechanism, supported by experimental in vitro evidence, refers to the inhibition by ANP of Proprotein convertase subtilisin/kexin type 9 (PCSK9), the enzyme responsible for Low Density Lipoprotein (LDL) receptor (LDLr) degradation [13]. Specifically, the original work by Bordicchia M. et al. demonstrates that ANP inhibits PCSK9 expression in human adipocytes, therefore reducing LDLr degradation [12]. It is known that the inhibition of PCSK9, through a specific antibody, allows the accumulation of LDLr and the decrease of LDL cholesterol level in the blood [13]. This strategy has represented a breakthrough of the current therapeutic approaches to treat hypercholesterolemia [14]. By blocking PCSK9 induction, ANP appears to mimick, although to a much lower extent, the action of PCSK9 inhibitors, evolocumab and alirocumab [15,16]. It will be interesting in the future to test LDL cholesterol levels in patients undertaking ARNi and presenting with higher ANP circulating levels [17].

Both the hemodynamic and cellular effects of NPs explain the pathogenetic involvement of NPs in hypertension and related target organ damage. In particular, as discussed in this issue of the Journal, the comprehension of the fine molecular mechanisms underlying hypertension has been largely improved through the dissection of the molecular genetics of the NPs system [18]. Nowadays, genetic variations of *Nppa*, *Nppb*, CNP gene (*Nppc*), *Npr-1*, NPRC gene (*Nprc*), *Corin*, and Proprotein convertase subtilisin/kexin type 6 gene (*PCSK6*) are known contributors to hypertension development in experimental models as well as in humans through a decreased function of the system and of its impact on BP regulation [18]. Furthermore, by dissecting molecular alterations of the NPs system components, we have been able to understand, at least in part, the pathogenesis of cardiovascular damage in hypertension. Most importantly, a harmful variant of human *Nppa* (the T2238C/ANP, rs5065), that is frequently encountered in the general population (14% frequency of the allele variant), has shown functional deleterious properties that completely diverge from those of the wild type form, which makes this molecular variant a significant contributor to cardiovascular acute events such as stroke and myocardial infarction [19]. On the other hand, a protective *Nppa* variant (rs5068) is able to reduce the cardiometabolic risk by increasing the circulating ANP level and its beneficial cardiovascular and metabolic properties [20]. Furthermore, a less frequent *Nppa* variant (rs5063) was associated to reduced left ventricular mass in hypertension [21]. Overall, the experience gained from several research groups with the studies on molecular variants of *Nppa* support the existence of genetic predictors of cardiovascular risk that contribute to the individual risk profile (as part of the emerging field of predictive medicine).

NPs represent today well established and useful diagnostic biomarkers in HF, being of particular help for the differential diagnosis of dyspnea in the emergency room [22]. The increase of amino-terminal (NT)-proBNP/BNP levels reflects the ventricular dysfunction characterizing the condition of HF with reduced ejection fraction (HFrEF), whereas their decrease reliably reflects functional cardiac improvement due to the therapeutic interventions [23]. ANP behaves in a similar manner, although it is not routinely used in clinical practice mainly due to its shorter half-life and lability. The mid-regional amino-terminal ANP (MR-proANP), detected through an immunoassay toward the segment including aminoacids 53-90 of the ANP amino-terminal portion, is a more stable form and offers more specific useful applications [24,25]. Both ANP and BNP also play a prognostic role in HF [26,27]. The accumulation of NPs is not sufficient to maintain a proper hemodynamic balance in cardiac failure, particularly with the progression of the disease. In fact, a state of "resistance to NPs" is described in HF patients, raising the need to increase further their plasma levels in order to achieve a better circulatory homeostasis in cardiac failure [28]. In this issue of the Journal, the role of NPs is discussed in the condition of HF with preserved ejection fraction (HFpEF) [29]. Although with some controversies, lower levels of BNP are found by the majority of the studies in HFpEF [29,30]. Of interest, the significance of MR-proANP in the context of HFpEF is growing as a more specific and more informative marker that parallels the trend of BNP [24]. Therefore, raising the NPs levels is expected to allow an improved hemodynamic profile in HFpEF as well. The upcoming results of the PARAGON trial (that tested the potential benefits of ARNi in HFpEF patients) could soon clarify this important question [31].

HF is often associated to atrial fibrillation (AF), a condition that on its own presents with higher BNP levels [32,33]. This combination raises the need to interpret correctly the level of NPs for both diagnostic and prognostic purposes. The BNP level may not differ between HF patients with AF and HF patients without AF [34]. In fact, higher cut-off levels of BNP need to be taken into consideration to improve the specificity and likelihood of correct diagnosis of HF in the presence of AF [35]. Moreover, as discussed in this issue, the role of NPs in the screening for the new onset of incident AF and for the prediction of AF recurrence after cardioversion and pulmonary vein isolation may reveal useful in the clinical setting [33,36,37].

In the context of HF, a renal dysfunction often develops (cardiorenal syndrome). As reviewed in this issue by Okamoto R. et al. [38], BNP is a major player in the heart–kidney connection and it plays important protective roles within the kidneys mainly through its inhibitory effect on the renin-angiotensin system and the sympathetic nervous system. Thus, by promoting diuresis, natriuresis, and vasorelaxation, it counteracts not only HF but also chronic kidney disease (CKD) development. In fact, BNP and NT-proBNP levels are higher in acute HF patients with renal dysfunction as compared to patients with normal renal function [39]. Importantly, it has been shown that BNP infusion may contribute to prevent development of CKD in HF [40].

The strength of the relevance of NPs in HF is reinforced by an interesting review article of this Special issue. Specifically, the article by Cao Z. et al. focuses on the role of NT-proBNP/BNP as valuable diagnostic biomarkers of cardiac dysfunction in deceased individuals [41]. This original observation extends the application of these HF biomarkers to forensic medicine apart from the standardized use in clinical practice. No other biomarker has ever been reported to diagnose cardiac dysfunction postmortem.

An important component of the NPs system is represented by CNP, which acts through either the NPRB or NPRC receptors. CNP is mainly synthetized by endothelial cells and also by cardiomyocytes and fibroblasts. It circulates in the blood at very low amounts, offering a clear example of an autocrine/paracrine mediator within the cardiovascular system [42]. The most recent discoveries regarding CNP functions have been reported in the review article by Moyes A. et al. [43]. These authors underscore novel functions of CNP, such as control of inflammation, angiogenesis, cell proliferation, and anti-atherosclerotic effect in the blood vessel; control of cardiomyocyte contractility, fibrosis, hypertrophy and even of electrophysiological activity of the heart [43]. These multiple functions make CNP a multifaceted paracrine regulator within the cardiovascular system. In the presence of a dysregulation of CNP, the development of CVDs is favored. For instance, since CNP controls BP levels through a potent vasodilation within the microvasculature, abnormalities of CNP function contribute to hypertension development [44]. CNP increases in HF, in parallel to ANP and BNP and to its receptor NPRC, and it correlates to disease severity and outcome [45]. In fact, these observations have focused the attention to CNP as a potential therapeutic target in both hypertension and HF.

Among the recent discoveries regarding the NPs system, the one that deserves particular attention is the potential involvement of NPRC signaling in the pathogenesis of pulmonary arterial hypertension [46]. The article by Egom E. provides a revision of the literature supporting the link between abnormalities of NPRC signaling and pulmonary vascular remodeling, pulmonary fibrosis, and chronic obstructive pulmonary disease [47]. The latter are explained by the disruption of the anti-proliferative effects of NPRC via the Gqα/mitogen-activated protein (MAP) kinase signaling pathway [48].

The main therapeutic approaches to treat CVDs involving the NPs system are based on either the development of peptide analogs or the blockade of peptides catabolism [3,4]. In this issue, Cannone V. et al. describe one of the most promising ANP analog, the MANP, a 40 amino acid peptide with a 12 amino acid extension to the carboxyl-terminus of ANP [49]. This peptide analog, that is more resistant to degradation, is progressively gaining more interest for its future application in clinical practice. In fact, it has been tested in both experimental and clinical settings with evidence of a significant prolonged anti-hypertensive effect. Its cardiometabolic protective properties are also being currently investigated in humans [49].

An overview of the strategies aimed at blocking the NPs catabolism through a NPRC blockade, and particularly through NEP inhibition, is presented by Volpe M. et al. [50]. The approach based on NEP inhibition led to the recent development of a new class of drug called ARNi, which currently represents a valuable therapeutic tool for the treatment of HRrEF and may become, in the near future, an essential tool for the treatment strategy toward many other CVDs, possibly also hypertension [5,50]. So far, the only available compound is sacubitril/valsartan.

Overall, the comprehension of the multiple functional roles of NPs, gained over the last 35 years, makes this hormonal system an essential contributor to the maintenance of the cardiovascular health. On the other hand, a deeper understanding of the complex molecular mechanisms underlying the functionality of NPs has opened a new way to relevant therapeutic innovations. Future years, through the continuous efforts of several research groups, will certainly reveal more insights on this multifaceted cardiovascular hormonal system.

Funding: This work was supported by a grant from the Italian Ministry of Health and by the "5 per mille" grant.

Conflicts of Interest: M.V. has received honoraria for participating to Advisory Boards of Novartis.

References

1. Levin, E.R.; Gardner, D.G.; Samson, W.K. Natriuretic peptides. *N. Engl. J. Med.* **1998**, *339*, 321–328. [PubMed]
2. Rubattu, S.; Sciarretta, S.; Valenti, V.; Stanzione, R.; Volpe, M. Natriuretic peptides: An update on bioactivity, potential therapeutic use and implication in cardiovascular diseases. *Am. J. Hypertens.* **2008**, *21*, 733–741. [CrossRef] [PubMed]
3. Volpe, M.; Rubattu, S.; Burnett, J., Jr. Natriuretic peptides in cardiovascular diseases: Current use and perspectives. *Eur. Heart J.* **2014**, *35*, 419–425. [CrossRef] [PubMed]
4. Das, B.B.; Solinger, R. Role of natriuretic peptide family in cardiovascular medicine. *Cardiovasc. Hematol. Agents Med. Chem.* **2009**, *7*, 29–42. [CrossRef] [PubMed]
5. Volpe, M.; Tocci, G.; Battistoni, A.; Rubattu, S. Angiotensin II receptor blocker nephrilysin inhibitor (ARNI): New avenues in cardiovascular therapy. *High Blood Press. Cardiovasc. Prev.* **2015**, *22*, 241–246. [CrossRef]
6. Kario, K. The Sacubitril/Valsartan, a first-in-class, angiotensin receptor neprilysin inhibitor (ARNI): Potential uses in hypertension, heart failure, and beyond. *Curr. Cardiol. Rep.* **2018**, *20*, 5. [CrossRef]

7. Forte, M.; Madonna, M.; Schiavon, S.; Valenti, V.; Versaci, F.; Biondi Zoccai, G.; Frati, G.; Sciarretta, S. Cardiovascular pleiotropic effects of atrial natriuretic peptide. *Int. J. Mol. Sci.* **2019**, *20*, 3874. [CrossRef]
8. Pandey, K.N. Genetic ablation and guanylyl cyclase/natriuretic peptide receptor A: Impact on the pathophysiology of cardiovascular dysfunction. *Int. J. Mol. Sci.* **2019**, *20*, 3496. [CrossRef]
9. Rubattu, S.; Sciarretta, S.; Volpe, M. Atrial natriuretic peptide gene variants and circulating levels: Implications in cardiovascular diseases. *Clin. Sci. Lond* **2014**, *127*, 1–13. [CrossRef]
10. Lafontan, M.; Moro, C.; Berlan, M.; Crampes, F.; Sengenes, C.; Galitzky, J. Control of lipolysis by natriuretic peptides and cyclic GMP. *Trends Endocrinol. Metab.* **2008**, *19*, 130–137. [CrossRef]
11. Bordicchia, M.; Liu, D.; Amri, E.Z.; Ailhaud, G.; Dessì-Fulgheri, P.; Zhang, C.; Takahashi, N.; Sarzani, R.; Collins, S. Cardiac natriuretic peptides act via p38 MAPK to induce the brown fat thermogenic program in mouse and human adipocytes. *J. Clin. Invest.* **2012**, *122*, 1022–1036. [CrossRef] [PubMed]
12. Bordicchia, M.; Spannella, F.; Ferretti, G.; Bacchetti, T.; Vignini, A.; Di Pentima, C.; Mazzanti, L.; Sarzani, R. PCSK9 is Expressed in Human Visceral Adipose Tissue and Regulated by Insulin and Cardiac Natriuretic Peptides. *Int. J. Mol. Sci.* **2019**, *20*, 245. [CrossRef] [PubMed]
13. Lagace, T.A. PCSK9 and LDLR degradation: Regulatory mechanisms in circulation and in cells. *Curr. Opin. Lipidol.* **2014**, *25*, 387–393. [CrossRef] [PubMed]
14. Urban, D.; Pöss, J.; Böhm, M.; Laufs, U. Targeting the proprotein convertase subtilisin/kexin type 9 for the treatment of dyslipidemia and atherosclerosis. *J. Am. Coll. Cardiol.* **2013**, *62*, 1401–1408. [CrossRef] [PubMed]
15. Sabatine, M.S.; Giugliano, R.P.; Keech, A.C.; Honarpour, N.; Wiviott, S.D.; Murphy, S.A.; Kuder, J.F.; Wang, H.; Liu, T.; Wasserman, S.M.; et al. Evolocumab and Clinical Outcomes in Patients with Cardiovascular Disease. *N. Engl. J. Med.* **2017**, *376*, 1713–1722. [CrossRef]
16. Tomlinson, B.; Hu, M.; Zhang, Y.; Chan, P.; Liu, Z.M. Alirocumab for the treatment of hypercholesterolemia. *Expert Opin. Biol. Ther.* **2017**, *17*, 633–643. [CrossRef]
17. Ibrahim, N.E.; McCarthy, C.P.; Shrestha, S.; Gaggin, H.K.; Mukai, R.; Szymonifka, J.; Apple, F.S.; Burnett, J.C., Jr.; Iyer, S.; Januzzi, J.L., Jr. Effect of neprilysin inhibition on various natriuretic peptide assays. *J. Am. Coll. Cardiol.* **2019**, *73*, 1273–1284. [CrossRef]
18. Rubattu, S.; Forte, M.; Marchitti, S.; Volpe, M. Molecular implications of natriuretic peptides in the protection from hypertension and target organ damage development. *Int. J. Mol. Sci.* **2019**, *20*, 798. [CrossRef]
19. Rubattu, S.; Sciarretta, S.; Marchitti, S.; Bianchi, F.; Forte, M.; Volpe, M. The T2238C atrial natriuretic peptide molecular variant and the risk of cardiovascular diseases. *Int. J. Mol. Sci.* **2018**, *19*, 540. [CrossRef]
20. Cannone, V.; Scott, C.G.; Decker, P.A.; Larson, N.B.; Palmas, W.; Taylor, K.D.; Wang, T.J.; Gupta, D.K.; Bielinski, S.J.; Burnett, J., Jr. A favorable cardiometabolic profile is associated with the G allele of the genetic variant rs5068 in African Americans: The Multi-Ethnic Study of Atherosclerosis (MESA). *PLoS ONE* **2017**, *12*, e0189858. [CrossRef]
21. Rubattu, S.; Bigatti, G.; Evangelista, A.; Lanzani, C.; Stanzione, R.; Zagato, L.; Manunta, P.; Marchitti, S.; Venturelli, V.; Bianchi, G.; et al. Association of atrial natriuretic and type-A natriuretic peptide receptor gene polymorphisms with left ventricular mass in human essential hypertension. *J. Am. Coll. Cardiol.* **2006**, *48*, 499–505. [CrossRef] [PubMed]
22. Maisel, A.S.; Krishnaswamy, P.; Nowak, R.M.; McCord, J.; Hollander, J.E.; Duc, P.; Omland, T.; Storrow, A.B.; Abraham, W.T.; Wu, A.H.; et al. Breathing Not Properly Multinational Study Investigators. Rapid measurement of B-type natriuretic peptide in the emergency diagnosis of heart failure. *N. Engl. J. Med.* **2002**, *347*, 161–167. [CrossRef] [PubMed]
23. Mukoyama, M.; Nnakao, K.; Saito, Y.; Ogawa, Y.; Hosoda, K.; Suga, S.; Shirakami, G.; Jougasaki, M.; Imura, H. Increased human brain natriuretic peptide in congestive heart failure. *N. Engl. J. Med.* **1990**, *323*, 757–758. [PubMed]
24. Cui, K.; Huang, W.; Fan, J.; Lei, H. Midregional pro-atrial natriuretic peptide is a superior biomarker to N-terminal pro-B-type natriuretic peptide in the diagnosis of heart failure patients with preserved ejection fraction. *Med. Baltim.* **2018**, *97*, e12277. [CrossRef] [PubMed]
25. Maisel, A.S.; Duran, J.M.; Wettersten, N. Natriuretic peptides in heart failure: Atrial and B-type natriuretic peptides. *Heart Fail. Clin.* **2018**, *14*, 13–25. [CrossRef] [PubMed]
26. Seronde, M.F.; Gayat, E.; Logeart, D.; Lassus, J.; Laribi, S.; Boukef, R.; Sibellas, F.; Launay, J.M.; Manivet, P.; Sadoune, M.; et al. Comparison of the diagnostic and prognostic values of B-type and atrial-type natriuretic peptides in acute heart failure. *Int. J. Cardiol.* **2013**, *168*, 3404–3411. [CrossRef] [PubMed]

27. Lam, C.S.P.; Gamble, G.D.; Ling, L.H.; Sim, D.; Leong, K.T.G.; Yeo, P.S.D.; Ong, H.Y.; Jaufeerally, F.; Ng, T.P.; Cameron, V.A.; et al. Mortality associated with heart failure with preserved vs reduced ejection fraction in a prospective multi-ethnic cohort study. *Eur. Heart J.* **2018**, *39*, 1770–1780. [CrossRef] [PubMed]
28. Volpe, M.; Rubattu, S. Natriuretic peptides. In *Hypertension and Heart Failure*; Updates in Hypertension and Cardiovascular Protection; Dorobantu, M., Mancia, G., Grassi, G., Voicu, V., Eds.; Springer: Cham, Switzerland, 2019; pp. 87–100.
29. Tanase, D.M.; Rdu, S.; Al Shurbaji, S.; Baroi, G.L.; Costea, C.L.; Turliuc, M.D.; Ouatu, A.; Floria, M. Natriuretic peptides in heart failure with preserved left ventricular ejection fraction: From molecular evidences to clinical implications. *Int. J. Mol. Sci.* **2019**, *20*, 2629. [CrossRef]
30. Brunner-La Rocca, H.P.; Sanders-van Wijk, S. Natriuretic peptides in chronic heart failure. *Card. Fail. Rev.* **2019**, *5*, 44–49. [CrossRef]
31. Solomon, S.D.; Rizkala, A.R.; Gong, J.; Wang, W.; Anand, I.S.; Ge, J.; Lam, C.S.P.; Maggioni, A.P.; Martinez, F.; Packer, M.; et al. Angiotensin receptor neprilysin inhibition in heart failure with preserved ejection fraction: Rationale and design of the PARAGON-HF trail. *JACC Heart Fail.* **2017**, *5*, 471–482. [CrossRef]
32. Ellinor, P.T.; Low, A.F.; Patton, K.K.; Shea, M.A.; Macrae, C.A. Discordant atrial natriuretic peptide and brain natriuretic peptide levels in lone atrial fibrillation. *J. Am. Coll. Cardiol.* **2005**, *45*, 82–86. [CrossRef] [PubMed]
33. Baba, M.; Yoshida, K.; Ieda, M. Clinical application of natriuretic peptides in heart failure and atrial fibrillation. *Int. J. Mol. Sci.* **2019**, *20*, 2824. [CrossRef] [PubMed]
34. Richards, M.; Di Somma, S.; Mueller, C.; Nowak, R.; Peacock, W.F.; Ponikowski, P.; Mockel, M.; Hogan, C.; Wu, A.H.; Clopton, P.; et al. Atrial fibrillation impairs the diagnostic performance of cardiac natriuretic peptides in dyspneic patients: Results from the BACH Study (Biomarkers in Acute Heart Failure). *JACC Heart Fail.* **2013**, *1*, 192–199. [CrossRef] [PubMed]
35. Knudsen, C.W.; Omland, T.; Clopton, P.; Westheim, A.; Wu, A.H.; Duc, P.; McCord, J.; Nowak, R.M.; Hollander, J.E.; Storrow, A.B.; et al. Impact of atrial fibrillation on the diagnostic performance of cardiac natriuretic peptide concentration in dyspneic patients: An analysis from the breathing not properly multinational study. *J. Am. Coll. Cardiol.* **2005**, *46*, 838–844. [CrossRef] [PubMed]
36. Beck-da-Silva, L.; de Bold, A.; Fraser, M.; Williams, K.; Haddad, H. Brain natriuretic peptide predicts successful cardioversion in patients with atrial fibrillation and maintenance of sinus rhythm. *Can. J. Cardiol.* **2004**, *20*, 1245–1248. [PubMed]
37. Deng, H.; Shantsila, A.; Guo, P.; Zhan, X.; Fang, X.; Liao, H.; Liu, Y.; Wei, W.; Fu, L.; Wu, S.; et al. Multiple biomarkers and arrhythmia outcome following catheter ablation of atrial fibrillation: The Guangzhou Atrial Fibrillation Project. *J. Arrhytm.* **2018**, *34*, 617–625. [CrossRef] [PubMed]
38. Okamoto, R.; Ali, Y.; Hashizume, R.; Suzuki, N.; Ito, M. BNP as a major player in the heart-kidney connection. *Int. J. Mol. Sci.* **2019**, *20*, 3581. [CrossRef] [PubMed]
39. Dos Reis, D.; Fraticelli, L.; Bassand, A.; Manzo-Silberman, S.; Peschanski, N.; Charpentier, S.; Elbaz, M.; Savary, D.; Bonnefoy-Cudraz, E.; Laribi, S.; et al. Impact of renal dysfunction on the management and outcome of acute heart failure: Results from the French prospective, multicenter, DeFSSICA survey. *BMJ Open* **2019**, *9*, e022776. [CrossRef]
40. McKie, P.M.; Schirger, J.A.; Benike, S.L.; Hastad, L.K.; Slusser, J.P.; Hodge, D.O.; Redfield, M.M.; Burnett, J.C., Jr.; Chen, H.H. Chronic subcutaneous brain natriuretic peptide therapy in asymptomatic systolic heart failure. *Eur. J. Heart Fail.* **2016**, *18*, 433–441. [CrossRef]
41. Cao, Z.; Jia, Y.; Zhu, B. BNP and NT-proBNP as Diagnostic Biomarkers for Cardiac Dysfunction in Both Clinical and Forensic Medicine. *Int. J. Mol. Sci.* **2019**, *20*, 1820. [CrossRef]
42. Horio, T.; Tokudome, T.; Maki, T.; Yoshihara, F.; Suga, S.; Nishikimi, T.; Kojima, M.; Kawano, Y.; Kangawa, K. Gene expression, secretion, and autocrine action of C-type natriuretic peptide in cultured adult rat cardiac fibroblasts. *Endocrinology* **2003**, *144*, 2279–2284. [CrossRef] [PubMed]
43. Moyes, A.; Hobbs, A. C-Type Natriuretic Peptide: A Multifaceted Paracrine Regulator in the Heart and Vasculature. *Int. J. Mol. Sci.* **2019**, *20*, 2281. [CrossRef] [PubMed]

44. Nakao, K.; Kuwahara, K.; Nishikimi, T.; Nakagawa, Y.; Kinoshita, H.; Minami, T.; Kuwabara, Y.; Yamada, C.; Yamada, Y.; Tokudome, T.; et al. Endothelium-Derived C-Type Natriuretic Peptide Contributes to Blood Pressure Regulation by Maintaining Endothelial Integrity. *Hypertension* **2017**, *69*, 286–296. [CrossRef] [PubMed]
45. Del Ry, S.; Passino, C.; Maltinti, M.; Emdin, M.; Giannessi, D. C-type natriuretic peptide plasma levels increase in patients with chronic heart failure as a function of clinical severity. *Eur. J. Heart Fail.* **2005**, *7*, 1145–1148. [CrossRef] [PubMed]
46. Egom, E.E.A.; Feridooni, T.; Pharithi, R.B.; Khan, B.; Shiwani, H.A.; Maher, V.; El Hiani, Y.; Rose, R.A.; Pasumarthi, K.B.S.; Ribama, H.A. New insights and new hope for pulmonary arterial hypertension. *Int. J. Physiol. Pathophysiol. Pharmacol.* **2017**, *9*, 112–118. [PubMed]
47. Egom, E.E.A. Pulmonary arterial hypertension due to NPR-C mutation: A novel paradigm for normal and pathologic remodeling? *Int. J. Mol. Sci.* **2019**, *20*, 3063. [CrossRef] [PubMed]
48. Li, Y.; Hashim, S.; Anand-Srivastava, M.B. Intracellular peptides of natriuretic peptide receptor-C inhibit vascular hypertrophy via Gqa/MAP kinase signaling pathways. *Cardiovasc. Res.* **2006**, *72*, 464–472. [CrossRef]
49. Cannone, V.; Cabassi, A.; Volpi, R.; Burnett, J.C., Jr. Atrial natriuretic peptide: A molecular target of novel therapeutic approaches to cardio-metabolic disease. *Int. J. Mol. Sci.* **2019**, *20*, 3265. [CrossRef]
50. Volpe, M.; Rubattu, S.; Battistoni, A. ARNi: A novel approach to counteract cardiovascular diseases. *Int. J. Mol. Sci.* **2019**, *20*, 2092. [CrossRef]

© 2019 by the authors. Licensee MDPI, Basel, Switzerland. This article is an open access article distributed under the terms and conditions of the Creative Commons Attribution (CC BY) license (http://creativecommons.org/licenses/by/4.0/).

Review

Cardiovascular Pleiotropic Effects of Natriuretic Peptides

Maurizio Forte [1,†], Michele Madonna [1,†], Sonia Schiavon [2], Valentina Valenti [3], Francesco Versaci [3], Giuseppe Biondi Zoccai [2,4], Giacomo Frati [1,2] and Sebastiano Sciarretta [1,2,*]

1. IRCCS NEUROMED, 86077 Pozzilli, Italy
2. Department of Medico-Surgical Sciences and Biotechnologies, Sapienza University of Rome, 04100 Latina, Italy
3. Department of Cardiology, Santa Maria Goretti Hospital, 04100 Latina, Italy
4. Mediterranea Cardiocentro, 80122 Napoli, Italy
* Correspondence: sebastiano.sciarretta@uniroma1.it
† These authors equally contributed to this work.

Received: 28 June 2019; Accepted: 7 August 2019; Published: 8 August 2019

Abstract: Atrial natriuretic peptide (ANP) is a cardiac hormone belonging to the family of natriuretic peptides (NPs). ANP exerts diuretic, natriuretic, and vasodilatory effects that contribute to maintain water–salt balance and regulate blood pressure. Besides these systemic properties, ANP displays important pleiotropic effects in the heart and in the vascular system that are independent of blood pressure regulation. These functions occur through autocrine and paracrine mechanisms. Previous works examining the cardiac phenotype of loss-of-function mouse models of ANP signaling showed that both mice with gene deletion of ANP or its receptor natriuretic peptide receptor A (NPR-A) developed cardiac hypertrophy and dysfunction in response to pressure overload and chronic ischemic remodeling. Conversely, ANP administration has been shown to improve cardiac function in response to remodeling and reduces ischemia-reperfusion (I/R) injury. ANP also acts as a pro-angiogenetic, anti-inflammatory, and anti-atherosclerotic factor in the vascular system. Pleiotropic effects regarding brain natriuretic peptide (BNP) and C-type natriuretic peptide (CNP) were also reported. In this review, we discuss the current evidence underlying the pleiotropic effects of NPs, underlying their importance in cardiovascular homeostasis.

Keywords: Atrial Natriuretic peptide; natriuretic peptides; cardiac remodelling; cardiac hypertrophy; vascular homeostasis

1. Introduction

Atrial natriuretic peptide (ANP) was the first member of the natriuretic peptides (NPs) family to be discovered, in 1981 [1–3]. The other two members, brain natriuretic peptide (BNP) and C-type natriuretic peptide (CNP) were identified a few years later [4,5]. NPs share some similarities: All three peptides are encoded by genes including three exons and display a 17-amino-acid ring structure in their active forms [6,7]. NPs are synthetized as pre-hormones and subsequently cleaved into the biological active carboxy terminal forms (α-ANP, BNP-32, CNP-22), together with their respective amino-terminal ends. The latter are the more stable circulating form of NPs [6–8]. ANP and BNP are synthesized prevalently in the heart, atria, and ventricles, respectively. However, a lesser expression of ANP and BNP has been reported in other areas [6–8]. Endothelium is the principal source of CNP production [6–8].

The systemic effects of NPs are well described in the literature [9,10]. They are prevalently secreted in response to the mechanical stretch of myocardial walls induced by volume or pressure overload and then, once secreted, they exert diuretic, natriuretic, and vasodilatory effects, thereby maintaining cardio–renal homeostasis and hemodynamic status through the regulation of water–salt balance and body fluid volume [9–11]. NP secretion is also induced by other hormones, such as endothelin 1, angiotensin II, and by the adrenergic system [12–14]. In turn, NPs inhibit the renin–angiotensin–aldosterone system (RAAS) and sympathetic nervous system (SNS) [3,15]. Based on all these effects, NPs may be considered as pivotal players in the pathophysiology of hypertension. Mice with homozygous genetic ablation of the ANP gene display absent levels of circulating ANP and develop salt-sensitive hypertension [16]. For these reasons, several therapeutic strategies for the management of hypertensive subjects have been created with the aim of increasing circulating levels of NPs [9,10]. NPs are also valid diagnostic and prognostic markers in cardiovascular diseases (CVDs), such as heart failure (HF), coronary artery diseases, valvular diseases, myocardial infarction, and stroke, also in apparently healthy individuals [17–21].

Among the NPs, both the active (BNP) and the inactive forms (either glycosylated or not-glycosylated proBNP and NTpro-BNP) of BNP are considered as the first line biomarkers for the diagnosis and progression of acute and chronic HF [22]. This may be due to the longer half-life and stability of BNP as compared to the other NPs (see below). Circulating levels of BNP and NTproBNP have been shown to directly correlate with the clinical outcomes in patients with HF and to be more accurate if compared to other markers used routinely, such as troponin [22]. However, the three forms of BNP detected in plasma shared different analytical properties, and to date there are no valid commercial kits able to detect only the active form of BNP, which is the form with the shorter half-life. The latter is in part attributable to the cross-reactivity of antibodies with glycosylated and not-glycosylated forms of proBNP. Of note, the inactive forms of BNP display a lower intra-individual biological variation and they have a higher ability to predict HF progression as compared to BNP. For these reasons, it is recommended to consider the level of BNP and proBNPs together in HF patients [22]. It is also strongly recommended that NPs levels should be evaluated in conjunction with other clinical parameters, such as renal function, body mass index, and cardiac imaging [23].

Besides their systemic effects, accumulating lines of evidence indicate that ANP has beneficial pleiotropic effects on the cardiovascular system at baseline and in response to stress. These pleiotropic functions are mediated by paracrine and autocrine mechanisms and are independent of blood pressure regulation [9,24]. ANP was shown to reduce maladaptive cardiac remodeling, exerting anti-hypertrophic and anti-fibrotic effects in response to pressure overload and chronic ischemic remodeling [25–28]. ANP administration reduces ischemia-reperfusion (I/R) injury. In addition, it was previously demonstrated that ANP induces anti-inflammatory and pro-angiogenetic effects in the vascular system [29–31]. On the other hand, similar effects do not seem to be induced by BNP, which is only able to exert anti-fibrotic functions [9,32].

In this review, we will discuss the pleiotropic beneficial effects of natriuretic peptides in the heart and vascular system, with a particular focus on ANP. We will summarize the results obtained in loss-of-function mouse models of ANP signaling undergoing cardiac and metabolic stress. We also discuss the available human evidence regarding the pleiotropic effects of ANP in the cardiovascular system.

2. Overview of ANP Metabolism

Once synthetized, the inactive form of ANP (pro-ANP) is stored in secretory granules of atrial cardiomyocytes [8]. Following this, pro-ANP is cleaved by the atrial natriuretic peptide-converting enzyme (Corin) in α-ANP and NT-proANP, which are the two detectable plasmatic forms [10]. Accordingly, corin knockout (KO) mice showed reduced level of active ANP and develop a mild hypertension [33]. The proprotein convertase subtilisin/kexin type 6 (PCSK6) is another important member of NP metabolic cascade. PCSK6 cleaves and activates the zymogen corin. PCSK6 KO mice develop salt-sensitive hypertension, with the consequent inhibition of ANP processing and corin activity in the heart. Notably, a genetic variant in the catalytic domain of PCSK6 was found in hypertensive patients. When transfected in cells, this variant of PCSK6 was unable to activate corin [34]. Similarly, genetic variants in corin gene that associate with hypertension and heart diseases were reported to impair PCSK6-mediated corin activation [35–37].

ANP exerts its biological effects by the interaction with the natriuretic peptide receptor type A (NPR-A). BNP also interacts with NPR-A, whereas CNP binds with a higher affinity to the type B receptor (NPR-B) [38]. NPR-A and NPR-B are widely distributed in the body. They are prevalently expressed in the kidney, brain, vascular system, heart, adrenal gland, and pancreas [39,40]. The adipose tissue is also a target of NPs. In this regard, NPRs are highly expressed in human adipose cells and both ANP and BNP were found to regulate lipid metabolism, by promoting lipolysis and adiponectin secretion [41–45]. This evidence suggests an important cross-talk between cardiac and adipose tissue [46] Both NPR-A and B are coupled to a guanylate cyclase (GC) and induce an increase of intracellular levels of cyclic guanosine monophosphate (cGMP) upon their activation [9,10,47]. Specifically, NPR-A and NPR-B are characterized by an external region, devoted to the interaction with NPs, a membrane region and an intracellular region containing the GC catalytic domain [48]. An additional receptor, NPR-C, does not show GC activity and it is instead coupled with an inhibitory G protein ($G_i\alpha$), which leads to the reduction of intracellular cyclic adenosine monophosphate (cAMP) levels, when NPR-C is activated. NPR-C contains an external region, which is the homologous of the external region of NPR-A and NPR-B. In contrast, its intracellular region is composed by only 37 amino acids. NPR-C is responsible for the clearance of NPs [47,49]. The expression of NPR-C has been reported in heart, vascular cells, pancreas, gastrointestinal tract, neurons, and chondrocytes [40]. The clearance of NPs mediated by NPR-C occurs through its internalization and delivery to lysosomes. However, known cytoplasmic international motifs have not been identified in NPR-C sequences [48]. It has been reported that the removal of ANP from circulation occurs mainly in the lung, liver, and kidney [50]. Clearance of ANP ranges from 0.5 min to 4 min [48], which is comparable with the clearance of other vasoactive hormones, such as angiotensin II and vasopressin [51,52]. BNP displays the longest half-life (about 23 min) among NPs, whereas CNP has the shortest half-life, which was very close to ANP (about 3 min) [53–55]. The differences between the half-life of NPs are attributable to their different binding affinity with NPR-C. In this case, ANP displays the major affinity with NPR-C. However, aside from its functions as a clearance receptor, it is now well established that NPR-C also mediates important cellular functions of NPs, especially CNP.

Besides NPR-C, the degradation of circulating NPs is also achieved by the neutral endopeptidase neprilysin (NEP) [56–58]. NEP is ubiquitously expressed in the body. It also cleaves other vasoactive peptides, such as angiotensin I and II, bradichinin [59]. The NPs degradation mediated by NEP occurs when NPR-C is saturated [60]. Recently, drugs combining NEP inhibition with angiotensin II receptor inhibition, named ARNi, have been introduced in the management of patients with HF, in order to maintain high levels of circulating NPs, as discussed in detail below [61,62].

3. ANP and Cardiac Pleiotropic Effects

The expression of NPR-A in cardiac cells, both in myocytes and in fibroblasts suggests a fundamental role played by ANP in the heart through autocrine and paracrine mechanisms (Figure 1) [63,64]. In fact, in vitro studies revealed that ANP inhibits cell growth and proliferation in cardiomyocytes and promotes apoptosis [65–67]. The observation underlying the anti-hypertrophic effects of ANP has been extensively investigated in KO models of ANP and NPR-A. For example, it was shown that mice with genetic disruption of ANP develop cardiac hypertrophy and hypertension in response to chronic hypoxia (3 to 5 weeks) and high salt diet administration [16,68–70]. Feng et al. demonstrated that the observed cardiac hypertrophy found in ANP KO models was independent of change in blood pressure. In fact, when the blood pressure was normalized by a low salt diet regimen ANP KO mice also developed cardiac hypertrophy, compared to wild-type animals [71]. In addition, ANP deficient mice undergoing 2 weeks of volume overload through an aorto-caval fistula showed left ventricular hypertrophy (LVH) and dysfunction when fed with both a regular and a low salt diet [27]. Franco et al. [24] further investigated the development of cardiac hypertrophy in heterozygous

ANP KO mouse after 1 week of pressure overload induced by transverse aortic constriction (TAC). The authors showed that the partial inhibition of ANP also leads to hypertrophy and adverse cardiac remodeling either at baseline and after hemodynamic stress without any significant effect on blood pressure. Conversely, the exogenous administration of ANP was found to be sufficient to reduce cardiac remodeling in a model of chronic myocardial infarction [72]. Kinoshita et al. demonstrated that ANP exerts its anti-hypertrophic actions on cardiac myocytes through the inhibition of the transient receptor potential subfamily C (TRPC-6), in a protein kinases G (PKG) mediated manner. TRPC-6 triggers hypertrophic stimuli by activating the calcineurin–nuclear factor of activated T cells (NFAT) signaling. In the same study, the authors found that inhibition of PKG, as well as modifications in the phosphorylation site of TRPC-6, blunts the anti-hypertrophic actions of ANP [73].

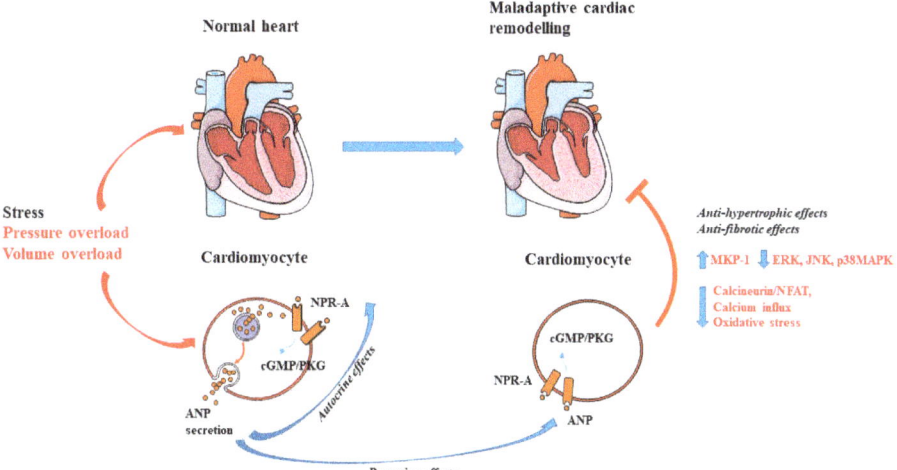

Figure 1. Local action of atrial natriuretic peptide (ANP) in cardiomyocytes. ANP is secreted by cardiomyocytes in response to cardiac stress. After secretion, ANP binds to the natriuretic peptide receptor (NPR-A) and activates the cGMP/PKG pathway. Anti-hypertrophic and anti-fibrotic effects of ANP occur through autocrine and paracrine mechanisms. Arrow-headed lines indicate activation, whereas bar-headed lines indicate inhibition. Legend: c-Jun N-terminal kinase (JNK); cyclic guanosine monophosphate (cGMP); extracellular signal-regulated kinase (ERK); mitogen-activated protein kinase phosphatase-1 (MKP-1); natriuretic peptide receptor type A (NPR-A); nuclear factor of activated T-cells (NFAT); p38 mitogen-activated protein kinase (MAPKs); protein kinases G (PKG). See text for further details. The figure was made using tools provided by Servier Medical Arts, amongst others.

Interesting findings have also been obtained in KO of NPR-A (NPR-A KO). NPR-A null mice are characterized by cardiac hypertrophy and chamber dilatation at three months of age. These effects were not exclusively attributable to the observed increase of blood pressure levels in these animals, since the magnitude of cardiac hypertrophy was exaggerated with respect to what could be expected based on the severity of hypertension. In addition, similar levels of cardiac hypertrophy could not be observed in other mouse models of hypertension [74]. Knowles et al. corroborated these data in NPR-A KO undergoing TAC. The authors showed the exacerbation of cardiac hypertrophy and cardiac dysfunction in NPR-A null mice compared to wild-type mice; they also reported that chronic administration of different anti-hypertensive drugs, such as enalapril, furosemide, hydralazine, propranolol, and losartan was not able to reduce cardiac mass in NPR-A KO [75]. Consistently, cardiac overexpression of NPR-A reduced cardiomyocyte size in both wild-type and NPR-A KO mice, along with the reduction of ANP mRNA levels [76]. These changes were not associated with hemodynamic alteration [76]. To test the local effect of ANP in determining cardiomyocytes size, Holtwick et al. generated a model with cardiomyocyte-specific gene deletion of NPR-A (NPR-AcKO). They demonstrated a hypertrophic cardiac response in NPR-AcKO at baseline, which was accelerated in mice subjected to aortic constriction and analyzed 10 days after surgery. Moreover, in NPR-AcKO mice undergoing surgery, the authors observed a reduction of blood pressure levels, probably attributable to the increase of plasma ANP levels [26]. These findings support the concept that ANP acts as an intrinsic regulator of cardiac hypertrophy [25]. Mechanistically, it was demonstrated that ANP reduces hypertrophy induced by angiotensin-II (Ang-II) and endothelin-1 (ET-1), by increasing mitogen-activated protein kinase phosphatase-1 (MKP-1) signaling in a cGMP-dependent manner in cultured cardiomyocytes [77]. MKP-1 activation inhibits extracellular signal-regulated kinases (ERKs), c-Jun N-terminal kinases (JNK), and p38MAPKs, known inducers of cell proliferation and hypertrophy [78]. ANP was also reported to reduce protein synthesis in cardiomyocytes and fibroblasts in vitro treated with adrenergic stimuli by the inhibition of calcium influx mediated by norepinephrine [79]. The involvement of the cardiac calcineurin-NFAT pathway in cardiac hypertrophy was also reported in the NPR-A KO mice [80]. In addition, exogenous administration of ANP was able to inhibit calcineurin/NFAT signaling in cultured cardiomyocytes treated with phenylephrine [73,80]. Other studies showed that the anti-hypertrophic effects of ANP are associated with the reduction of oxidative stress in isolated cardiomyocytes treated with Ang-II and ET-1 [81]. However, most of the previous studies dissecting the mechanisms of the antihypertrophic effects of ANP were associative and further investigations are warranted to address this issue.

Human evidence also underlines the association between ANP and cardiac mass. In this regard, Rubattu et al. found that reduced levels of circulating ANP were associated with an increase in left ventricular mass in individuals with essential hypertension. Similar data were obtained in the same study in subjects carrying an allelic variant of ANP gene promoter, which is associated with reduced circulating ANP levels [82]. Similarly, a deletion mutation in the NPR-A gene was found to be associated with the development of LVH without hypertension [83]. ANP levels were found to be lower in the presence of obesity and metabolic syndrome, because of either increased clearance or reduced synthesis. [84]. Accordingly, plasma levels of ANP were found to be inversely correlated with the increase of cardiac mass in hypertensive subjects with metabolic syndrome or obesity [85,86].

4. ANP and Ischemia/Riperfusion Injury

Pre-clinical and clinical studies showed that ANP is able to attenuate I/R injury. ANP was found to protect isolated rat hearts from I/R injury and to increase post-ischemic cGMP level when administered at the time of reperfusion [87]. Similar findings were obtained in isolated rabbit hearts, in which infusion of ANP prior to reperfusion was reported to significantly decrease infarction area. Similarly, administration of a cell-permeable cGMP analogue was able to mimic the protection exerted by ANP. In contrast, the mitochondrial ATP sensitive potassium channel (mKATP) inhibitor 5-Hydroxyde-canoate (5-HD) blunted it [88]. Other studies revealed that the pre-ischemic infusion of ANP in isolated rat hearts was unable to limit I/R injury when the hearts were co-treated with N-nitro l-arginine methyl ester (L-NAME) or a protein kinase C (PKC) synthetase inhibitor or with a mKATP channel blocker. The latter findings suggest that the nitric oxide (NO)-PKC pathway and the mKATP channel activation are likely involved in the beneficial effects of ANP during I/R [89].

The role of ANP was also observed in models of I/R in vivo. For example, in dogs undergoing 30-min of ischemia followed by 60 min of reperfusion, ANP was shown to decrease ventricular extrasystoles and atrial fibrillation when administered either during artery occlusion or during reperfusion. An increase of myocardial ATP was found in the ischemic myocardium of ANP-treated animals. On the other hand, no differences were observed in hemodynamic parameters, suggesting that the protective effects of ANP were mediated prevalently by the elevation of cGMP [90]. In another study, ANP administration during ischemia or immediately before reperfusion was found to limit cardiac injury in pigs [91]. In pigs subjected to 30 min of ischemia followed by 4 h of reperfusion, ANP was reported to decrease myocardial injury, in association with the increase of myocardial expression of peroxisome proliferator activated receptor γ, a transcription factor involved in myocardial protection during I/R [92,93]. Of interest, Charan et al. showed that ANP was able to restore the cardioprotection conferred by ischemic pre-conditioning (four cycles of 5 min of ischemia followed by 5 min of reperfusion) in diabetic hearts, likely through an improvement of NO metabolism [94]. NPR-A KO subjected to myocardial infarction by permanent ligation of left coronary artery showed higher mortality within 1 week, as compared to wild-type mice and also a reduced water and sodium excretion. In addition, NPR-A KO mice showed exacerbated cardiac hypertrophy, fibrosis and dysfunction 4 weeks after myocardial infarction. Notably, cardiac fibrosis was absent in NPR-A mice carrying a deletion in Ang-II type 1A receptor whereas the higher mortality and cardiac hypertrophy remained unaltered [95].

Similar effects were observed in human patients with acute myocardial infarction (AMI). In fact, Hayashi et al. reported that ANP improves left ventricular ejection fraction (LVEF) and prevents left ventricular enlargement in patients with anterior AMI receiving reperfusion therapy. In the same study, the authors showed the suppression of the renin-angiotensin-aldosterone system and endothelin-1 (ET-1) pathways, known mediators of left ventricular remodeling [96]. Moreover, ANP administered immediately after coronary angioplasty limited I/R injury, reduced ST-segment elevation and increased LVEF in AMI patients [97]. Kasama et al. studied the effects of ANP on left ventricular remodeling in patients with first anterior AMI. In this study, ANP was continuously infused before and after primary coronary angioplasty. ANP drastically reduced I/R injury, inhibited LV remodeling, and ameliorated LV function, together with the reduction of cardiac sympathetic nerve activity [98]. In the J-WIND (Japan-Working Groups of Acute Myocardial Infarction for the Reduction of Necrotic Damage) clinical trial, ANP infusion was shown to decrease infarct size and to limit reperfusion injury in patients affected by AMI undergoing reperfusion therapy [99].

5. ANP and Vascular Pleiotropic Effects

Previous evidence indicates that ANP can be expressed and secreted by aortic endothelial cells [100], suggesting a local action of ANP also in the endothelium. Mice with endothelial-specific deletion of NPR-A gene (NPR-A-ecKO) undergoing limb ischemia displayed impaired angiogenesis until 5 weeks post-surgery. In addition, NPR-A-ecKO observed 10 days after TAC showed a decreased capillary density in the heart, which was associated with the development of cardiac hypertrophy and fibrosis. In the same study, ANP was also able to induce endothelial cell proliferation and angiogenesis in vitro [101]. Other evidence has demonstrated that physiological doses of ANP are able to induce endothelial cell proliferation and migration along with increase of phospho-Akt and phospho-ERK1/2. In contrast, excessive doses of ANP leads to opposite effects [29]. Of interest, we reported that a molecular variant of ANP (C2238-αANP), which is associated with increased cardiovascular risk, severely affects endothelial function in vitro and ex vivo in isolated mouse arteries [102–105]. C2238-αANP interacts with NPR-C receptors, leading to the inhibition of the cAMP/Akt/protein kinase A (PKA) pathway and activation of NADPH oxidase. The latter contributes to the increase of reactive oxygen species (ROS) and promotes endothelial dysfunction [102]. We also found that subjects carrying the C2238-αANP gene variant showed endothelial dysfunction [102].

Kiemer et al. reported that ANP reduces inflammation in endothelial cells by inhibiting the TNF-α-induced expression of adhesion molecules, such as E-selectin and ICAM-1. The latter is achieved by the activation of the nuclear factor of kappa light polypeptide gene enhancer in B-cells inhibitor (IκB) and the consequent inhibition of the nuclear factor kappa-light-chain-enhancer of activated B cells (NF-κB), in a cGMP-dependent manner [106]. In separate studies, the same group found that ANP is able to inhibit the macrophage expression of ciclooxygenases-2 (Cox-2) induced by lipopolysaccharides and to reduce inducible nitric oxide synthase (iNOS) expression [107,108]. Of note, inhibition of neprilysin was shown to potentiate the effects of ANP and to limit polymorphonuclear neutrophil-vascular cell interactions in vitro under hypoxia [109]. Although the evidence described above suggests that ANP may preserve vascular function through autocrine and paracrine mechanisms, it should be better clarified whether these effects are prominently attributable to endothelial-derived or also to cardiac-derived ANP (Figure 2). ANP increases systemic endothelial permeability through endothelial NPR-A, thereby maintaining intravascular volume homeostasis. Cardiac ANP reduces lung endothelial permeability in pathological conditions [110]. It should also be noted that most of the evidence regarding the endothelial effects of ANP were observed in vitro. In addition, most of the studies only investigated the contribution of exogenous ANP, without exploring the impact of endogenous endothelial-derived ANP.

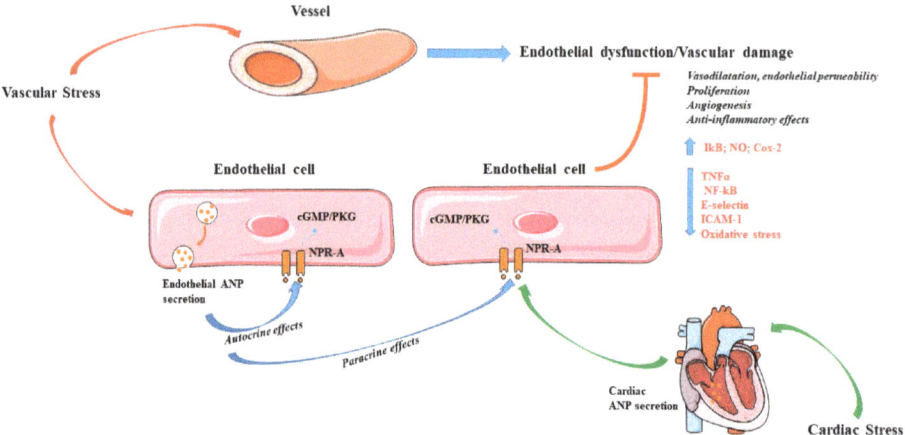

Figure 2. Effects of atrial natriuretic peptide (ANP) in the vascular system. The contribution of either endothelial and cardiac-derived ANP in response to stress prevents endothelial dysfunction and vascular damage. ANP increases vasodilatation and endothelial permeability, stimulates angiogenesis, proliferation, nitric oxide production, and exerts anti-inflammatory effects. Arrow-headed lines indicate activation whereas bar-headed lines indicate inhibition. Legend: cyclic guanosine monophosphate (cGMP); ciclooxygenases-2 (Cox-2); intercellular Adhesion Molecule 1 (ICAM-1); natriuretic peptide receptor type A (NPR-A); nitric oxide (NO); nuclear factor kappa-light-chain-enhancer of activated B cells (NFkB); protein kinases G (PKG); tumour necrosis factor alpha (TNFα). See text for further details. The figure was made using tools provided by Servier Medical Arts, amongst others.

6. Local Actions of BNP and CNP in the Cardiovascular System

Accumulating lines of evidence also suggest a local action of BNP and CNP in the cardiovascular system, although to a lesser extent than ANP. For example, in mice lacking BNP gene (BNP KO) cardiac fibrosis was observed 7 days after pressure overload. However, no signs of ventricular hypertrophy were evident in BNP KO mice [111]. Infusion of BNP was found to inhibit TGF-β-induced cardiac fibroblast proliferation and to reduce the expression of genes involved in fibrosis, myofibroblast conversion, and inflammation. These effects were mediated by the activation of PKG and mitogen-activated protein kinase (MEK)/extracellular signal-regulated kinase (ERK) pathway [112].

CNP was shown to be secreted by cardiac fibroblasts exposed to pro-fibrotic stimuli. CNP secretion by cardiac fibroblasts leads to the reduction of collagen synthesis in a cGMP dependent manner [32]. These findings were supported in vivo in rats subjected to myocardial infarction induced by coronary ligation. CNP administered for 2 weeks starting 4 days after myocardial infarction was found to decrease left ventricular enlargement, cross-sectional area of cardiomyocytes, and markers of cardiac fibrosis and hypertrophy, independently of changes in blood pressure [113]. Similarly, mice with cardiac overexpression of CNP undergoing myocardial infarction showed the decrease in cardiac hypertrophy and an amelioration of cardiac function after 3 weeks of permanent ligation of left coronary artery [114]. Consistently, rats overexpressing a dominant-negative form of NPR-B developed cardiac hypertrophy at 6 months of age, which was accelerated in animals undergoing chronic volume overload by infrarenal aortocaval shunt and studied 8 weeks after surgery. These data suggest the role of NPR-B in the regulation of hypertrophy [115]. In addition, Izumiya et al. found that chronic CNP administration (2 weeks) concomitantly with Ang-II treatment attenuated Ang-II-induced cardiac hypertrophy without affecting blood pressure in mice. In this case, the effects of CNP were associated with the decrease of ROS and NOX4 expression [116].

Other works suggest that endothelial-derived CNP acts as an important regulator of blood pressure. In fact, endothelial specific CNP KO mice (CNP-ecKO) develop hypertension, atherogenesis, aneurysm, and showed an impaired endothelial-dependent vasorelaxation [117,118]. Conversely, no effects on blood pressure levels and acetylcholine-induced vasorelaxation in isolated arteries were reported in vascular smooth muscle cell (VSMC)-specific NPR-B KO model, which also preserved. The results indicate that CNP is required for the maintenance of blood pressure and endothelial function independently of VSMC NPR-B. However, the effect of CNP in the induction of vasodilation was abolished in VSMC NPR-B KO [117]. Spiranek et al. further demonstrated that the action of CNP on vasodilatation is strictly dependent on vessel diameter. In fact, administration of CNP induced a vasodilatation on precapillary arterioles and capillaries and did not affect proximal arterioles. The vasodilatory effects of CNP was preserved in mice lacking NPR-B in endothelial cells, whereas it was abolished when NPR-B was deleted in microcirculatory VSMCs and in pericytes. In these models, a peripheral resistance and chronic arterial hypertension was observed, with a preserved renal function [119].

The involvement of NPR-C in mediating the vascular protection exerted by CNP was also investigated. For example, Moyes et al. reported that NPR-C agonists were able to induce vasorelaxation and to lower blood pressure in wild-type mice [118].

CNP also plays an important role in vascular remodeling. For example, it was previously reported that CNP induces vasodilatation of human forearm resistance vessels in an endothelium-independent manner [120]. Of note, CNP was also found to reduce VSMC proliferation induced by platelet-derived growth factor (PDGF) and to reduce intimal lesions in rat common carotid arteries after vascular injury [121,122]. Finally, CNP suppresses inflammation and fibrosis in rabbit carotid arteries subjected to shear stress, by the enhancement of NO production [123].

7. Therapeutic Interventions Targeting NPs

The evidence described here suggests that NPs act as critical regulators of cardiac and vascular homeostasis. As a consequence, the modulation of NPs may represent a useful strategy for the prevention or treatment of cardiovascular diseases. In this regard, intravenous administration of ANP was reported to improve cardiac remodeling in patients with AMI undergoing primary coronary angioplasty [98]. ANP was also proven to improve cardiac function in patients with AMI, without affecting blood pressure [99]. However, a renal resistance to ANP was reported in patients with chronic HF (CHF). In this regard, both in preclinical models and in patients, ANP treatment was found to attenuate diuresis and natriuresis [124–129]. Moreover, patients with congestive heart failure showed sodium retention and oedema despite high levels of circulating cardiac NPs [130]. Renal resistance to NPs, defined as the "endocrine heart paradox", may be explained by three main mechanisms, as reviewed in detail by Clerico et al. [128]: (i) inactivation of circulating NPs, (ii) downregulation of NPRs, and (iii) by mechanisms acting downstream to NPRs.

Apart from ANP administration, synthetic NPs have been developed and tested in different clinical trials. Anaritide and carperitide are synthetic forms of ANP, whereas neseritide and cenderitide are the recombinant forms of BNP and CNP, respectively [131,132]. It was reported that neseritide improves hemodynamic function and global clinical status, such as dyspnea and fatigue in patients with congestive HF [132,133]. However, it is now known that safety and efficacious of neseritide appear to be low. A previous large clinical trial conducted in patients with acute HF showed that neseritide was not able to significantly reduce mortality and alleviate symptoms in these subjects [134]. In the same study, neseritide increased hypotension. In addition, two meta-analyses revealed that neseritide impairs renal function and increases the risk of short-term mortality in acutely decompensated HF patients [135,136]. These detrimental effects appear to be mediated also by hypotension induced by the drug.

ARNi, angiotensin receptor–neprilysin inhibitors, are able to enhance circulating levels of NPs, as previously reported. To date, ARNi have been shown to be efficacious in patients with HF with reduced ejection fraction (HF-REF) [61]. In the PARADIGM-HF trial, LCZ696 (sacubitril/valsartan), also known as ENTRESTO, was shown to reduce mortality and morbidity in patients with HF-REF in a manner more efficacious than enalapril alone [137,138]. However, the effects of LCZ696 on NEP inhibition in acute heart failure requires further investigations. In fact, Vodovar et al. reported that elevated levels of BNP (916 pg/mL) correlated with a decrease of NEP activity in patients with acute decompensated HF. BNP also inhibits NEP in vitro [139]. Thus, it is not clear the potential impact of high BNP levels on the efficacy of ENTRESTO in patients with acute HF [59].

The effects of ARNi are being testing also in ongoing clinical trials, in the setting of cardiac remodeling [140] and endothelial function (ClinicalTrials.gov Identifier: NCT03119623). Pre-clinical evidence about the use of ARNi in cardiac and vascular remodeling are promising. For example, LCZ696 used at a dose that does not lower blood pressure, was recently reported to reduce cardiac rupture and survival in a mouse model of myocardial infarction. The beneficial effect of LCZ696 was associated with the suppression of pro-inflammatory cytokines interleukin (IL)-1β and IL-6 and extracellular matrix degradation. The latter was achieved by the reduction of metalloprotease-9 (MMP-9) activity and expression [141]. In addition, sacubitril/valsartan was recently demonstrated to improve cardiac remodeling and to reduce infarct size in experimental models of myocardial infarction [142,143]. Besides ARNi, omapatrilat is an anti-hypertensive agent that combines neprylisin and angiotensin-converting enzyme (ACE) inhibition. Omapatrilat was shown to be an efficacious antihypertensive agent and to be a promising drug in a phase II and phase III HF trial [144,145]. However, it failed to reduce death and hospitalization in the OVERTURE clinical trial, in patients with CHF [146]. Adverse effects were also reported for omapatrilat. For example, it was shown that omapatrilat causes symptomatic hypotension and increases the risk of angioedema if compared to the administration of the only ACE inhibitor [145,146].

8. Conclusions and Perspectives

In this review, we summarized the relevant literature about the role of NPs, and in particular of ANP, in the improvement of cardiac and vascular remodeling in different stress conditions. In vivo and in vitro studies revealed that local ANP secretion by cardiac and endothelial cells play important pleiotropic functions through autocrine and paracrine effects. This evidence suggests that ANP acts as a main regulator of cardiovascular homeostasis in an autocrine and paracrine manner. Overall, the enhancement of ANP levels may be a promising target for the prevention and treatment of cardiovascular diseases, in both hypertensive and not hypertensive patients. In this regard, inhibition of NEP may be the appropriate strategy to potentiate the effects of NPs. However, some issues regarding pleiotropic effects of NPs in the cardiovascular system should be elucidated. First of all, the molecular mechanisms underlying the autocrine and paracrine functions of ANP need to be dissected. Secondly, downstream effectors of the cGMP/PKG pathway involved in the anti-remodeling actions of NPs should be identified. Thirdly, additional studies are needed to elucidate the effects of NPs in the vascular system, especially those related to metabolic alterations. Finally, it should be assessed in the future whether the modulation of NPs should contribute to the improvement of the clinical management of individuals at high risk of developing cardiac and vascular damage.

Funding: This work was partially supported by a grant from the Italian Ministry of Health to SS (GR-2013-02355401) and from a Sapienza University Grant of Italian Ministry of Research to SS (RG11816433FC0145).

Conflicts of Interest: The authors declare no conflict of interest.

References

1. de Bold, A.J.; Borenstein, H.B.; Veress, A.T.; Sonnenberg, H. A rapid and potent natriuretic response to intravenous injection of atrial myocardial extract in rats. *Life Sci.* **1981**, *28*, 89–94. [CrossRef]
2. Kangawa, K.; Matsuo, H. Purification and complete amino acid sequence of alpha-human atrial natriuretic polypeptide (alpha-hANP). *Biochem. Biophys. Res. Commun.* **1984**, *118*, 131–139. [CrossRef]
3. Levin, E.R.; Gardner, D.G.; Samson, W.K. Natriuretic peptides. *N. Engl. J. Med.* **1998**, *339*, 321–328. [PubMed]
4. Sudoh, T.; Minamino, N.; Kangawa, K.; Matsuo, H. C-type natriuretic peptide (CNP): a new member of natriuretic peptide family identified in porcine brain. *Biochem. Biophys. Res. Commun.* **1990**, *168*, 863–870. [CrossRef]
5. Sudoh, T.; Kangawa, K.; Minamino, N.; Matsuo, H. A new natriuretic peptide in porcine brain. *Nature* **1988**, *332*, 78–81. [CrossRef] [PubMed]
6. Potter, L.R.; Yoder, A.R.; Flora, D.R.; Antos, L.K.; Dickey, D.M. Natriuretic peptides: their structures, receptors, physiologic functions and therapeutic applications. *Handb. Exp. Pharmacol.* **2009**, 341–366.
7. He, X.L.; Dukkipati, A.; Garcia, K.C. Structural determinants of natriuretic peptide receptor specificity and degeneracy. *J. Mol. Biol.* **2006**, *361*, 698–714. [CrossRef] [PubMed]
8. Matsuo, A.; Nagai-Okatani, C.; Nishigori, M.; Kangawa, K.; Minamino, N. Natriuretic peptides in human heart: Novel insight into their molecular forms, functions, and diagnostic use. *Peptides* **2019**, *111*, 3–17. [CrossRef] [PubMed]
9. Rubattu, S.; Sciarretta, S.; Valenti, V.; Stanzione, R.; Volpe, M. Natriuretic peptides: an update on bioactivity, potential therapeutic use, and implication in cardiovascular diseases. *Am. J. Hypertens* **2008**, *21*, 733–741. [CrossRef]
10. Volpe, M.; Rubattu, S.; Burnett, J., Jr. Natriuretic peptides in cardiovascular diseases: current use and perspectives. *Eur. Heart J.* **2014**, *35*, 419–425. [CrossRef]
11. Rubattu, S.; Forte, M.; Marchitti, S.; Volpe, M. Molecular Implications of Natriuretic Peptides in the Protection from Hypertension and Target Organ Damage Development. *Int. J. Mol. Sci.* **2019**, *20*, 798. [CrossRef] [PubMed]
12. Soualmia, H.; Barthelemy, C.; Masson, F.; Maistre, G.; Eurin, J.; Carayon, A. Angiotensin II-induced phosphoinositide production and atrial natriuretic peptide release in rat atrial tissue. *J. Cardiovasc Pharmacol.* **1997**, *29*, 605–611. [CrossRef] [PubMed]
13. Stasch, J.P.; Hirth-Dietrich, C.; Kazda, S.; Neuser, D. Endothelin stimulates release of atrial natriuretic peptides in vitro and in vivo. *Life Sci.* **1989**, *45*, 869–875. [CrossRef]
14. Thibault, G.; Amiri, F.; Garcia, R. Regulation of natriuretic peptide secretion by the heart. *Annu. Rev. Physiol.* **1999**, *61*, 193–217. [CrossRef] [PubMed]
15. Lee, C.Y.; Burnett, J.C., Jr. Natriuretic peptides and therapeutic applications. *Heart Fail. Rev.* **2007**, *12*, 131–142. [CrossRef] [PubMed]
16. John, S.W.; Krege, J.H.; Oliver, P.M.; Hagaman, J.R.; Hodgin, J.B.; Pang, S.C.; Flynn, T.G.; Smithies, O. Genetic decreases in atrial natriuretic peptide and salt-sensitive hypertension. *Science* **1995**, *267*, 679–681. [CrossRef] [PubMed]
17. Seronde, M.F.; Gayat, E.; Logeart, D.; Lassus, J.; Laribi, S.; Boukef, R.; Sibellas, F.; Launay, J.M.; Manivet, P.; Sadoune, M.; et al. Comparison of the diagnostic and prognostic values of B-type and atrial-type natriuretic peptides in acute heart failure. *Int. J. Cardiol.* **2013**, *168*, 3404–3411. [CrossRef] [PubMed]
18. Paget, V.; Legedz, L.; Gaudebout, N.; Girerd, N.; Bricca, G.; Milon, H.; Vincent, M.; Lantelme, P. N-terminal pro-brain natriuretic peptide: a powerful predictor of mortality in hypertension. *Hypertension* **2011**, *57*, 702–709. [CrossRef] [PubMed]
19. Sabatine, M.S.; Morrow, D.A.; de Lemos, J.A.; Omland, T.; Sloan, S.; Jarolim, P.; Solomon, S.D.; Pfeffer, M.A.; Braunwald, E. Evaluation of multiple biomarkers of cardiovascular stress for risk prediction and guiding medical therapy in patients with stable coronary disease. *Circulation* **2012**, *125*, 233–240. [CrossRef]
20. Daniels, L.B. Natriuretic Peptides and Assessment of Cardiovascular Disease Risk in Asymptomatic Persons. *Curr. Cardiovasc. Risk Rep.* **2010**, *4*, 120–127. [CrossRef]
21. Volpe, M.; Battistoni, A.; Rubattu, S. Natriuretic peptides in heart failure: Current achievements and future perspectives. *Int. J. Cardiol.* **2019**, *281*, 186–189. [CrossRef] [PubMed]

22. Clerico, A.; Passino, C.; Franzini, M.; Emdin, M. Cardiac biomarker testing in the clinical laboratory: where do we stand? General overview of the methodology with special emphasis on natriuretic peptides. *Clin. Chim. Acta.* **2015**, *443*, 17–24. [CrossRef] [PubMed]
23. Mueller, C.; McDonald, K.; de Boer, R.A.; Maisel, A.; Cleland, J.G.F.; Kozhuharov, N.; Coats, A.J.S.; Metra, M.; Mebazaa, A.; Ruschitzka, F.; et al. Heart Failure Association of the European Society of, C. Heart Failure Association of the European Society of Cardiology practical guidance on the use of natriuretic peptide concentrations. *Eur. J. Heart Fail.* **2019**, *21*, 715–731. [CrossRef] [PubMed]
24. Franco, V.; Chen, Y.F.; Oparil, S.; Feng, J.A.; Wang, D.; Hage, F.; Perry, G. Atrial natriuretic peptide dose-dependently inhibits pressure overload-induced cardiac remodeling. *Hypertension* **2004**, *44*, 746–750. [CrossRef] [PubMed]
25. Molkentin, J.D. A friend within the heart: natriuretic peptide receptor signaling. *J. Clin. Investig.* **2003**, *111*, 1275–1277. [CrossRef] [PubMed]
26. Holtwick, R.; van Eickels, M.; Skryabin, B.V.; Baba, H.A.; Bubikat, A.; Begrow, F.; Schneider, M.D.; Garbers, D.L.; Kuhn, M. Pressure-independent cardiac hypertrophy in mice with cardiomyocyte-restricted inactivation of the atrial natriuretic peptide receptor guanylyl cyclase-A. *J. Clin. Investig.* **2003**, *111*, 1399–1407. [CrossRef]
27. Mori, T.; Chen, Y.F.; Feng, J.A.; Hayashi, T.; Oparil, S.; Perry, G.J. Volume overload results in exaggerated cardiac hypertrophy in the atrial natriuretic peptide knockout mouse. *Cardiovasc. Res.* **2004**, *61*, 771–779. [CrossRef] [PubMed]
28. Calvieri, C.; Rubattu, S.; Volpe, M. Molecular mechanisms underlying cardiac antihypertrophic and antifibrotic effects of natriuretic peptides. *J. Mol. Med. (Berl.)* **2012**, *90*, 5–13. [CrossRef] [PubMed]
29. Kook, H.; Itoh, H.; Choi, B.S.; Sawada, N.; Doi, K.; Hwang, T.J.; Kim, K.K.; Arai, H.; Baik, Y.H.; Nakao, K. Physiological concentration of atrial natriuretic peptide induces endothelial regeneration in vitro. *Am. J. Physiol. Heart Circ. Physiol.* **2003**, *284*, H1388–H1397. [CrossRef]
30. Moro, C.; Klimcakova, E.; Lolmede, K.; Berlan, M.; Lafontan, M.; Stich, V.; Bouloumie, A.; Galitzky, J.; Arner, P.; Langin, D. Atrial natriuretic peptide inhibits the production of adipokines and cytokines linked to inflammation and insulin resistance in human subcutaneous adipose tissue. *Diabetologia* **2007**, *50*, 1038–1047. [CrossRef]
31. Alexander, M.R.; Knowles, J.W.; Nishikimi, T.; Maeda, N. Increased atherosclerosis and smooth muscle cell hypertrophy in natriuretic peptide receptor A-/-apolipoprotein E-/- mice. *Arterioscler. Thromb. Vasc. Biol.* **2003**, *23*, 1077–1082. [CrossRef] [PubMed]
32. Horio, T.; Tokudome, T.; Maki, T.; Yoshihara, F.; Suga, S.; Nishikimi, T.; Kojima, M.; Kawano, Y.; Kangawa, K. Gene expression, secretion, and autocrine action of C-type natriuretic peptide in cultured adult rat cardiac fibroblasts. *Endocrinology* **2003**, *144*, 2279–2284. [CrossRef] [PubMed]
33. Chan, J.C.; Knudson, O.; Wu, F.; Morser, J.; Dole, W.P.; Wu, Q. Hypertension in mice lacking the proatrial natriuretic peptide convertase corin. *Proc. Natl. Acad. Sci. USA* **2005**, *102*, 785–790. [CrossRef] [PubMed]
34. Chen, S.; Cao, P.; Dong, N.; Peng, J.; Zhang, C.; Wang, H.; Zhou, T.; Yang, J.; Zhang, Y.; Martelli, E.E.; et al. PCSK6-mediated corin activation is essential for normal blood pressure. *Nat. Med.* **2015**, *21*, 1048–1053. [CrossRef] [PubMed]
35. Zhang, Y.; Zhou, T.; Niu, Y.; He, M.; Wang, C.; Liu, M.; Yang, J.; Zhang, Y.; Zhou, J.; Fukuda, K.; et al. Identification and functional analysis of CORIN variants in hypertensive patients. *Hum. Mutat.* **2017**, *38*, 1700–1710. [CrossRef] [PubMed]
36. Rame, J.E.; Drazner, M.H.; Post, W.; Peshock, R.; Lima, J.; Cooper, R.S.; Dries, D.L. Corin I555(P568) allele is associated with enhanced cardiac hypertrophic response to increased systemic afterload. *Hypertension* **2007**, *49*, 857–864. [CrossRef] [PubMed]
37. Rame, J.E.; Tam, S.W.; McNamara, D.; Worcel, M.; Sabolinski, M.L.; Wu, A.H.; Dries, D.L. Dysfunctional corin i555(p568) allele is associated with impaired brain natriuretic peptide processing and adverse outcomes in blacks with systolic heart failure: results from the Genetic Risk Assessment in Heart Failure substudy. *Circ. Heart Fail.* **2009**, *2*, 541–548. [CrossRef] [PubMed]
38. Suga, S.; Nakao, K.; Hosoda, K.; Mukoyama, M.; Ogawa, Y.; Shirakami, G.; Arai, H.; Saito, Y.; Kambayashi, Y.; Inouye, K.; et al. Receptor selectivity of natriuretic peptide family, atrial natriuretic peptide, brain natriuretic peptide, and C-type natriuretic peptide. *Endocrinology* **1992**, *130*, 229–239. [CrossRef]

39. Vanderheyden, M.; Bartunek, J.; Goethals, M. Brain and other natriuretic peptides: molecular aspects. *Eur. J. Heart Fail.* **2004**, *6*, 261–268. [CrossRef]
40. Zois, N.E.; Bartels, E.D.; Hunter, I.; Kousholt, B.S.; Olsen, L.H.; Goetze, J.P. Natriuretic peptides in cardiometabolic regulation and disease. *Nat. Rev. Cardiol.* **2014**, *11*, 403–412. [CrossRef]
41. Sengenes, C.; Berlan, M.; De Glisezinski, I.; Lafontan, M.; Galitzky, J. Natriuretic peptides: a new lipolytic pathway in human adipocytes. *FASEB J.* **2000**, *14*, 1345–1351. [CrossRef] [PubMed]
42. Dessi-Fulgheri, P.; Sarzani, R.; Rappelli, A. Role of the natriuretic peptide system in lipogenesis/lipolysis. *Nutr. Metab. Cardiovasc. Dis.* **2003**, *13*, 244–249. [CrossRef]
43. Bordicchia, M.; Spannella, F.; Ferretti, G.; Bacchetti, T.; Vignini, A.; Di Pentima, C.; Mazzanti, L.; Sarzani, R. PCSK9 is Expressed in Human Visceral Adipose Tissue and Regulated by Insulin and Cardiac Natriuretic Peptides. *Int. J. Mol. Sci.* **2019**, *20*, 245. [CrossRef] [PubMed]
44. Bordicchia, M.; Ceresiani, M.; Pavani, M.; Minardi, D.; Polito, M.; Wabitsch, M.; Cannone, V.; Burnett, J.C., Jr.; Dessi-Fulgheri, P.; Sarzani, R. Insulin/glucose induces natriuretic peptide clearance receptor in human adipocytes: a metabolic link with the cardiac natriuretic pathway. *Am. J. Physiol. Regul. Integr. Comp. Physiol.* **2016**, *311*, R104–R114. [CrossRef] [PubMed]
45. Sarzani, R.; Marcucci, P.; Salvi, F.; Bordicchia, M.; Espinosa, E.; Mucci, L.; Lorenzetti, B.; Minardi, D.; Muzzonigro, G.; Dessi-Fulgheri, P.; et al. Angiotensin II stimulates and atrial natriuretic peptide inhibits human visceral adipocyte growth. *Int. J. Obes. (Lond.)* **2008**, *32*, 259–267. [CrossRef] [PubMed]
46. Jordan, J.; Birkenfeld, A.L.; Melander, O.; Moro, C. Natriuretic Peptides in Cardiovascular and Metabolic Crosstalk: Implications for Hypertension Management. *Hypertension* **2018**, *72*, 270–276. [CrossRef]
47. Kuhn, M. Molecular Physiology of Membrane Guanylyl Cyclase Receptors. *Physiol. Rev.* **2016**, *96*, 751–804. [CrossRef] [PubMed]
48. Potter, L.R. Natriuretic peptide metabolism, clearance and degradation. *FEBS J.* **2011**, *278*, 1808–1817. [CrossRef] [PubMed]
49. Rubattu, S.; Sciarretta, S.; Morriello, A.; Calvieri, C.; Battistoni, A.; Volpe, M. NPR-C: a component of the natriuretic peptide family with implications in human diseases. *J. Mol. Med. (Berl.)* **2010**, *88*, 889–897. [CrossRef]
50. Hollister, A.S.; Rodeheffer, R.J.; White, F.J.; Potts, J.R.; Imada, T.; Inagami, T. Clearance of atrial natriuretic factor by lung, liver, and kidney in human subjects and the dog. *J. Clin. Investig.* **1989**, *83*, 623–628. [CrossRef]
51. Nakao, K.; Sugawara, A.; Morii, N.; Sakamoto, M.; Yamada, T.; Itoh, H.; Shiono, S.; Saito, Y.; Nishimura, K.; Ban, T.; et al. The pharmacokinetics of alpha-human atrial natriuretic polypeptide in healthy subjects. *Eur. J. Clin. Pharmacol.* **1986**, *31*, 101–103. [CrossRef] [PubMed]
52. Yandle, T.G.; Richards, A.M.; Nicholls, M.G.; Cuneo, R.; Espiner, E.A.; Livesey, J.H. Metabolic clearance rate and plasma half life of alpha-human atrial natriuretic peptide in man. *Life Sci.* **1986**, *38*, 1827–1833. [CrossRef]
53. Mukoyama, M.; Nakao, K.; Hosoda, K.; Suga, S.; Saito, Y.; Ogawa, Y.; Shirakami, G.; Jougasaki, M.; Obata, K.; Yasue, H.; et al. Brain natriuretic peptide as a novel cardiac hormone in humans. Evidence for an exquisite dual natriuretic peptide system, atrial natriuretic peptide and brain natriuretic peptide. *J. Clin. Investig.* **1991**, *87*, 1402–1412. [CrossRef] [PubMed]
54. Holmes, S.J.; Espiner, E.A.; Richards, A.M.; Yandle, T.G.; Frampton, C. Renal, endocrine, and hemodynamic effects of human brain natriuretic peptide in normal man. *J. Clin. Endocrinol. Metab.* **1993**, *76*, 91–96. [PubMed]
55. Hunt, P.J.; Richards, A.M.; Espiner, E.A.; Nicholls, M.G.; Yandle, T.G. Bioactivity and metabolism of C-type natriuretic peptide in normal man. *J. Clin. Endocrinol. Metab.* **1994**, *78*, 1428–1435. [PubMed]
56. Kerr, M.A.; Kenny, A.J. The purification and specificity of a neutral endopeptidase from rabbit kidney brush border. *Biochem. J.* **1974**, *137*, 477–488. [CrossRef] [PubMed]
57. Vanneste, Y.; Michel, A.; Dimaline, R.; Najdovski, T.; Deschodt-Lanckman, M. Hydrolysis of alpha-human atrial natriuretic peptide in vitro by human kidney membranes and purified endopeptidase-24.11. Evidence for a novel cleavage site. *Biochem. J.* **1988**, *254*, 531–537. [CrossRef] [PubMed]
58. Yandle, T.G.; Brennan, S.O.; Espiner, E.A.; Nicholls, M.G.; Richards, A.M. Endopeptidase-24.11 in human plasma degrades atrial natriuretic factor (ANF) to ANF(99-105/106-126). *Peptides* **1989**, *10*, 891–894. [CrossRef]
59. Bayes-Genis, A. Neprilysin in Heart Failure: From Oblivion to Center Stage. *JACC Heart Fail.* **2015**, *3*, 637–640. [CrossRef]

60. Hashimoto, Y.; Nakao, K.; Hama, N.; Imura, H.; Mori, S.; Yamaguchi, M.; Yasuhara, M.; Hori, R. Clearance mechanisms of atrial and brain natriuretic peptides in rats. *Pharm. Res.* **1994**, *11*, 60–64. [CrossRef]
61. Volpe, M.; Rubattu, S.; Battistoni, A. ARNi: A Novel Approach to Counteract Cardiovascular Diseases. *Int. J. Mol. Sci.* **2019**, *20*, 2092. [CrossRef] [PubMed]
62. Mogensen, U.M.; Gong, J.; Jhund, P.S.; Shen, L.; Kober, L.; Desai, A.S.; Lefkowitz, M.P.; Packer, M.; Rouleau, J.L.; Solomon, S.D.; et al. Effect of sacubitril/valsartan on recurrent events in the Prospective comparison of ARNI with ACEI to Determine Impact on Global Mortality and morbidity in Heart Failure trial (PARADIGM-HF). *Eur. J. Heart Fail.* **2018**, *20*, 760–768. [CrossRef] [PubMed]
63. Lin, X.; Hanze, J.; Heese, F.; Sodmann, R.; Lang, R.E. Gene expression of natriuretic peptide receptors in myocardial cells. *Circ. Res.* **1995**, *77*, 750–758. [CrossRef] [PubMed]
64. Singh, G.; Kuc, R.E.; Maguire, J.J.; Fidock, M.; Davenport, A.P. Novel snake venom ligand dendroaspis natriuretic peptide is selective for natriuretic peptide receptor-A in human heart: downregulation of natriuretic peptide receptor-A in heart failure. *Circ. Res.* **2006**, *99*, 183–190. [CrossRef] [PubMed]
65. Horio, T.; Nishikimi, T.; Yoshihara, F.; Matsuo, H.; Takishita, S.; Kangawa, K. Inhibitory regulation of hypertrophy by endogenous atrial natriuretic peptide in cultured cardiac myocytes. *Hypertension* **2000**, *35*, 19–24. [CrossRef] [PubMed]
66. Wu, C.F.; Bishopric, N.H.; Pratt, R.E. Atrial natriuretic peptide induces apoptosis in neonatal rat cardiac myocytes. *J. Biol. Chem.* **1997**, *272*, 14860–14866. [CrossRef] [PubMed]
67. Cao, L.; Gardner, D.G. Natriuretic peptides inhibit DNA synthesis in cardiac fibroblasts. *Hypertension* **1995**, *25*, 227–234. [CrossRef] [PubMed]
68. Sun, J.Z.; Chen, S.J.; Majid-Hasan, E.; Oparil, S.; Chen, Y.F. Dietary salt supplementation selectively downregulates NPR-C receptor expression in kidney independently of ANP. *Am. J. Physiol. Renal. Physiol.* **2002**, *282*, F220–F227. [CrossRef] [PubMed]
69. Klinger, J.R.; Warburton, R.R.; Pietras, L.A.; Smithies, O.; Swift, R.; Hill, N.S. Genetic disruption of atrial natriuretic peptide causes pulmonary hypertension in normoxic and hypoxic mice. *Am. J. Physiol.* **1999**, *276*, L868–L874. [CrossRef] [PubMed]
70. Sun, J.Z.; Chen, S.J.; Li, G.; Chen, Y.F. Hypoxia reduces atrial natriuretic peptide clearance receptor gene expression in ANP knockout mice. *Am. J. Physiol. Lung Cell Mol. Physiol.* **2000**, *279*, L511–L519. [CrossRef] [PubMed]
71. Feng, J.A.; Perry, G.; Mori, T.; Hayashi, T.; Oparil, S.; Chen, Y.F. Pressure-independent enhancement of cardiac hypertrophy in atrial natriuretic peptide-deficient mice. *Clin. Exp. Pharmacol. Physiol.* **2003**, *30*, 343–349. [CrossRef] [PubMed]
72. Kasama, S.; Furuya, M.; Toyama, T.; Ichikawa, S.; Kurabayashi, M. Effect of atrial natriuretic peptide on left ventricular remodelling in patients with acute myocardial infarction. *Eur. Heart J.* **2008**, *29*, 1485–1494. [CrossRef] [PubMed]
73. Kinoshita, H.; Kuwahara, K.; Nishida, M.; Jian, Z.; Rong, X.; Kiyonaka, S.; Kuwabara, Y.; Kurose, H.; Inoue, R.; Mori, Y.; et al. Inhibition of TRPC6 channel activity contributes to the antihypertrophic effects of natriuretic peptides-guanylyl cyclase-A signaling in the heart. *Circ. Res.* **2010**, *106*, 1849–1860. [CrossRef] [PubMed]
74. Oliver, P.M.; Fox, J.E.; Kim, R.; Rockman, H.A.; Kim, H.S.; Reddick, R.L.; Pandey, K.N.; Milgram, S.L.; Smithies, O.; Maeda, N. Hypertension, cardiac hypertrophy, and sudden death in mice lacking natriuretic peptide receptor A. *Proc. Natl. Acad. Sci. USA* **1997**, *94*, 14730–14735. [CrossRef] [PubMed]
75. Knowles, J.W.; Esposito, G.; Mao, L.; Hagaman, J.R.; Fox, J.E.; Smithies, O.; Rockman, H.A.; Maeda, N. Pressure-independent enhancement of cardiac hypertrophy in natriuretic peptide receptor A-deficient mice. *J. Clin. Investig.* **2001**, *107*, 975–984. [CrossRef] [PubMed]
76. Kishimoto, I.; Rossi, K.; Garbers, D.L. A genetic model provides evidence that the receptor for atrial natriuretic peptide (guanylyl cyclase-A) inhibits cardiac ventricular myocyte hypertrophy. *Proc. Natl. Acad. Sci. USA* **2001**, *98*, 2703–2706. [CrossRef]
77. Hayashi, D.; Kudoh, S.; Shiojima, I.; Zou, Y.; Harada, K.; Shimoyama, M.; Imai, Y.; Monzen, K.; Yamazaki, T.; Yazaki, Y.; et al. Atrial natriuretic peptide inhibits cardiomyocyte hypertrophy through mitogen-activated protein kinase phosphatase-1. *Biochem. Biophys. Res. Commun.* **2004**, *322*, 310–319. [CrossRef]
78. Nemoto, S.; Sheng, Z.; Lin, A. Opposing effects of Jun kinase and p38 mitogen-activated protein kinases on cardiomyocyte hypertrophy. *Mol. Cell Biol.* **1998**, *18*, 3518–3526. [CrossRef]

79. Calderone, A.; Thaik, C.M.; Takahashi, N.; Chang, D.L.; Colucci, W.S. Nitric oxide, atrial natriuretic peptide, and cyclic GMP inhibit the growth-promoting effects of norepinephrine in cardiac myocytes and fibroblasts. *J. Clin. Investig.* **1998**, *101*, 812–818. [CrossRef]
80. Tokudome, T.; Horio, T.; Kishimoto, I.; Soeki, T.; Mori, K.; Kawano, Y.; Kohno, M.; Garbers, D.L.; Nakao, K.; Kangawa, K. Calcineurin-nuclear factor of activated T cells pathway-dependent cardiac remodeling in mice deficient in guanylyl cyclase A, a receptor for atrial and brain natriuretic peptides. *Circulation* **2005**, *111*, 3095–3104. [CrossRef]
81. Laskowski, A.; Woodman, O.L.; Cao, A.H.; Drummond, G.R.; Marshall, T.; Kaye, D.M.; Ritchie, R.H. Antioxidant actions contribute to the antihypertrophic effects of atrial natriuretic peptide in neonatal rat cardiomyocytes. *Cardiovasc. Res.* **2006**, *72*, 112–123. [CrossRef] [PubMed]
82. Rubattu, S.; Bigatti, G.; Evangelista, A.; Lanzani, C.; Stanzione, R.; Zagato, L.; Manunta, P.; Marchitti, S.; Venturelli, V.; Bianchi, G.; et al. Association of atrial natriuretic peptide and type a natriuretic peptide receptor gene polymorphisms with left ventricular mass in human essential hypertension. *J. Am. Coll. Cardiol.* **2006**, *48*, 499–505. [CrossRef] [PubMed]
83. Nakayama, T.; Soma, M.; Takahashi, Y.; Rehemudula, D.; Kanmatsuse, K.; Furuya, K. Functional deletion mutation of the 5′-flanking region of type A human natriuretic peptide receptor gene and its association with essential hypertension and left ventricular hypertrophy in the Japanese. *Circ. Res.* **2000**, *86*, 841–845. [CrossRef] [PubMed]
84. Bartels, E.D.; Nielsen, J.M.; Bisgaard, L.S.; Goetze, J.P.; Nielsen, L.B. Decreased expression of natriuretic peptides associated with lipid accumulation in cardiac ventricle of obese mice. *Endocrinology* **2010**, *151*, 5218–5225. [CrossRef] [PubMed]
85. Rubattu, S.; Sciarretta, S.; Ciavarella, G.M.; Venturelli, V.; De Paolis, P.; Tocci, G.; De Biase, L.; Ferrucci, A.; Volpe, M. Reduced levels of N-terminal-proatrial natriuretic peptide in hypertensive patients with metabolic syndrome and their relationship with left ventricular mass. *J. Hypertens.* **2007**, *25*, 833–839. [CrossRef] [PubMed]
86. Cuspidi, C.; Meani, S.; Fusi, V.; Severgnini, B.; Valerio, C.; Catini, E.; Leonetti, G.; Magrini, F.; Zanchetti, A. Metabolic syndrome and target organ damage in untreated essential hypertensives. *J. Hypertens.* **2004**, *22*, 1991–1998. [CrossRef] [PubMed]
87. Sangawa, K.; Nakanishi, K.; Ishino, K.; Inoue, M.; Kawada, M.; Sano, S. Atrial natriuretic peptide protects against ischemia-reperfusion injury in the isolated rat heart. *Ann. Thorac. Surg.* **2004**, *77*, 233–237. [CrossRef]
88. Yang, X.M.; Philipp, S.; Downey, J.M.; Cohen, M.V. Atrial natriuretic peptide administered just prior to reperfusion limits infarction in rabbit hearts. *Basic Res. Cardiol.* **2006**, *101*, 311–318. [CrossRef]
89. Okawa, H.; Horimoto, H.; Mieno, S.; Nomura, Y.; Yoshida, M.; Shinjiro, S. Preischemic infusion of alpha-human atrial natriuretic peptide elicits myoprotective effects against ischemia reperfusion in isolated rat hearts. *Mol. Cell. Biochem.* **2003**, *248*, 171–177. [CrossRef]
90. Takata, Y.; Hirayama, Y.; Kiyomi, S.; Ogawa, T.; Iga, K.; Ishii, T.; Nagai, Y.; Ibukiyama, C. The beneficial effects of atrial natriuretic peptide on arrhythmias and myocardial high-energy phosphates after reperfusion. *Cardiovasc. Res.* **1996**, *32*, 286–293. [CrossRef]
91. Wakui, S.; Sezai, A.; Tenderich, G.; Hata, M.; Osaka, S.; Taniguchi, Y.; Koerfer, R.; Minami, K. Experimental investigation of direct myocardial protective effect of atrial natriuretic peptide in cardiac surgery. *J. Thorac. Cardiovasc. Surg.* **2010**, *139*, 918–925. [CrossRef] [PubMed]
92. Suzuki, T.; Saiki, Y.; Horii, A.; Fukushige, S.; Kawamoto, S.; Adachi, O.; Akiyama, M.; Ito, K.; Masaki, N.; Saiki, Y. Atrial natriuretic peptide induces peroxisome proliferator activated receptor gamma during cardiac ischemia-reperfusion in swine heart. *Gen. Thorac Cardiovasc. Surg.* **2017**, *65*, 85–95. [CrossRef] [PubMed]
93. Abdelrahman, M.; Sivarajah, A.; Thiemermann, C. Beneficial effects of PPAR-gamma ligands in ischemia-reperfusion injury, inflammation and shock. *Cardiovasc. Res.* **2005**, *65*, 772–781. [CrossRef] [PubMed]
94. Charan, K.; Goyal, A.; Gupta, J.K.; Yadav, H.N. Role of atrial natriuretic peptide in ischemic preconditioning-induced cardioprotection in the diabetic rat heart. *J. Surg. Res.* **2016**, *201*, 272–278. [CrossRef] [PubMed]
95. Nakanishi, M.; Saito, Y.; Kishimoto, I.; Harada, M.; Kuwahara, K.; Takahashi, N.; Kawakami, R.; Nakagawa, Y.; Tanimoto, K.; Yasuno, S.; et al. Role of natriuretic peptide receptor guanylyl cyclase-A in myocardial infarction evaluated using genetically engineered mice. *Hypertension* **2005**, *46*, 441–447. [CrossRef] [PubMed]

96. Hayashi, M.; Tsutamoto, T.; Wada, A.; Maeda, K.; Mabuchi, N.; Tsutsui, T.; Horie, H.; Ohnishi, M.; Kinoshita, M. Intravenous atrial natriuretic peptide prevents left ventricular remodeling in patients with first anterior acute myocardial infarction. *J. Am. Coll. Cardiol.* **2001**, *37*, 1820–1826. [CrossRef]
97. Kuga, H.; Ogawa, K.; Oida, A.; Taguchi, I.; Nakatsugawa, M.; Hoshi, T.; Sugimura, H.; Abe, S.; Kaneko, N. Administration of atrial natriuretic peptide attenuates reperfusion phenomena and preserves left ventricular regional wall motion after direct coronary angioplasty for acute myocardial infarction. *Circ. J.* **2003**, *67*, 443–448. [CrossRef]
98. Kasama, S.; Toyama, T.; Hatori, T.; Sumino, H.; Kumakura, H.; Takayama, Y.; Ichikawa, S.; Suzuki, T.; Kurabayashi, M. Effects of intravenous atrial natriuretic peptide on cardiac sympathetic nerve activity and left ventricular remodeling in patients with first anterior acute myocardial infarction. *J. Am. Coll. Cardiol.* **2007**, *49*, 667–674. [CrossRef]
99. Kitakaze, M.; Asakura, M.; Kim, J.; Shintani, Y.; Asanuma, H.; Hamasaki, T.; Seguchi, O.; Myoishi, M.; Minamino, T.; Ohara, T.; et al. Human atrial natriuretic peptide and nicorandil as adjuncts to reperfusion treatment for acute myocardial infarction (J-WIND): two randomised trials. *Lancet* **2007**, *370*, 1483–1493. [CrossRef]
100. Brandt, R.R.; Heublein, D.M.; Mattingly, M.T.; Pittelkow, M.R.; Burnett, J.C., Jr. Presence and secretion of atrial natriuretic peptide from cultured human aortic endothelial cells. *Am. J. Physiol.* **1995**, *268*, H921–H925. [CrossRef]
101. Kuhn, M.; Volker, K.; Schwarz, K.; Carbajo-Lozoya, J.; Flogel, U.; Jacoby, C.; Stypmann, J.; van Eickels, M.; Gambaryan, S.; Hartmann, M.; et al. The natriuretic peptide/guanylyl cyclase—A system functions as a stress-responsive regulator of angiogenesis in mice. *J. Clin. Investig.* **2009**, *119*, 2019–2030. [CrossRef] [PubMed]
102. Sciarretta, S.; Marchitti, S.; Bianchi, F.; Moyes, A.; Barbato, E.; Di Castro, S.; Stanzione, R.; Cotugno, M.; Castello, L.; Calvieri, C.; et al. C2238 atrial natriuretic peptide molecular variant is associated with endothelial damage and dysfunction through natriuretic peptide receptor C signaling. *Circ. Res.* **2013**, *112*, 1355–1364. [CrossRef]
103. Barbato, E.; Bartunek, J.; Mangiacapra, F.; Sciarretta, S.; Stanzione, R.; Delrue, L.; Cotugno, M.; Marchitti, S.; Iaccarino, G.; Sirico, G.; et al. Influence of rs5065 atrial natriuretic peptide gene variant on coronary artery disease. *J. Am. Coll. Cardiol.* **2012**, *59*, 1763–1770. [CrossRef] [PubMed]
104. Scarpino, S.; Marchitti, S.; Stanzione, R.; Evangelista, A.; Di Castro, S.; Savoia, C.; Quarta, G.; Sciarretta, S.; Ruco, L.; Volpe, M.; et al. Reactive oxygen species-mediated effects on vascular remodeling induced by human atrial natriuretic peptide T2238C molecular variant in endothelial cells in vitro. *J. Hypertens.* **2009**, *27*, 1804–1813. [CrossRef] [PubMed]
105. Rubattu, S.; Sciarretta, S.; Marchitti, S.; Bianchi, F.; Forte, M.; Volpe, M. The T2238C Human Atrial Natriuretic Peptide Molecular Variant and the Risk of Cardiovascular Diseases. *Int. J. Mol. Sci.* **2018**, *19*, 540. [CrossRef] [PubMed]
106. Kiemer, A.K.; Weber, N.C.; Vollmar, A.M. Induction of IkappaB: atrial natriuretic peptide as a regulator of the NF-kappaB pathway. *Biochem. Biophys. Res. Commun.* **2002**, *295*, 1068–1076. [CrossRef]
107. Kiemer, A.K.; Lehner, M.D.; Hartung, T.; Vollmar, A.M. Inhibition of cyclooxygenase-2 by natriuretic peptides. *Endocrinology* **2002**, *143*, 846–852. [CrossRef]
108. Kiemer, A.K.; Vollmar, A.M. Autocrine regulation of inducible nitric-oxide synthase in macrophages by atrial natriuretic peptide. *J. Biol. Chem.* **1998**, *273*, 13444–13451. [CrossRef]
109. Mtairag, E.M.; Houard, X.; Rais, S.; Pasquier, C.; Oudghiri, M.; Jacob, M.P.; Meilhac, O.; Michel, J.B. Pharmacological potentiation of natriuretic peptide limits polymorphonuclear neutrophil-vascular cell interactions. *Arterioscler. Thromb. Vasc. Biol.* **2002**, *22*, 1824–1831. [CrossRef]
110. Kuhn, M. Endothelial actions of atrial and B-type natriuretic peptides. *Br. J. Pharmacol.* **2012**, *166*, 522–531. [CrossRef]
111. Tamura, N.; Ogawa, Y.; Chusho, H.; Nakamura, K.; Nakao, K.; Suda, M.; Kasahara, M.; Hashimoto, R.; Katsuura, G.; Mukoyama, M.; et al. Cardiac fibrosis in mice lacking brain natriuretic peptide. *Proc. Natl. Acad. Sci. USA* **2000**, *97*, 4239–4244. [CrossRef]

112. Kapoun, A.M.; Liang, F.; O'Young, G.; Damm, D.L.; Quon, D.; White, R.T.; Munson, K.; Lam, A.; Schreiner, G.F.; Protter, A.A. B-type natriuretic peptide exerts broad functional opposition to transforming growth factor-beta in primary human cardiac fibroblasts: fibrosis, myofibroblast conversion, proliferation, and inflammation. *Circ. Res.* **2004**, *94*, 453–461. [CrossRef]
113. Soeki, T.; Kishimoto, I.; Okumura, H.; Tokudome, T.; Horio, T.; Mori, K.; Kangawa, K. C-type natriuretic peptide, a novel antifibrotic and antihypertrophic agent, prevents cardiac remodeling after myocardial infarction. *J. Am. Coll. Cardiol.* **2005**, *45*, 608–616. [CrossRef] [PubMed]
114. Wang, Y.; de Waard, M.C.; Sterner-Kock, A.; Stepan, H.; Schultheiss, H.P.; Duncker, D.J.; Walther, T. Cardiomyocyte-restricted over-expression of C-type natriuretic peptide prevents cardiac hypertrophy induced by myocardial infarction in mice. *Eur. J. Heart Fail.* **2007**, *9*, 548–557. [CrossRef] [PubMed]
115. Langenickel, T.H.; Buttgereit, J.; Pagel-Langenickel, I.; Lindner, M.; Monti, J.; Beuerlein, K.; Al-Saadi, N.; Plehm, R.; Popova, E.; Tank, J.; et al. Cardiac hypertrophy in transgenic rats expressing a dominant-negative mutant of the natriuretic peptide receptor B. *Proc. Natl. Acad. Sci. USA* **2006**, *103*, 4735–4740. [CrossRef] [PubMed]
116. Izumiya, Y.; Araki, S.; Usuku, H.; Rokutanda, T.; Hanatani, S.; Ogawa, H. Chronic C-Type Natriuretic Peptide Infusion Attenuates Angiotensin II-Induced Myocardial Superoxide Production and Cardiac Remodeling. *Int. J. Vasc. Med.* **2012**, *2012*, 246058. [CrossRef] [PubMed]
117. Nakao, K.; Kuwahara, K.; Nishikimi, T.; Nakagawa, Y.; Kinoshita, H.; Minami, T.; Kuwabara, Y.; Yamada, C.; Yamada, Y.; Tokudome, T.; et al. Endothelium-Derived C-Type Natriuretic Peptide Contributes to Blood Pressure Regulation by Maintaining Endothelial Integrity. *Hypertension* **2017**, *69*, 286–296. [CrossRef]
118. Moyes, A.J.; Khambata, R.S.; Villar, I.; Bubb, K.J.; Baliga, R.S.; Lumsden, N.G.; Xiao, F.; Gane, P.J.; Rebstock, A.S.; Worthington, R.J.; et al. Endothelial C-type natriuretic peptide maintains vascular homeostasis. *J. Clin. Investig.* **2014**, *124*, 4039–4051. [CrossRef]
119. Spiranec, K.; Chen, W.; Werner, F.; Nikolaev, V.O.; Naruke, T.; Koch, F.; Werner, A.; Eder-Negrin, P.; Dieguez-Hurtado, R.; Adams, R.H.; et al. Endothelial C-Type Natriuretic Peptide Acts on Pericytes to Regulate Microcirculatory Flow and Blood Pressure. *Circulation* **2018**, *138*, 494–508. [CrossRef]
120. Honing, M.L.; Smits, P.; Morrison, P.J.; Burnett, J.C., Jr.; Rabelink, T.J. C-type natriuretic peptide-induced vasodilation is dependent on hyperpolarization in human forearm resistance vessels. *Hypertension* **2001**, *37*, 1179–1183. [CrossRef]
121. Hutchinson, H.G.; Trindade, P.T.; Cunanan, D.B.; Wu, C.F.; Pratt, R.E. Mechanisms of natriuretic-peptide-induced growth inhibition of vascular smooth muscle cells. *Cardiovasc. Res.* **1997**, *35*, 158–167. [CrossRef]
122. Furuya, M.; Miyazaki, T.; Honbou, N.; Kawashima, K.; Ohno, T.; Tanaka, S.; Kangawa, K.; Matsuo, H. C-type natriuretic peptide inhibits intimal thickening after vascular injury. *Ann. N. Y. Acad. Sci.* **1995**, *748*, 517–523. [CrossRef] [PubMed]
123. Qian, J.Y.; Haruno, A.; Asada, Y.; Nishida, T.; Saito, Y.; Matsuda, T.; Ueno, H. Local expression of C-type natriuretic peptide suppresses inflammation, eliminates shear stress-induced thrombosis, and prevents neointima formation through enhanced nitric oxide production in rabbit injured carotid arteries. *Circ. Res.* **2002**, *91*, 1063–1069. [CrossRef] [PubMed]
124. Cody, R.J.; Atlas, S.A.; Laragh, J.H.; Kubo, S.H.; Covit, A.B.; Ryman, K.S.; Shaknovich, A.; Pondolfino, K.; Clark, M.; Camargo, M.J.; et al. Atrial natriuretic factor in normal subjects and heart failure patients. Plasma levels and renal, hormonal, and hemodynamic responses to peptide infusion. *J. Clin. Investig.* **1986**, *78*, 1362–1374. [CrossRef] [PubMed]
125. Scriven, T.A.; Burnett, J.C., Jr. Effects of synthetic atrial natriuretic peptide on renal function and renin release in acute experimental heart failure. *Circulation* **1985**, *72*, 892–897. [CrossRef] [PubMed]
126. Wambach, G.; Schittenhelm, U.; Bonner, G.; Kaufmann, W. Renal and adrenal resistance against atrial natriuretic peptide in congestive heart failure: effect of angiotensin I-converting-enzyme inhibition. *Cardiology* **1989**, *76*, 418–427. [CrossRef] [PubMed]
127. Riegger, G.A.; Elsner, D.; Kromer, E.P.; Daffner, C.; Forssmann, W.G.; Muders, F.; Pascher, E.W.; Kochsiek, K. Atrial natriuretic peptide in congestive heart failure in the dog: plasma levels, cyclic guanosine monophosphate, ultrastructure of atrial myoendocrine cells, and hemodynamic, hormonal, and renal effects. *Circulation* **1988**, *77*, 398–406. [CrossRef]

128. Clerico, A.; Recchia, F.A.; Passino, C.; Emdin, M. Cardiac endocrine function is an essential component of the homeostatic regulation network: physiological and clinical implications. *Am. J. Physiol. Heart Circ. Physiol.* **2006**, *290*, H17–H29. [CrossRef]
129. Charloux, A.; Piquard, F.; Doutreleau, S.; Brandenberger, G.; Geny, B. Mechanisms of renal hyporesponsiveness to ANP in heart failure. *Eur. J. Clin. Investig.* **2003**, *33*, 769–778. [CrossRef]
130. Goetze, J.P.; Kastrup, J.; Rehfeld, J.F. The paradox of increased natriuretic hormones in congestive heart failure patients: does the endocrine heart also fail in heart failure? *Eur. Heart J.* **2003**, *24*, 1471–1472. [CrossRef]
131. Rubattu, S.; Calvieri, C.; Pagliaro, B.; Volpe, M. Atrial natriuretic peptide and regulation of vascular function in hypertension and heart failure: implications for novel therapeutic strategies. *J. Hypertens.* **2013**, *31*, 1061–1072. [CrossRef] [PubMed]
132. Colucci, W.S.; Elkayam, U.; Horton, D.P.; Abraham, W.T.; Bourge, R.C.; Johnson, A.D.; Wagoner, L.E.; Givertz, M.M.; Liang, C.S.; Neibaur, M.; et al. Intravenous nesiritide, a natriuretic peptide, in the treatment of decompensated congestive heart failure. Nesiritide Study Group. *N. Engl. J. Med.* **2000**, *343*, 246–253. [CrossRef] [PubMed]
133. Publication Committee for the VMAC Investigators (Vasodilatation in the Management of Acute CHF). Intravenous nesiritide vs nitroglycerin for treatment of decompensated congestive heart failure: a randomized controlled trial. *JAMA* **2002**, *287*, 1531–1540. [CrossRef]
134. O'Connor, C.M.; Starling, R.C.; Hernandez, A.F.; Armstrong, P.W.; Dickstein, K.; Hasselblad, V.; Heizer, G.M.; Komajda, M.; Massie, B.M.; McMurray, J.J.; et al. Effect of nesiritide in patients with acute decompensated heart failure. *N. Engl. J. Med.* **2011**, *365*, 32–43. [CrossRef] [PubMed]
135. Sackner-Bernstein, J.D.; Skopicki, H.A.; Aaronson, K.D. Risk of worsening renal function with nesiritide in patients with acutely decompensated heart failure. *Circulation* **2005**, *111*, 1487–1491. [CrossRef] [PubMed]
136. Sackner-Bernstein, J.D.; Kowalski, M.; Fox, M.; Aaronson, K. Short-term risk of death after treatment with nesiritide for decompensated heart failure: a pooled analysis of randomized controlled trials. *JAMA* **2005**, *293*, 1900–1905. [CrossRef] [PubMed]
137. McMurray, J.J.; Packer, M.; Desai, A.S.; Gong, J.; Lefkowitz, M.P.; Rizkala, A.R.; Rouleau, J.L.; Shi, V.C.; Solomon, S.D.; Swedberg, K.; et al. Committees, Angiotensin-neprilysin inhibition versus enalapril in heart failure. *N. Engl. J. Med.* **2014**, *371*, 993–1004. [CrossRef] [PubMed]
138. Packer, M.; McMurray, J.J.; Desai, A.S.; Gong, J.; Lefkowitz, M.P.; Rizkala, A.R.; Rouleau, J.L.; Shi, V.C.; Solomon, S.D.; Swedberg, K.; et al. Coordinators, Angiotensin receptor neprilysin inhibition compared with enalapril on the risk of clinical progression in surviving patients with heart failure. *Circulation* **2015**, *131*, 54–61. [CrossRef] [PubMed]
139. Vodovar, N.; Seronde, M.F.; Laribi, S.; Gayat, E.; Lassus, J.; Januzzi, J.L., Jr.; Boukef, R.; Nouira, S.; Manivet, P.; Samuel, J.L.; et al. Elevated Plasma B-Type Natriuretic Peptide Concentrations Directly Inhibit Circulating Neprilysin Activity in Heart Failure. *JACC Heart Fail.* **2015**, *3*, 629–636. [CrossRef] [PubMed]
140. Januzzi, J.L.; Butler, J.; Fombu, E.; Maisel, A.; McCague, K.; Pina, I.L.; Prescott, M.F.; Riebman, J.B.; Solomon, S. Rationale and methods of the Prospective Study of Biomarkers, Symptom Improvement, and Ventricular Remodeling During Sacubitril/Valsartan Therapy for Heart Failure (PROVE-HF). *Am. Heart J.* **2018**, *199*, 130–136. [CrossRef]
141. Ishii, M.; Kaikita, K.; Sato, K.; Sueta, D.; Fujisue, K.; Arima, Y.; Oimatsu, Y.; Mitsuse, T.; Onoue, Y.; Araki, S.; et al. Cardioprotective Effects of LCZ696 (Sacubitril/Valsartan) After Experimental Acute Myocardial Infarction. *JACC Basic Transl. Sci.* **2017**, *2*, 655–668. [CrossRef]
142. Torrado, J.; Cain, C.; Mauro, A.G.; Romeo, F.; Ockaili, R.; Chau, V.Q.; Nestler, J.A.; Devarakonda, T.; Ghosh, S.; Das, A.; et al. Sacubitril/Valsartan Averts Adverse Post-Infarction Ventricular Remodeling and Preserves Systolic Function in Rabbits. *J. Am. Coll. Cardiol.* **2018**, *72*, 2342–2356. [CrossRef] [PubMed]
143. Pfau, D.; Thorn, S.L.; Zhang, J.; Mikush, N.; Renaud, J.M.; Klein, R.; deKemp, R.A.; Wu, X.; Hu, X.; Sinusas, A.J.; et al. Angiotensin Receptor Neprilysin Inhibitor Attenuates Myocardial Remodeling and Improves Infarct Perfusion in Experimental Heart Failure. *Sci. Rep.* **2019**, *9*, 5791. [CrossRef] [PubMed]
144. Rouleau, J.L.; Pfeffer, M.A.; Stewart, D.J.; Isaac, D.; Sestier, F.; Kerut, E.K.; Porter, C.B.; Proulx, G.; Qian, C.; Block, A.J. Comparison of vasopeptidase inhibitor, omapatrilat, and lisinopril on exercise tolerance and morbidity in patients with heart failure: IMPRESS randomised trial. *Lancet* **2000**, *356*, 615–620. [CrossRef]

145. Kostis, J.B.; Packer, M.; Black, H.R.; Schmieder, R.; Henry, D.; Levy, E. Omapatrilat and enalapril in patients with hypertension: The Omapatrilat Cardiovascular Treatment vs. Enalapril (OCTAVE) trial. *Am. J. Hypertens.* **2004**, *17*, 103–111. [CrossRef] [PubMed]
146. Packer, M.; Califf, R.M.; Konstam, M.A.; Krum, H.; McMurray, J.J.; Rouleau, J.L.; Swedberg, K. Comparison of omapatrilat and enalapril in patients with chronic heart failure: The Omapatrilat Versus Enalapril Randomized Trial of Utility in Reducing Events (OVERTURE). *Circulation* **2002**, *106*, 920–926. [CrossRef]

© 2019 by the authors. Licensee MDPI, Basel, Switzerland. This article is an open access article distributed under the terms and conditions of the Creative Commons Attribution (CC BY) license (http://creativecommons.org/licenses/by/4.0/).

Review

Genetic Ablation and Guanylyl Cyclase/Natriuretic Peptide Receptor-A: Impact on the Pathophysiology of Cardiovascular Dysfunction

Kailash N. Pandey

Department of Physiology, Tulane University Health Sciences Center, School of Medicine, New Orleans, LA 70112, USA; kpandey@tulane.edu; Tel.: +1-(504)-988-1628; Fax: +1-(504)-988-2675

Received: 25 July 2019; Accepted: 10 August 2019; Published: 14 August 2019

Abstract: Mice bearing targeted gene mutations that affect the functions of natriuretic peptides (NPs) and natriuretic peptide receptors (NPRs) have contributed important information on the pathogenesis of hypertension, kidney disease, and cardiovascular dysfunction. Studies of mice having both complete gene disruption and tissue-specific gene ablation have contributed to our understanding of hypertension and cardiovascular disorders. These phenomena are consistent with an oligogenic inheritance in which interactions among a few alleles may account for genetic susceptibility to hypertension, renal insufficiency, and congestive heart failure. In addition to gene knockouts conferring increased risks of hypertension, kidney disorders, and cardiovascular dysfunction, studies of gene duplications have identified mutations that protect against high blood pressure and cardiovascular events, thus generating the notion that certain alleles can confer resistance to hypertension and heart disease. This review focuses on the intriguing phenotypes of *Npr1* gene disruption and gene duplication in mice, with emphasis on hypertension and cardiovascular events using mouse models carrying *Npr1* gene knockout and/or gene duplication. It also describes how *Npr1* gene targeting in mice has contributed to our knowledge of the roles of NPs and NPRs in dose-dependently regulating hypertension and cardiovascular events.

Keywords: atrial natriuretic peptide; guanylyl cyclase/natriuretic peptide receptor-A; gene-knockout; gene-duplication; hypertension; congestive heart failure

1. Introduction

Almost 40 years ago, de Bold and coworkers established that the extracts of atria contain natriuretic, diuretic, and vasorelaxant activity [1]. This pioneering discovery led them to identify and characterize atrial natriuretic peptide (ANP) in the heart [1,2]. Natriuretic peptides (NPs) exhibit critical functions in the control of renal, cardiovascular, endocrine, neural, and skeletal homeostasis [2–9]. The natriuretic, diuretic, and vasorelaxant effects of ANP, the first member of the NP hormone family to be discovered, are largely directed at lowering blood pressure (BP) and cardiovascular homeostasis [5,6,10]. Subsequently, two other NPs, brain natriuretic peptide (BNP) and C-type natriuretic peptide (CNP), were discovered. Each of these peptide hormones is encoded by a separate gene [6,11–13]. The three natriuretic peptides (ANP, BNP, and CNP) have highly homologous structure, but distinct sites of synthesis and secretion. Both ANP and BNP are predominantly synthesized in the heart. ANP concentrations range from 50- to 100-fold higher than BNP [14]. CNP, which is largely synthesized in endothelial and neuronal cells is not much released into the circulation [15]. All three natriuretic peptide hormones have highly homologous structure, bind to specific cell-surface cognate receptors, and exert distinct biological functions [4,6,16].

Endogenous peptide hormones, including ANP, BNP, CNP, and urodilatin (Uro), are considered to have integral roles in hypertension and cardiovascular regulation [2,4,6,7,17]. ANP and BNP not only regulate BP, but also maintain antagonistic action in response to the renin–angiotensin–aldosterone system (RAAS) [4,6,17–19]. They also have effects on endothelial cell function, cartilage growth, immunity, and mitochondrial biogenesis [7,10,20–23]. ANP and BNP are used in hospitals to diagnose the etiologies of shortness of breath, cardiovascular dysfunction, and congestive heart failure in patients with emergency conditions [24]. The discovery of structurally related natriuretic peptides indicated that the physiological control of BP and cardiovascular homeostasis is complex and multifactorial.

Three subtypes of NP receptors have been identified and characterized: natriuretic peptide receptor-A (NPRA), receptor-B (NPRB), and receptor-C (NPRC). NPRA and NPRB both contain intrinsic cytosolic guanylyl cyclase (GC) catalytic domains, produce intracellular second messenger cGMP. These receptors are also referred to, respectively, as guanylyl cyclase-A (GC-A) and guanylyl cyclase-B (GC-B) [3,6,16,25,26]. Thus, both NPRA and NPRB have also been respectively designated as guanylyl cyclase/natriuretic peptide receptor-A (GC-A/NPRA) and guanylyl cyclase/natriuretic peptide receptor-B (GC-B/NPRB). GC-A/NPRA is a principal locus involved in the regulatory action of ANP and BNP [6,10,27,28].

Biochemical, immunohistochemical, and molecular studies have suggested that ANP, BNP, CNP and their three distinct receptor subtypes (NPRA, NPRB, and NPRC) have widespread tissue and cellular distributions, indicating the occurrence of pleotropic actions at both systemic and local levels (Figure 1). Both ANP and BNP activate NPRA, which produces second-messenger cGMP in response to hormone binding. CNP activates NPRB, which also generates cGMP, but all three NPs indiscriminately bind to NPRC, which lacks a GC domain and does not produce cGMP [29–31]. The ligand–receptor complexes of NPRA, NPRB, and NPRC are rapidly internalized in the intracellular compartments and a majority of ligand-bound receptors are degraded in the lysosomes; however, a small population of receptors recycle back to the plasma membrane [6,32–36]. The combined cellular, biochemical, molecular, and pharmacological aspects of NPs and their prototype cognate receptors have demonstrated hallmark physiological and pathophysiological effects, including renal, cardiovascular, neuronal, and immunological effects that are important in health and disease [6,9,10,23,37–39].

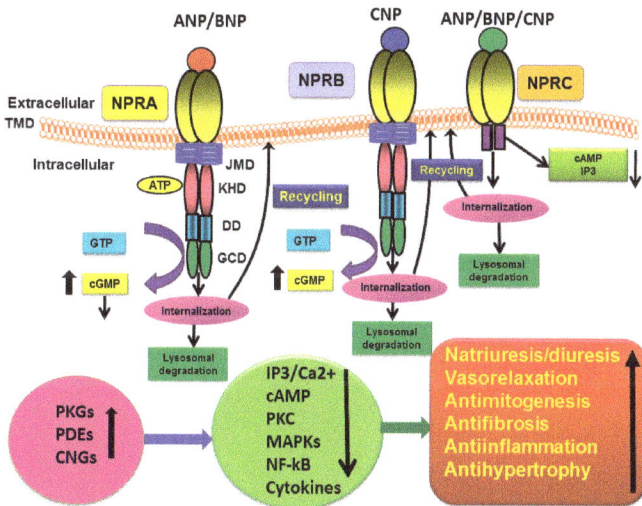

Figure 1. Diagrammatic representation of ligand-stimulation of ligand–receptor complex activation and the physiological functions of natriuretic peptide receptor-A (NPRA), receptor-B (NPRB), and receptor-C (NPRC): Ligand-binding activates NPRA and NPRB, which leads to enhanced production of

the intracellular second messenger cGMP, as well as with stimulation and activation of ligand-dependent cellular and physiological responsiveness. CNP activates NPRB and all three NPs activate NPRC. The bound-ligand receptor complexes of NPRA, NPRB, and NPRC are rapidly internalized and a large population of receptors is degraded in lysosomes. However, a small population of ligand-bound receptors is recycled back to the plasma membranes. ANP, atrial natriuretic peptide; BNP, brain natriuretic peptide; CNP, C-type natriuretic peptide; LBD, ligand binding domain; TMD, transmembrane domain; JMD, juxtamembrane domain; KHD, kinase like homology domain; DD, dimerization domain; GCD, guanylyl cyclase catalytic domain; IP$_3$, inositol trisphosphate; PKG, cGMP-dependent protein kinase or protein kinase G; PDE, phosphodiesterase; CNG, gated-ion channel; MAPKs, mitogen-activated protein kinase; NF-κB, nuclear factor kappa B.

Gaining insight into the intricacies of the ANP-BNP/NPRA/cGMP signaling system is of pivotal importance for understanding both receptor biology and the pathology of disease conditions arising from abnormal hormone–receptor interplay. It has been demonstrated that binding of ANP and BNP to the extracellular ligand-binding domain of NPRA causes a conformational change, thereby transmitting the signal to the intracellular carboxyl-terminal GC catalytic region, which produces second-messenger cGMP from substrate GTP in target cells and tissues [16,25,30]. Earlier studies focused on elucidating the biochemical and molecular nature of the function of GC-A/NPRA [10,38–40]. Both cultured cells in vitro and gene-targeted (gene-knockout and gene-duplication) mouse models in vivo have been used to gain understanding of the NPs and their receptors to delineate their normal and abnormal control of pathophysiological processes in hypertension and cardiovascular disease states. Although there has been a strong focus on the functional roles of ANP, BNP, and NPRA in cardiovascular, endocrine, and renal homeostasis, in-depth studies are still needed to identify their potential molecular targets in cardiovascular diseases states. It is expected that ongoing and future studies on natriuretic peptides and their receptors will provide with new therapeutic targets and novel loci for the control and treatment of hypertension, cardiovascular events, and neurological dysfunction.

2. Historical Perspectives and Background

An increase in atrial stretch triggers the release of atrial natriuretic peptide (ANP) and brain natriuretic peptide (BNP), which largely control homeostasis of extracellular fluid, blood volume, and BP, and cardiovascular homeostasis under both normal and pathophysiological conditions [1,2,4,6,7,17,41–44]. BNP exhibits a pattern of action similar to that of ANP, but there are differences between ANP and BNP with regard to pro-hormone processing, storage, and release [17,45–49]. Although circulating levels of both ANP and BNP are elevated in patients with cardiovascular and renal disease, the infusion of BNP in patients with congestive heart failure has shown more beneficial effects [50,51]. NPRA is considered to be a primary ANP- and BNP-signaling molecule because major cellular and physiological responsiveness of these peptide hormones is mimicked by cGMP analogs [4,6,14,43,52].

Efforts to design chimeric natriuretic peptides has led to the synthesis of biologically active peptide molecules, which represent single-chemical entities that combine the structural and functional properties of two different natriuretic peptides [53]. These chimeric natriuretic peptides, which exhibit the actions of two different natriuretic peptides, often reduce adverse biological effects. One such chimeric peptide molecule, which is derived from CNP and *Dendroaspis* natriuretic peptide (DNP) is known as CD-NP; it is produced by the fusion of 22-amino acid residues of CNP with the 15-amino-acid residues of C-terminus of DNP [53]. The chimeric hormone CD-NP exhibits vasorelaxant properties, effectively reduces cardiac volume overload, and exerts renal and cardiovascular protective effects. At the same time, the 15-amino-acid C-terminus of DNP is extremely resistant to neutral endopeptidase (NEP), making CD-NP a more stable and desirable peptide molecule than are naturally occurring natriuretic peptides hormones [53,54].

Experimental evidence suggests that the ANP-BNP/NPRA system has an important role in renal function, which is directed toward lowering BP to reduce the heart load [9,10,17,38,55]. Determining the physiological role of ANP by studies using large infusions of hormone has been difficult because of the interdependent nature of many organ-system responses and the counter-regulatory effect that is often evoked. Earlier, several approaches were used in attempts to define the physiological responses of ANP. For example, cardiac appendectomy was used to prevent ANP release. However, the missing normal cardiac function resulted in a lack of the physiological reflexes normally induced by atria [56]. Another approach used the monoclonal antibody against circulating ANP, but problems occurred because of the nonspecific effects of the antigen–antibody complexes. One classical approach was to use receptor antagonists specifically to inhibit the signaling pathway of NPRA and block cGMP production. Although two compounds, A71915 and HS-142-1, were shown to diminish the effect of ANP, neither completely inhibited NPRA [6,17,57–60].

The advent of molecular biology opened the way for genetic analysis, which is a faster and more accurate way to evaluate physiological and pathophysiological functions. Genetic studies of live mice have been accomplished by gene targeting (gene-knockout and gene-duplication), in which the desired genetic changes are specially made in mouse germ lines. This allows examination of the effect of genes of interest, such as those controlling hypertension and cardiovascular dysfunction [61]. Gene titration (gene-targeting) in mice is a novel method for use in investigations that have either ablated (gene-knockout/gene-disruption) or increased the number of gene copies (gene-duplication) at the normal chromosome location [62]. Gene targeting, including gene-ablation and gene-duplication, can lead to the production of offspring having zero-copy (0/0; −/−), one-copy (1/0; +/−), two-copies (1/1; +/+), three-copies (2/1; ++/+), or four-copies (2/2; ++/++) of the target gene of interest. Genetically modified animals have provided excellent models for the study of gene-dose-dependent physiological responses in vivo [63].

Studies with *Npr1* (encoding NPRA) gene-knockout mice have demonstrated that a deficiency in NPRA increases BP by 35–45 mmHg and causes renal, vascular, and cardiac dysfunction that leads to hypertensive heart disease in mice, much like that observed in untreated hypertensive patients [64–69]. On the other hand, increased expression (gene-duplication) of NPRA in mice significantly reduces BP and increases second-messenger cGMP corresponding to the level of *Npr1* gene copy number in both male and female mice [70–77].

3. Ablation of *Nppa* and *Nppb* Triggers Hypertension and Cardiovascular Dysfunction

Gene-targeting strategies in mice have provided novel approaches to the study of physiological responses corresponding to gene dosages in vivo [10,38,78,79]. Experimental evidence has led to the notion that the ANP and BNP system is important in regulating BP by its direct vasodilatory effect, as well as by natriuretic and diuretic responses and the antagonistic actions of the RAAS [4,6,9,17,52]. Studies using gene disruption of *Nppa* (coding pro-ANP) and *Nppb* (coding for pro-BNP) have shown that ANP- and BNP-deficient genetic strains of mice exhibit a defect in ANP and BNP synthesis that can cause hypertension and cardiovascular dysfunction in homozygous null mutant mice with no circulating or cardiac ANP and BNP, respectively [80,81]. Therefore, genetic defects that reduce the activity of the NPs system can be considered important candidates as contributors to essential hypertension and cardiovascular dysfunction. Previous findings showing elevated BP in genetic mouse models with loss of ANP or BNP expression and circulating hormones have provided strong support for a physiological and pathophysiological role of the NPs hormone system in regulating BP and cardiovascular homeostasis [80–82].

3.1. Gene Ablation of Nppa Increases Blood Pressure

Gene-targeting studies of *Nppa* and *Nppb* in mice carrying gene-knockout have presented strong evidence of the physiological roles of ANP and BNP hormones and their signaling system in hypertension and cardiovascular dysfunction [80–82]. In *Nppa* homozygous null mutant ($Nppa^{-/-}$) mice on standard- or intermediate-salt diets, BP was elevated by 8–12 mmHg. Heterozygous ($Nppa^{+/-}$) mice on a standard-salt diet showed a normal amount of circulating ANP and normal BP. On the other hand, heterozygous $Nppa^{+/-}$ mice on a high-salt diet became hypertensive; their BP was elevated by 20–28 mmHg [80]. These studies demonstrated that a mutation in the *Nppa* allele could lead to salt-sensitive hypertension even though the plasma ANP level was not significantly reduced. Thus, genetically reduced production of ANP could cause salt-sensitive hypertension. Transgenic mice overexpressing ANP developed sustained hypotension. Also, their mean arterial pressure (MAP) was 25–30 mmHg lower than that in nontransgenic control animals [82,83]. Overexpression of ANP in hypertensive mice significantly lowered systolic BP; similarly the somatic delivery of *Nppa* in SHR exhibited a sustained reduction in BP, suggesting that ANP gene therapy could be used to treat hypertensive humans [84,85].

A significant positive correlation between plasma ANP levels and pulmonary vascular resistance in normal individuals and patients with heart and lung disease has also been found [86,87]. Moreover, it has been demonstrated that after chronic hypoxic exposure transgenic mice overexpressing ANP, as compared to nontransgenic control mice, develop very low right ventricular hypertension and vascular remodeling [88]. As compared to control animals, mice lacking functional *Nppa* ($Nppa^{-/-}$) developed more severe pulmonary hypertension and right ventricular hypertrophy with chronic hypoxia [89–92]. These early findings supported the notion that *Nppa* has a critical role in modulating hypertension and right ventricular hypertrophy in hypoxic conditions.

ANP transgene in mice has been shown to be capable of maintaining sodium balance in mice on a very low-salt diet without evidence of salt depletion, suggesting an independent action of ANP on BP and renal hemodynamic functions [82,93]. A similar conclusion was advanced based on findings in *Nppa* gene-knockout mice, in which relative hypertension was observed after mice had been on a high-salt diet for one week [80]. Further studies indicated that the lack of ANP action in *Nppa* gene-knockout mice might be due to the inability of ANP to properly regulate plasma renin and angiotensin II (Ang II) levels, which result in the development of an additional component of salt-sensitive hypertension [94]. A deficiency in endogenous ANP might play a regulatory role in the hypertensive disease state; however, the genetic basis of salt-sensitive variants of hypertension still remain controversial. It is possible that sensitization of arterial BP to dietary salt may develop as a consequence of defective functional alterations in both salt- and BP-regulating mechanisms. As a result, the genetic deficiency in ANP synthesis might play a critical role in the regulation of BP homeostasis. Monoclonal antibodies against ANP were often used to block the endogenous action of ANP in spontaneously hypertensive rats (SHR), which resulted in accelerated development of hypertension in these animals compared with Wistar Kyoto (WKY) rats [95].

3.2. Role of Nppa/Nppb in Cardiac Remodeling and Dysfunction

ANP and BNP exert cardioprotective effects not only as circulating endocrine hormones, but also as local autocrine and paracrine factors. Plasma levels of both ANP and BNP are markedly elevated under the pathophysiological conditions of cardiac dysfunction, fibrosis, hypertrophy, pulmonary embolism, and congestive heart failure [66,67,96–101]. The expression of both ANP and BNP is greatly increased in proportion to the severity of the pathophysiology of cardiac remolding and disorders in the experimental animal models and humans [7,97,100,102,103]. Levels of ANP and BNP are greatly increased in the cardiac tissues and plasma of patients with hypertensive heart disease and congestive heart failure. In humans, high plasma levels of ANP and BNP tend to predict congestive heart failure and both of these peptide hormones appear to reduce the pre- and after-load of the heart [96,97,102,104]. In patients with severe congestive heart failure, both ANP and BNP levels increase, but BNP levels rises 25-fold to 50-fold higher than do ANP levels [9,105]. Both *Nppa* and *Nppb* genes are overexpressed in hypertrophied hearts, suggesting that both the autocrine and/or paracrine effects of ANP and BNP predominate and endogenously protect against the maladaptive factors of pathological cardiac hypertrophy and dysfunction [10,38,66,67,73,77,96,100,106,107].

Ventricular expression of *Nppa* and *Nppb* is more closely associated with local cardiac hypertrophy and fibrosis [66,67]. Since the half-life of plasma BNP is longer than that of ANP, diagnostic evaluations of NPs have predicted and favored BNP as a critical indicator of cardiovascular dysfunction and congestive heart failure in emergency-room patients with chest pain [97]. However, NT-proBNP also seems to be a stronger indicator of the risk of congestive heart failure and cardiovascular disorders [108–111]. The expression levels of *Nppa* and *Nppb* are greatly stimulated in hypertrophied and failing hearts [66,106,112]. Interestingly, mutation in the *Nppa* promoter has been shown to trigger cardiac hypertrophy in WKY and Wistar Kyoto-derived hypertensive (WKYH) rats [113].

Earlier studies suggested that expression of *Nppa* is one of the critical responders to different hypertrophic stimuli in the heart [114]. Thus, the induction of *Nppa* gene expression can be considered to be a pathognomonic indicator of cardiac hypertrophy that seems to be conserved across species [115]. Similarly, *Nppb* gene expression and BNP levels are most robustly increased in patients with cardiac hypertrophy [116,117]. Increased cardiac mass and hypertrophic growth are characterized by increased synthesis of contractile and matrix proteins and reactivation of embryonic genes [115,118]. It is thought that the function of ANP and BNP is to reduce load-specific changes in the heart by increasing vasodilatory responses and enhancing natriuretic and diuretic effects, leading to decreased body fluid and blood volume [4,7,9,17]. It has been found that ANP transgene overexpression in mice results in decreased heart weight and prevents hypertrophic responses [88]. Mice lacking *Nppb* did not develop cardiac hypertrophy or increased BP, but did show an anti-fibrotic role for BNP [81]. It is thought that the direct effect of ANP and BNP can be mimicked by the use of agents such as 8-Br-cGMP or zaprinast, which increase intracellular second-messenger cGMP levels and have anti-hypertrophic and apoptotic effects in the heart [38,55,119–122].

4. Genetic Disruption of *Npr1* and Pathophysiology of Hypertension and Cardiovascular Events

Much of the early work on NPRA was directed at determining the biochemical character and cellular signaling of the receptor molecule; later studies were aimed at delineating the structure–function relationship and physiological and pathophysiological significance of NPRA in regulating hypertension and cardiovascular events. Although the early studies provided a wealth of information about the biochemical and molecular properties of NPRA, the mechanisms regulating the physiological functions of the ANP-BNP/NPRA system in vivo remained less understood. Development and analyses of gene-targeted (gene-knockout and gene-duplication) mutant mouse models of *Npr1* allowed a more direct assessment of the physiological function of *Npr1* in intact animals in vivo (Table 1). Disruption of *Npr1* encoding NPRA increases BP and causes hypertensive heart disease in null mutant mice; this heart disease is similar to that seen in patients with untreated hypertensive heart disease [6,9,71,123]. In contrast, increased expression of *Npr1* in gene-duplicated mice significantly reduces BP and cardiovascular events and increases GC activity and intracellular second-messenger cGMP, which corresponds to *Npr1* gene-copy numbers in vivo [64,71–74,77]. Thus, complete absence of NPRA protein in mice causes hypertension and leads to cardiac hypertrophy; also, particularly in males, the absence of NPRA protein leads to lethal vascular complications and aortic dissection [64,66,100,106]. The resultant *Npr1* gene-knockout phenotypes in mice reflect the critical roles of NPRA in physiological and pathophysiological processes occurring in hypertension and cardiovascular disorders.

Table 1. The gene-knockout disease-specific phenotypes of mice with gene-disruption of natriuretic peptides and their cognate receptors.

Gene Nomenclature	Peptide/Protein Nomenclature	Gene-Knockout Phenotype in Mouse
Nppa	ANP	Volume overload, high blood pressure and hypertension, salt sensitivity, fibrosis, and cardiac disorders [80,82–84]
Nppb	BNP	Hypertension, sodium excretion, vascular complication, and fibrosis [81,124]
Nppc	CNP	Dwarfism, reduced bone growth, and impaired endochondral ossification [125–127]
Npr1	NPRA	High blood pressure and hypertension, salt sensitivity, volume overload, and cardiac hypertrophy, fibrosis, and inflammation [64–66,73,77,97,100,128]
Npr2	NPRB	Dwarfism, seizures, female sterility, and decreased adiposity [129,130]
Npr3	NPRC	Bone deformation and long bone overgrowth [62]

Nomenclature of gene and peptide or protein with gene-knockout phenotype. The gene nomenclature is indicated in italic and peptide or protein is indicated in capital letters. *Nppa*, coding for pro-ANP; *Nppb*, coding for pro-BNP; *Nppc*, coding for pro-CNP; ANP, atrial natriuretic peptide; BNP, brain natriuretic peptide; CNP, C-Type natriuretic peptide; *Npr1* (coding for NPRA); *Npr2*, (coding for NPRB); *Npr3* (coding for NPRC). NPRA, natriuretic peptide receptor-A; NPRB, natriuretic peptide receptor-B; and NPRC, natriuretic peptide receptor-C.

4.1. Effect of Npr1 Gene Ablation on Hypertension

The ANP/NPRA system plays important roles in the pathophysiology of hypertension and cardiovascular regulation. However, the precise mechanism of action of ANP/NPRA is still not well understood. Genetic studies of *Npr1* gene-deficient mice have provided evidence that NPRA deficiency increases BP in mutant mice as compared with wild-type animals [18,64,65,68,101,128]. Moreover, ablation of *Npr1* has indicated that the BP of homozygous mutant mice (0-copy) remains elevated and largely unchanged in response to either minimal- or high-salt diet, suggesting that mutations in *Npr1* may also explain salt-resistant form of hypertension [19,71,101,128]. Increase in circulating ANP/BNP is an important marker of cardiac hypertrophy and heart failure in both animal models and humans. Thus, increases in ANP and BNP are considered one of the most robust responses to hypertrophic factors in the heart.

To examine whether the heart releases other natriuretic substances in response to volume expansion, *Npr1* gene-disrupted mice were volume-expanded by the infusion of an iso-oncotic solution [131]. In these mice, the release of ANP from the heart was shown by marked elevation in plasma ANP levels. Wild-type control mice responded to the volume expansion with diuresis and natriuresis; however, gene-disrupted mutant mice showed diminished diuresis and natriuresis in response to saline infusions [38,131,132]. Although early studies raised the question of whether or not saline infusion would contribute to hemodilution, it was possible that an exaggerated response might have been produced in saline-infused mice. In our studies, the administration of pure blood to anesthetized *Npr1* homozygous null mutant (0-copy), wild-type (2-copy), and gene-duplicated (4-copy) mice to produce intravascular volume expansion was not accompanied by hemodilution [65]. There was a moderate increase in plasma protein and hematocrit in both wild-type controls and *Npr1* null mutant mice, indicating that no hemodilution occurred with pure blood volume expansion in these animals. In wild-type control mice, urine flow and sodium excretion were associated with rises in glomerular filtration rate (GFR), but the MAP did not increase. In contrast, *Npr1* 0-copy mutant mice exhibited only a small change in urine flow and sodium excretion, in spite of the greater increases in MAP [65]. These results established that NPRA is essential for natriuretic and diuretic action in kidneys, mediating renal hemodynamic changes in response to acute pure blood volume expansion.

The ablation of *Npr1* increases BP by 35–45 mmHg in $Npr1^{-/-}$ (0-copy) mutant mice as compared with $Npr1^{+/+}$ (2-copy) wild-type mice [18,64,66,68]. However, increased copy number of *Npr1* in gene-duplicated $Npr1^{++/+}$ and $Npr1^{++/++}$ (3-copy and 4-copy) mice significantly reduced BP, corresponding to the increasing number of *Npr1* gene copies [19,65,71,73,77]. Our early studies examined the mechanisms mediating the responsiveness of varying numbers of *Npr1* gene copies by determining urine flow, sodium excretion, renal plasma flow (RPF), GFR, and BP in *Npr1* 0-copy, 2-copy, and 4-copy mice in a *Npr1* gene-dose-dependent manner [19,65,69,73,74,77]. Volume expansion with whole blood infusion significantly increased MAP in *Npr1* gene-disrupted (0-copy) mice compared with wild-type (2-copy) and gene-duplicated (4-copy) mice [65]. In 0-copy mice, the GFR was lower by almost 35% and higher by almost 45% in 4-copy mice than in 2-copy control mice. Similarly, RPF was lower by 30% in 0-copy and higher by almost 60% in gene-duplicated 4-copy mice as compared with 2-copy control animals. The 0-copy mice retained significantly higher levels of sodium and water than did other mice; however, gene-duplicated 4-copy mice showed drastically lower levels of both sodium and water than did 2-copy mice. Significant increases in plasma creatinine concentrations and urinary albumin excretion, together with reduced creatinine clearance rates, suggested the onset of renal insufficiency in *Npr1* 0-copy mutant mice. These results demonstrated that the ANP/NPRA/cGMP signaling axis is predominantly responsible for mediating the renal hemodynamic and sodium excretory in responses to intravascular blood volume expansion. Our early studies showed that ablation of *Npr1* in mice provokes kidney fibrosis, remodeling, and significant expression of pro-inflammatory and pro-fibrotic cytokines, as well as several genes participating in the nuclear factor kappa B (NF-κB) pathway [68,69]. The treatment of *Npr1* 0-copy mice with NF-κB inhibitors significantly reversed the renal hypertrophic and fibrotic responses in *Npr1* 0-copy animals [69].

4.2. Effect of Npr1 Gene Disruption on the Pathophysiology of Cardiac Dysfunction

Gene-disrupted *Npr1* mutant mice lacking NPRA showed marked cardiac hypertrophy and chamber dilation disproportionate to their increased BP [64,66,73]. Echocardiographic analysis of mice showed a compensated state of systemic hypertension in which cardiac hypertrophy and chamber dilation were evident, but there was no reduction in ventricular performance. Overall, the heart weight (HW) and echocardiographic data of *Npr1* gene-knockout mutant mice were accompanied by marked cardiac hypertrophy and ventricular enlargement [66,67,73,77]. Extensive cardiac hypertrophy in human seems to be associated with local ischemia that leads to the death of myocytes and cardiac fibrosis. Histological examination and morphometric comparisons has shown concentric hypertrophy with an increase in myocyte cross-sectional area in 0-copy null mutant male mice as compared with wild-type 2-copy animals [64,66,73]. Perivascular fibrosis has often been seen in male null mutant 0-copy mice; however, the extent of this pathology was somewhat variable [64,67]. The degree of cardiac hypertrophy and fibrosis was lower in female 0-copy mutant mice than in males [64,101]. Cardiac output and basal stroke calculations from echocardiography and hemodynamic data analysis suggested that the hearts of *Npr1* 0-copy mutant mice seem to work at least double as hard as those of 2-copy mice.

The signaling mechanisms that mediate the development of cardiac hypertrophy in vivo still are not fully understood. To examine whether the ANP-BNP/NPRA pathway directly modulates hypertrophic responses in the heart, both *Npr1* gene-knockout and wild-type mice were subjected to pressure overload (60 mmHg) induced by transverse aortic constriction (TAC) method [106]. In 0-copy mutant mice, TAC induced left ventricular weight/body weight (LV/BW) ratio, left ventricular end diastolic dimension (LVEDD) and significant reduction in cardiac function as compared with 2-copy mice [106]. These authors also observed that chronic treatment of *Npr1* 0-copy mice with enalapril, which is an angiotensin-converting enzyme (ACE) inhibitor, furosemide, which is a diuretic, or losartan, an angiotensin II type 1 receptor (AT1R) blocker, effectively reduced BP to normal levels, although their HW/BW ratios were not decreased. We observed that ACE and AT1R are upregulated in *Npr1* 0-copy mice as compared with 2-copy wild-type mice [67]. These findings suggested that the ANP-BNP/NPRA signaling mechanism exerts direct antihypertrophic effects in the heart, independent of the BP control. The deletion mutations in *Npr1* gene have been suggested to reduce the receptor activity of NPRA and were considered as a potential genetic factor for hypertension and cardiac hypertrophy in humans [133].

4.3. Effect of Npr1 Ablation on the Renin-Angiotensin System and Cardiovascular Disorders

There is evidence that the chronic hypotensive effect of ANP/NPRA is partly mediated by suppressing RAAS and that ANP/NPRA opposes ANG II-mediated vascular and renal effects [4] Pandey 2008; Pandey 2018. In the absence of the counter-regulatory effect of ANP-BNP/NPRA signaling, sensitization of BP might result, at least in part, from failure to overcome the effects of RAAS. Early findings led to the notion that ANP plays an important role in regulating renal function by its direct vasodilatory effect and natriuretic response, as well as the antagonistic action of RAAS [6,9,17,19]. Our previous studies showed that renin and Ang II levels were significantly elevated in newborn male *Npr1* gene-knockout mice, while both renal and circulating renin levels were drastically reduced in adult male knockout mice [18]. However, the adrenal renin contents in adult male mice remained significantly elevated, suggesting that inhibition of renin and Ang II levels is a compensatory response to increased BP in adult homozygous null mutant mice. Thus, NPRA may differentially regulate adrenal versus renal renin and Ang II levels, resulting in increased absorption of salt and water through the kidneys, which is critical for the development of hypertension and cardiac dysfunction in *Npr1* null mutant mice [18,19].

Ablation of the ANP/NPRA signaling system inhibits aldosterone synthesis and its release from adrenal glomerulosa cells [18,19]. This may account for this system's ANP-dependent renal natriuretic and diuretic effects. Our studies of *Npr1* gene-knockout mice suggested that at birth, the absence of NPRA allowed greater renin and Ang II levels and increased mRNA expression of renin [18,19]. However, when 0-copy mice reached 4–12 weeks of age, their circulating renin and Ang II levels were greatly reduced as compared to levels in 2-copy wild-type mice. We predicted that the decrease in renin content in adult *Npr1* 0-copy mice is most likely a result of progressive elevation of BP leading to inhibition of the synthesis of renin and its release from renal juxtaglomerular cells [65]. On the other hand, renin mRNA levels and renin contents, as well as both Ang II and aldosterone levels, were elevated in adult 2-copy mice compared with 0-copy mice [18,19].

Evidence suggests that the ANP/NPRA/cGMP system opposes the physiological and pathological actions of Ang II [6,134]. Our studies have shown that adrenal Ang II and aldosterone levels are decreased in *Npr1* gene-duplicated mice fed a high-salt diet, whereas a low-salt diet stimulated Ang II and aldosterone levels in the adrenal glands of both 0-copy and 4-copy mice [19,101]. On the other hand, a high-salt diet suppressed adrenal Ang II and aldosterone levels in 0-copy and 2-copy mice, but not in *Npr1* gene-duplicated 3-copy and 4-copy mice. These findings indicated that ANP/NPRA signaling is protective against high salt levels in *Npr1* gene-duplicated mice as compared with *Npr1* gene-knockout mice [19,101].

5. Consequences of Genetic Duplication of *Npr1*

Gene-duplication strategy by homologous recombination produces offspring, including 2-copies (1/1; wild-type), 3-copies (2/1), or 4-copies (2/2) of the target gene [61,71,72]. Method for duplicating genes by double-strand gap repair has also been reported [63]. In double-strand gap repair, the target gene acts as a template to fill in a gap between two regions of homology of the gene with a single crossover event to produce duplication of the target gene. Mice with 2, 3, or 4 copies of *Npr1* were produced to establish a direct gene-dose-dependent effect of *Npr1* on BP, cardiac hypertrophy, and inflammatory responses [19,65,73,77]. To determine whether mice with different *Npr1* copies are able to maintain stable physiological functions, *Npr1* gene-duplicated mice were used to examine the effect of changes in expression of the *Npr1* gene copy numbers on BP and cardiovascular disorders [19,71,73,101].

Our results also showed that changes in the *Npr1* copy number, with corresponding changes in expression of NPRA, causes progressive changes in BP in mice kept for three weeks on a low-salt diet (0.05% NaCl), intermediate-salt diet (2.0% NaCl), or high-salt diet (8% NaCl) [19,71,101]. Salt had a significant effect on BP in 1-copy mice, but only a modest effect in mice with 2-copies of *Npr1*. However, BP in animals having 3- and 4-copies of *Npr1* were negatively affected by increasing dietary salt levels. These results demonstrated that below-normal *Npr1* expression leads to salt-sensitive increases in BP, whereas above-normal *Npr1* expression lowers BP and protects against high dietary salt concentrations [19,71].

6. Genetic Disruption of *Npr1* and Cardiac Hypertrophy, Fibrosis, and Inflammation

Genetic ablation of *Npr1* in mice increases cardiac mass and the incidence of hypertrophy, fibrosis, and inflammation [64,66,67,73,77,100,101,135–137]. *Npr1* gene-knockout mice develop cardiac hypertrophy, fibrosis, and inflammatory responses independent of BP [66,67,138]. We have demonstrated that *Npr1* gene disruption in mice not only increases the expression of cardiac hypertrophic markers, pro-inflammatory cytokines, matrix metalloproteinases (MMP-2, MMP-9), and markers of fibrosis, but also increases the expression and activation of two transcription factors, NF-κB, and activating protein-1 (AP-1) [66,67,73,77,101]. Both of these transcription factors have been found to be associated with cardiac hypertrophy, fibrosis, and extracellular matrix remodeling in *Npr1* gene-disrupted mouse models [7,10,66,67,73]. Moreover, *Npr1* gene-disruption activates NF-κB and AP-1, leading to cardiovascular disease conditions.

ANP/NPRA signaling antagonizes Ang II-induced collagen synthesis by suppressing the activities of MMP-2, MMP-9, and nuclear translocation of NF-κB [73,139]. The expression of angiotensin-converting enzyme (ACE) and AT1R is significantly increased in *Npr1* null mutant mice compared with wild-type mice [67,138]. However, the expression of sarcolemal/endoplasmic reticulum Ca^{2+}-ATPase-2a (SERCA-2a) was progressively decreased in the hypertrophied hearts of *Npr1* gene-ablated mice [66,138].

ANP/NPRA signaling antagonizes Ang II and AT1R-mediated cardiac remodeling and defects; it also provides a protective mechanism in hypertrophied and failing hearts [67,73,77,101,140,141]. Selective inactivation of *Npr1* by homologous lox/Cre-mediated recombination led to only mild cardiac hypertrophy, but ANP levels were significantly increased [55,142–145]. In *Npr1* gene-knockout mice, both ANP and BNP levels were markedly increased. These mice showed a higher incidence of congestive heart failure and significantly greater mortality than did wild-type mice [64,66,67,135].

7. Association of *Nppa*, *Nppb*, and *Npr1* Polymorphisms with Cardiovascular Dysfunction

Genetic and clinical studies have demonstrated the association of single nucleotide polymorphisms (SNPs) of *Nppa*, *Nppb*, and *Npr1* polymorphisms with cardiovascular dysfunction in humans [107,133,146–151]. It is known that patients with monogenic forms of hypertension have rare genetic mutations [152,153]. Various genes among the pathways, including natriuretic peptides, RAAS, and the adrenergic systems, have been found to regulate inherited multigenic disease traits like hypertension and cardiovascular disorders. Nevertheless, the genetic determinants in these pathways that contribute to inter-individual differences in BP regulation have been linked only with the NPs and their receptor systems [146]. Moreover, SNPs of *Nppa*, *Nppb*, and *Npr1* have been found to be associated with increased circulating ANP and BNP, lower systolic and diastolic BP, and decreased prevalence of cardiometabolic syndrome [146,154,155]. Interestingly, an association between *Nppa* promoter polymorphism and cardiac hypertrophy has been demonstrated in hypertensive Italian patients, suggesting that patients carrying SNPs of *Nppa* have marked decreases in pro-ANP levels and left ventricular hypertrophy [147].

The genetic association between the microsatellite marker in *Npr1* promoter and left ventricular hypertrophy indicated that the ANP-BNP/NPRA signaling system contributes to cardiac remodeling and congestive heart failure in patients with essential hypertension [147]. The relationship between high BP and cardiovascular risk exists; thus, in the absence of ANP-BNP/NPRA/cGMP signaling, a small increase in BP may cause the detrimental consequences of cardiac events, including congestive heart failure.

It has been suggested that the underlying mechanism of high BP and cardiovascular events could be linked with *Npr1* mRNA instability, and that this leads to a decrease in translational products of receptor protein [156]. Consequently, a feedback mechanism could be elicited whereby diminished function of ANP-BNP/NPRA/cGMP signaling caused by the defect in the *Npr1* gene could, in compensation, trigger increased expression and release of ANP and/or BNP into the circulation. Earlier studies also indicated that substantial heritability of BP and cardiovascular risks might play a role for genetic factors [157]. It has been shown that SNP in human *Npr1* significantly decreases the expression of *Npr1*, suggesting that a single allele mutation may be associated with increased susceptibility to hypertension and cardiac hypertrophy [133]. In humans, a "4-minus" haplotype in the 3'-non-coding region of the *Npr1* promoter was found to be linked with elevated NT-proBNP levels [133,148]. Thus, it is thought that in humans there is a positive association between *Nppa*, *Nppb*, and *Npr1* gene polymorphisms and high BP, cardiac hypertrophy, and heart failure. The variations in SNPs of *Nppa*, *Nppb*, and *Npr1* in human patients seem to be associated with a family history of high BP and cardiovascular disorders, including myocardial infarction, cardiac mass, septal wall thickness, and congestive heart failure. Because common variants associated with BP and heart failure have not yet been fully identified, further studies are needed to characterize the most functionally significant markers of *Nppa*, *Nppb*, and *Npr1* variants in a large patient population. Interestingly, it has been shown that the minor allele of the ANP genetic variant rs5068 is associated with high circulating levels of ANP and BNP, low risk of hypertension, high plasma levels of high-density lipoprotein (HDL) cholesterol, and low prevalence of metabolic syndrome and obesity [155].

8. Therapeutic Use of Natriuretic Peptides

The physiological and pathophysiological functions of ANP, BNP, and NPRA signaling are associated with protective mechanisms in various organ systems, including the heart, kidneys, central nervous system, lung, and vasculature. The injection of recombinant ANP (Carperitide) was found to accelerate the recovery of blood flow with increased capillary density in ischemic and hyperglycemic conditions [158,159]. Recombinant BNP (nesiritide) also elicited natriuretic and vasodilatory action in heart failure patients [160]. Both ANP and BNP have been shown to exert antihypertrophic and antifibrotic functions [161]. Laser Doppler perfusion studies of *Npr1* gene-knockout mice have shown that blood flow recovery in ischemic limbs is markedly inhibited [162]. Critical hind limb ischemia also severely impaired these animals [163].

The key elements regulating BP and cardiovascular homeostasis in humans and mice are very similar in all of their main attributes related end-organ damage. Given the apparent action of ANP-BNP/NPRA system in antagonizing the RAAS, there is great potential in harnessing the ANP-BNP/NPRA axis as a novel therapeutic intervention in hypertension and cardiovascular diseases. Thus far, clinical trials have identified both benefits and risks of synthetic ANP (anaritide, carperitide) and BNP (nesiritide) for treating hypertension, renal insufficiency, and heart failure in humans; however, these drugs have not yet joined the mainstream of therapeutic treatments in the US [164–166]. Alternatively, enhancing endogenous levels of ANP-BNP/NPRA signaling by inhibiting neprilysin has recently been proven to be a critical component in the first new therapeutic targets for hypertension and heart failure to have been developed in decades [167].

9. Current and Future Perspectives

In about the last four decades, a large body of information has provided a unique perspective on delineating cellular, biochemical, molecular, and genetic information on the regulation and function of natriuretic peptides and their receptor system in the regulation of both the physiology and pathophysiology of cardiovascular homeostasis. Gene-targeting methods (knockout and gene duplication) have been used to delineate the genetic functions dictated by decreasing and/or increasing numbers of *Npr1* gene copies in mice in a gene-dose-dependent manner. Analyses of the molecular genetics and physiological and pathological phenotypes of *Npr1* gene-targeted mutant mice have defined the pivotal roles of ANP/NPRA/cGMP signaling in disease states by genetically modified *Npr1* gene copy numbers and their protein product levels in intact animals. An intriguing finding was that a common genetic variant at the *Nppa* and *Nppb* locus is associated with circulating concentrations of ANP and BNP concentrations, contributing to inter-individual variations in cardiovascular homeostasis. Future investigations should provide a better understanding of the genetic basis of *Npr1* in regulating hypertension, strokes, and cardiovascular disorders.

ANP and BNP are considered to be critical markers of congestive heart failure, but their therapeutic potential in the treatment of cardiovascular diseases such as high BP, renal failure, congestive heart failure, and stroke remains to be established. Ongoing molecular and genetic investigations of *Nppa*, *Nppb*, and *Npr1* should be of great value in resolving the genetic complexities related to hypertension and cardiovascular events, including congestive heart failure. Overall, current and ongoing and future studies should be directed toward providing a unique perspective on delineating the genetic, molecular, and physiological basis of *Nppa*, *Nppb*, and *Npr1* gene expression and regulation under physiological and pathological conditions in both normal and disease states in both experimental animals and humans.

The results of future investigations should provide new therapeutic targets for preventing and treating cardiovascular diseases, stroke, and other abnormal cardiac conditions. Above all, clinical studies are needed to characterize more functionally significant markers of *Nppa*, *Nppb*, and *Npr1* variants in large human populations. In the clinical environment, human recombinant ANP and BNP could possibly be used for heart failure therapy, but first more careful molecular and genetic investigations are needed. Future progress in this field will significantly strengthen and advance our knowledge of genetic and molecular approaches to evaluate the diverse physiological and pathophysiological functions related to cardiovascular disorders. In the future, we should expect to define the potential clinical implications of ongoing investigations of the pharmacogenomics of *Nppa*, *Nppb*, and *Npr1*. Right now we are just at the initial stage of the exponential phase of molecular therapeutics and genomic advancement of the functional aspects of natriuretic peptides and their receptor systems for use in the treatment, prevention, and control of hypertension, heart failure, and other cardiovascular and neurological disorders. More investigations are needed to extend this possibility.

Funding: The research work in the author's laboratory was supported by the National Institutes of Health grant ((R01 HL062147)) and Tulane University Carol Levin Bernick Research Grant Award.

Acknowledgments: I thank Kamala Pandey for her assistance in preparing this manuscript.

Conflicts of Interest: The authors declare no conflict of interest. The funders had no role in the design of the study; in the collection, analyses, or interpretation of data; in the writing of the manuscript, or in the decision to publish the results.

References

1. De Bold, A.J.; Borenstein, H.B.; Veress, A.T.; Sonnenberg, H. A rapid and potent natriuretic response to intravenous injection of atrial myocardial extract in rats. *Life Sci.* **1981**, *28*, 89–94. [CrossRef]
2. de Bold, A.J. Atrial natriuretic factor: A hormone produced by the heart. *Science* **1985**, *230*, 767–770. [CrossRef] [PubMed]
3. Drewett, J.G.; Garbers, D.L. The family of guanylyl cyclase receptors and their ligands. *Endocr. Rev.* **1994**, *15*, 135–162. [CrossRef] [PubMed]
4. Brenner, B.M.; Ballermann, B.J.; Gunning, M.E.; Zeidel, M.L. Diverse biological actions of atrial natriuretic peptide. *Physiol. Rev.* **1990**, *70*, 665–699. [CrossRef] [PubMed]
5. McGrath, M.F.; de Bold, M.L.; de Bold, A.J. The endocrine function of the heart. *Trends Endocrinol. Metab.* **2005**, *16*, 469–477. [CrossRef] [PubMed]
6. Pandey, K.N. Biology of natriuretic peptides and their receptors. *Peptides* **2005**, *26*, 901–932. [CrossRef] [PubMed]

7. Pandey, K.N. Emerging Roles of Natriuretic Peptides and their Receptors in Pathophysiology of Hypertension and Cardiovascular Regulation. *J. Am. Soc. Hypertens* **2008**, *2*, 210–226. [CrossRef] [PubMed]
8. de Bold, A.J.; Ma, K.K.; Zhang, Y.; de Bold, M.L.; Bensimon, M.; Khoshbaten, A. The physiological and pathophysiological modulation of the endocrine function of the heart. *Can. J. Physiol. Pharmacol.* **2001**, *79*, 705–714. [CrossRef]
9. Pandey, K.N. Molecular and genetic aspects of guanylyl cyclase natriuretic peptide receptor-A in regulation of blood pressure and renal function. *Physiol. Genom.* **2018**, *50*, 913–928. [CrossRef]
10. Pandey, K.N. The functional genomics of guanylyl cyclase/natriuretic peptide receptor-A: Perspectives and paradigms. *FEBS J.* **2011**, *278*, 1792–1807. [CrossRef]
11. LaPointe, M.C. Molecular regulation of the brain natriuretic peptide gene. *Peptides* **2005**, *26*, 944–956. [CrossRef] [PubMed]
12. Schulz, S. C-type natriuretic peptide and guanylyl cyclase B receptor. *Peptides* **2005**, *26*, 1024–1034. [CrossRef] [PubMed]
13. Rosenzweig, A.; Seidman, C.E. Atrial natriuretic factor and related peptide hormones. *Annu. Rev. Biochem.* **1991**, *60*, 229–255. [CrossRef] [PubMed]
14. Kojima, M.; Minamino, N.; Kangawa, K.; Matsuo, H. Cloning and sequence analysis of cDNA encoding a precursor for rat brain natriuretic peptide. *Biochem. Biophys. Res. Commun.* **1989**, *159*, 1420–1426. [CrossRef]
15. Suga, S.; Nakao, K.; Hosoda, K.; Mukoyama, M.; Ogawa, Y.; Shirakami, G.; Arai, H.; Saito, Y.; Kambayashi, Y.; Inouye, K. Phenotype-related alteration in expression of natriuretic peptide receptors in aortic smooth muscle cells. *Circ. Res.* **1992**, *71*, 34–39. [CrossRef] [PubMed]
16. Koller, K.J.; de Sauvage, F.J.; Lowe, D.G.; Goeddel, D.V. Conservation of the kinaselike regulatory domain is essential for activation of the natriuretic peptide receptor guanylyl cyclases. *Mol. Cell. Biol.* **1992**, *12*, 2581–2590. [CrossRef] [PubMed]
17. Levin, E.R.; Gardner, D.G.; Samson, W.K. Natriuretic peptides. *NEJM* **1998**, *339*, 321–328. [PubMed]
18. Shi, S.J.; Nguyen, H.T.; Sharma, G.D.; Navar, L.G.; Pandey, K.N. Genetic disruption of atrial natriuretic peptide receptor-A alters renin and angiotensin II levels. *Am. J. Physiol. Ren. Physiol.* **2001**, *281*, F665–F673. [CrossRef]
19. Zhao, D.; Vellaichamy, E.; Somanna, N.K.; Pandey, K.N. Guanylyl cyclase/natriuretic peptide receptor-A gene disruption causes increased adrenal angiotensin II and aldosterone levels. *Am. J. Physiol. Ren. Physiol.* **2007**, *293*, F121–F127. [CrossRef]
20. Garbers, D.L.; Chrisman, T.D.; Wiegn, P.; Katafuchi, T.; Albanesi, J.P.; Bielinski, V.; Barylko, B.; Redfield, M.M.; Burnett, J.C., Jr. Membrane guanylyl cyclase receptors: An update. *Trends Endocrinol. Metab.* **2006**, *17*, 251–258. [CrossRef]
21. Gardner, D.G. Natriuretic peptides: Markers or modulators of cardiac hypertrophy? *Trends Endocrinol. Metab.* **2003**, *14*, 411–416. [CrossRef]
22. Richards, A.M. Natriuretic peptides: Update on Peptide release, bioactivity, and clinical use. *Hypertension* **2007**, *50*, 25–30. [CrossRef]
23. Vollmer, A.M. The role of atrial natriuretic peptide in the immune system. *Peptides* **2005**, *26*, 1087–1094. [CrossRef]
24. Vasan, R.S.; Benjamin, E.J.; Larson, M.G.; Leip, E.P.; Wang, T.J.; Wilson, P.W.; Levy, D. Plasma natriuretic peptides for community screening for left ventricular hypertrophy and systolic dysfunction: The Framingham heart study. *JAMA* **2002**, *288*, 1252–1259. [CrossRef]
25. Pandey, K.N.; Singh, S. Molecular cloning and expression of murine guanylate cyclase/atrial natriuretic factor receptor cDNA. *J. Biol. Chem.* **1990**, *265*, 12342–12348.
26. Schulz, S.; Singh, S.; Bellet, R.A.; Singh, G.; Tubb, D.J.; Chin, H.; Garbers, D.L. The primary structure of a plasma membrane guanylate cyclase demonstrates diversity within this new receptor family. *Cell* **1989**, *58*, 1155–1162. [CrossRef]
27. Lucas, K.A.; Pitari, G.M.; Kazerounian, S.; Ruiz-Stewart, I.; Park, J.; Schulz, S.; Chepenik, K.P.; Waldman, S.A. Guanylyl cyclases and signaling by cyclic GMP. *Pharmacol. Rev.* **2000**, *52*, 375–414.
28. Tremblay, J.; Desjardins, R.; Hum, D.; Gutkowska, J.; Hamet, P. Biochemistry and physiology of the natriuretic peptide receptor guanylyl cyclases. *Mol. Cell. Biochem.* **2002**, *230*, 31–47. [CrossRef]
29. Koller, K.J.; Goddel, D.V. Molecular biology of the natriuretic peptides and their receptors. *Circulation* **1992**, *86*, 1081–1088. [CrossRef]

30. Garbers, D.L. Guanylyl cyclase receptors and their endocrine, paracrine, and autocrine ligands. *Cell* **1992**, *71*, 1–4. [CrossRef]
31. Fuller, F.; Porter, J.G.; Arfsten, A.E.; Miller, J.; Schilling, J.W.; Scarborough, R.M.; Lewicki, J.A.; Schenk, D.B. Atrial natriuretic peptide clearance receptor. Complete sequence and functional expression of cDNA clones. *J. Biol. Chem.* **1988**, *263*, 9395–9401.
32. Mani, I.; Garg, R.; Pandey, K.N. Role of FQQI motif in the internalization, trafficking, and signaling of guanylyl-cyclase/natriuretic peptide receptor-A in cultured murine mesangial cells. *Am. J. Physiol. Ren. Physiol.* **2016**, *310*, F68–F84. [CrossRef]
33. Mani, I.; Pandey, K.N. Emerging concepts of receptor endocytosis and concurrent intracellular signaling: Mechanisms of guanylyl cyclase/natriuretic peptide receptor-A activation and trafficking. *Cell Signal.* **2019**, *60*, 17–30. [CrossRef]
34. Pandey, K.N. Stoichiometric analysis of internalization, recycling, and redistribution of photoaffinity-labeled guanylate cyclase/atrial natriuretic factor receptors in cultured murine Leydig tumor cells. *J. Biol. Chem.* **1993**, *268*, 4382–4390.
35. Brackmann, M.; Schuchmann, S.; Anand, R.; Braunewell, K.H. Neuronal Ca2+ sensor protein VILIP-1 affects cGMP signalling of guanylyl cyclase B by regulating clathrin-dependent receptor recycling in hippocampal neurons. *J. Cell Sci.* **2005**, *118*, 2495–2505. [CrossRef]
36. Koh, G.Y.; Nussenzveig, D.R.; Okolicany, J.; Price, D.A.; Maack, T. Dynamics of atrial natriuretic factor-guanylate cyclase receptors and receptor-ligand complexes in cultured glomerular mesangial and renomedullary interstitial cells. *J. Biol. Chem.* **1992**, *267*, 11987–11994.
37. Kuhn, M. Cardiac and intestinal natriuretic peptides: Insights from genetically modified mice. *Peptides* **2005**, *26*, 1078–1085. [CrossRef]
38. Kishimoto, I.; Tokudome, T.; Nakao, K.; Kangawa, K. Natriuretic peptide system: An overview of studies using genetically engineered animal models. *FEBS J.* **2011**, *278*, 1830–1841. [CrossRef]
39. Misono, K.S.; Philo, J.S.; Arakawa, T.; Ogata, C.M.; Qiu, Y.; Ogawa, H.; Young, H.S. Structure, signaling mechanism and regulation of the natriuretic peptide receptor guanylate cyclase. *FEBS J.* **2011**, *278*, 1818–1829. [CrossRef]
40. Potter, L.R. Natriuretic peptide metabolism, clearance and degradation. *FEBS J.* **2011**, *278*, 1808–1817. [CrossRef]
41. Goetz, K.L. Evidence that atriopeptin is not a physiological regulator of sodium excretion. *Hypertension* **1990**, *15*, 9–19. [CrossRef]
42. Goetz, K.L. Renal natriuretic peptide (urodilatin?) and atriopeptin: Evolving concepts. *Am. J. Physiol.* **1991**, *261*, F921–F932. [CrossRef]
43. Espiner, E.A. Physiology of natriuretic peptides. *J. Intern. Med.* **1994**, *235*, 527–541. [CrossRef]
44. Sagnella, G.A. Measurement and significance of circulating natriuretic peptides in cardiovascular disease. *Clin. Sci. (Lond.)* **1998**, *95*, 519–529. [CrossRef]
45. Blaine, E. Atrial natriuretic factor plays a significant role in blody fluid hemostasis. *Hypertension* **1990**, *15*, 2–8. [CrossRef]
46. McGrath, M.F.; de Bold, A.J. Determinants of natriuretic peptide gene expression. *Peptides* **2005**, *26*, 933–943. [CrossRef]
47. Holmes, S.J.; Espiner, E.A.; Richards, A.M.; Yandle, T.G.; Frampton, C. Renal, endocrine, and hemodynamic effects of human brain natriuretic peptide in normal man. *J. Clin. Endocrinol. Metab.* **1993**, *76*, 91–96.
48. Lang, C.C.; McAlpine, H.M.; Choy, A.M.; Pringle, T.H.; Coutie, W.J.; Struthers, A.D. Effect of pericardiocentesis on plasma levels of brain natriuretic peptide in cardiac tamponade. *Am. J. Cardiol.* **1992**, *70*, 1628–1629. [CrossRef]
49. Lang, C.C.; Choy, A.M.; Struthers, A.D. Atrial and brain natriuretic peptides: A dual natriuretic peptide system potentially involved in circulatory homeostasis. *Clin. Sci. (Lond.)* **1992**, *83*, 519–527. [CrossRef]
50. Yoshimura, M.; Yasue, H.; Morita, E.; Sakaino, N.; Jougasaki, M.; Kurose, M.; Mukoyama, M.; Saito, Y.; Nakao, K.; Imura, H. Hemodynamic, renal, and hormonal responses to brain natriuretic peptide infusion in patients with congestive heart failure. *Circulation* **1991**, *84*, 1581–1588. [CrossRef]
51. Darbar, D.; Davidson, N.C.; Gillespie, N.; Choy, A.M.; Lang, C.C.; Shyr, Y.; McNeill, G.P.; Pringle, T.H.; Struthers, A.D. Diagnostic value of B-type natriuretic peptide concentrations in patients with acute myocardial infarction. *Am. J. Cardiol.* **1996**, *78*, 284–287. [CrossRef]

52. Anand-Srivastava, M.B.; Trachte, G.J. Atrial natriuretic factor receptors and signal transduction mechanisms. *Pharmacol. Rev.* **1993**, *45*, 455–497.
53. Lisy, O.; Huntley, B.K.; McCormick, D.J.; Kurlansky, P.A.; Burnett, J.C., Jr. Design synthesis, and actions of a novel chimeric natriuretic peptide: CD-NP. *J. Am. Coll. Cardiol.* **2008**, *52*, 60–68. [CrossRef]
54. Lee, C.Y.; Lieu, H.; Burnett, J.C., Jr. Designer natriuretic peptides. *J. Investig. Med.* **2009**, *57*, 18–21. [CrossRef]
55. Kuhn, M. Molecular Physiology of Membrane Guanylyl Cyclase Receptors. *Physiol. Rev.* **2016**, *96*, 751–804. [CrossRef]
56. Schwab, T.R.; Edwards, B.S.; Heublein, D.M.; Burnett, J.C., Jr. Role of atrial natriuretic peptide in volume-expansion natriuresis. *Am. J. Physiol.* **1986**, *251*, R310–R313. [CrossRef]
57. Delporte, C.; Poloczek, P.; Tastenoy, M.; Winand, J.; Christophe, J. Atrial natriuretic peptide binds to ANP-R1 receptors in neuroblastoma cells or is degraded extracellularly at the Ser-Phe bond. *Eur. J. Pharmacol.* **1992**, *227*, 247–256. [CrossRef]
58. Delporte, C.; Winand, J.; Poloczek, P.; von Geldern, T.; Christophe, J. Discovery of a potent atrial natriuretic peptide antagonist for ANPA receptors in the human neuroblastoma NB-OK-1 cell line. *Eur. J. Pharmacol.* **1992**, *224*, 183–188. [CrossRef]
59. Ohyama, Y.; Miyamoto, K.; Morishita, Y.; Matsuda, Y.; Saito, Y.; Minamino, N.; Kangawa, K.; Matsuo, H. Stable expression of natriuretic peptide receptors: Effects of HS-142-1, a non-peptide ANP antagonist. *Biochem. Biophys. Res. Commun.* **1992**, *189*, 336–342. [CrossRef]
60. Khurana, M.L.; Pandey, K.N. Receptor-mediated stimulatory effect of atrial natriuretic factor, brain natriuretic peptide, and C-type natriuretic peptide on testosterone production in purified mouse Leydig cells: Activation of cholesterol side-chain cleavage enzyme. *Endocrinology* **1993**, *133*, 2141–2149. [CrossRef]
61. Smithies, O.; Maeda, N. Gene targeting approaches to complex genetic diseases: Atherosclerosis and essential hypertension. *Proc. Natl. Acad. Sci. USA* **1995**, *92*, 5266–5272. [CrossRef]
62. Matsukawa, N.; Grzesik, W.J.; Takahashi, N.; Pandey, K.N.; Pang, S.; Yamauchi, M.; Smithies, O. The natriuretic peptide clearance receptor locally modulates the physiological effects of the natriuretic peptide system. *Pro. Natl. Acad. Sci. USA* **1999**, *96*, 7403–7408. [CrossRef]
63. Kim, H.S.; Krege, J.H.; Kluckman, K.D.; Hagaman, J.R.; Hodgin, J.B.; Best, C.F.; Jennette, J.C.; Coffman, T.M.; Maeda, N.; Smithies, O. Genetic control of blood pressure and the angiotensinogen locus. *Proc. Natl. Acad. Sci. USA* **1995**, *92*, 2735–2739. [CrossRef]
64. Oliver, P.M.; Fox, J.E.; Kim, R.; Rockman, H.A.; Kim, H.S.; Reddick, R.L.; Pandey, K.N.; Milgram, S.L.; Smithies, O.; Maeda, N. Hypertension, cardiac hypertrophy, and sudden death in mice lacking natriuretic peptide receptor A. *Proc. Natl. Acad. Sci. USA* **1997**, *94*, 14730–14735. [CrossRef]
65. Shi, S.J.; Vellaichamy, E.; Chin, S.Y.; Smithies, O.; Navar, L.G.; Pandey, K.N. Natriuretic peptide receptor A mediates renal sodium excretory responses to blood volume expansion. *Am. J. Physiol. Ren. Physiol.* **2003**, *285*, F694–F702. [CrossRef]
66. Vellaichamy, E.; Khurana, M.L.; Fink, J.; Pandey, K.N. Involvement of the NF-kappa B/matrix metalloproteinase pathway in cardiac fibrosis of mice lacking guanylyl cyclase/natriuretic peptide receptor A. *J. Biol. Chem.* **2005**, *280*, 19230–19242. [CrossRef]
67. Vellaichamy, E.; Zhao, D.; Somanna, N.; Pandey, K.N. Genetic disruption of guanylyl cyclase/natriuretic peptide receptor-A upregulates ACE and AT1 receptor gene expression and signaling: Role in cardiac hypertrophy. *Physiol. Genom.* **2007**, *31*, 193–202. [CrossRef]
68. Das, S.; Au, E.; Krazit, S.T.; Pandey, K.N. Targeted disruption of guanylyl cyclase-A/natriuretic peptide receptor-A gene provokes renal fibrosis and remodeling in null mutant mice: Role of proinflammatory cytokines. *Endocrinology* **2010**, *151*, 5841–5850. [CrossRef]
69. Das, S.; Periyasamy, R.; Pandey, K.N. Activation of IKK/NF-kappaB provokes renal inflammatory responses in guanylyl cyclase/natriuretic peptide receptor-A gene-knockout mice. *Physiol. Genom.* **2012**, *44*, 430–442. [CrossRef]
70. Kumar, P.; Periyasamy, R.; Das, S.; Neerukonda, S.; Mani, I.; Pandey, K.N. All-trans retinoic acid and sodium butyrate enhance natriuretic peptide receptor a gene transcription: Role of histone modification. *Mol. Pharmacol.* **2014**, *85*, 946–957. [CrossRef]
71. Oliver, P.M.; John, S.W.; Purdy, K.E.; Kim, R.; Maeda, N.; Goy, M.F.; Smithies, O. Natriuretic peptide receptor 1 expression influences blood pressures of mice in a dose-dependent manner. *Proc. Natl. Acad. Sci. USA* **1998**, *95*, 2547–2551. [CrossRef]

72. Pandey, K.N.; Oliver, P.M.; Maeda, N.; Smithies, O. Hypertension associated with decreased testosterone levels in natriuretic peptide receptor-A gene-knockout and gene-duplicated mutant mouse models. *Endocrinology* **1999**, *140*, 5112–5119. [CrossRef]
73. Vellaichamy, E.; Das, S.; Subramanian, U.; Maeda, N.; Pandey, K.N. Genetically altered mutant mouse models of guanylyl cyclase/natriuretic peptide receptor-A exhibit the cardiac expression of proinflammatory mediators in a gene-dose-dependent manner. *Endocrinology* **2014**, *155*, 1045–1056. [CrossRef]
74. Kumar, P.; Gogulamudi, V.R.; Periasamy, R.; Raghavaraju, G.; Subramanian, U.; Pandey, K.N. Inhibition of HDAC enhances STAT acetylation, blocks NF-kappaB, and suppresses the renal inflammation and fibrosis in Npr1 haplotype male mice. *Am. J. Physiol. Ren. Physiol.* **2017**, *313*, F781–F795. [CrossRef]
75. Gogulamudi, V.R.; Mani, I.; Subramanian, U.; Pandey, K.N. Genetic disruption of Npr1 depletes regulatory T cells and provokes high levels of proinflammatory cytokines and fibrosis in the kidneys of female mutant mice. *Am. J. Physiol. Ren. Physiol.* **2019**, *316*, F1254–F1272. [CrossRef]
76. Periyasamy, R.; Das, S.; Pandey, K.N. Genetic disruption of guanylyl cyclase/natriuretic peptide receptor-A upregulates renal (pro) renin receptor expression in Npr1 null mutant mice. *Peptides* **2019**, *114*, 17–28. [CrossRef]
77. Subramanian, U.; Kumar, P.; Mani, I.; Chen, D.; Kessler, I.; Periyasamy, R.; Raghavaraju, G.; Pandey, K.N. Retinoic acid and sodium butyrate suppress the cardiac expression of hypertrophic markers and proinflammatory mediators in Npr1 gene-disrupted haplotype mice. *Physiol. Genom.* **2016**, *48*, 477–490. [CrossRef]
78. Kim, H.S.; Lu, G.; John, S.W.M.; Maeda, N.; Smithies, O. Molecular phenotyping for analyzing subtle genetic effects in mice application to an angiotensinogen gene titration. *Proc. Natl. Acad. Sci. USA* **2002**, *99*, 4602–4607. [CrossRef]
79. Takahashi, N.; Smithies, O. Gene targeting approaches to analyzing hypertension. *J. Am. Soc. Nephrol.* **1999**, *10*, 1598–1605.
80. John, S.W.; Krege, J.H.; Oliver, P.M.; Hagaman, J.R.; Hodgin, J.B.; Pang, S.C.; Flynn, T.G.; Smithies, O. Genetic decreases in atrial natriuretic peptide and salt-sensitive hypertension. *Science* **1995**, *267*, 679–681. [CrossRef]
81. Tamura, N.; Ogawa, Y.; Chusho, H.; Nakamura, K.; Nakao, K.; Suda, M.; Kasahara, M.; Hashimoto, R.; Katsuura, G.; Mukoyama, M.; et al. Cardiac fibrosis in mice lacking brain natriuretic peptide. *Proc. Natl. Acad. Sci. USA* **2000**, *97*, 4239–4244. [CrossRef]
82. Melo, L.G.; Veress, A.T.; Ackermann, U.; Steinhelper, M.E.; Pang, S.C.; Tse, Y.; Sonnenberg, H. Chronic regulation of arterial blood pressure in ANP transgenic and knockout mice: Role of cardiovascular sympathetic tone. *Cardiovasc. Res.* **1999**, *43*, 437–444. [CrossRef]
83. Steinhelper, M.E.; Cochrane, K.L.; Field, L.J. Hypotension in transgenic mice expressing atrial natriuretic factor fusion genes. *Hypertension* **1990**, *16*, 301–307. [CrossRef]
84. Lin, K.F.; Chao, J.; Chao, L. Human atrial natriuretic peptide gene delivery reduces blood pressure in hypertensive rats. *Hypertension* **1995**, *26*, 847–853. [CrossRef]
85. Schillinger, K.J.; Tsai, S.Y.; Taffet, G.E.; Reddy, A.K.; Marian, A.J.; Entman, M.L.; Oka, K.; Chan, L.; O'Malley, B.W. Regulatable atrial natriuretic peptide gene therapy for hypertension. *Proc. Natl. Acad. Sci. USA* **2005**, *102*, 13789–13794. [CrossRef]
86. Klinger, J.R.; Cutaia, M. The natriuretic peptides. Clinical applications in patients with COPD. *Chest* **1996**, *110*, 1136–1138. [CrossRef]
87. Naruse, M.; Takeyama, Y.; Tanabe, A.; Hiroshige, J.; Naruse, K.; Yoshimoto, T.; Tanaka, M.; Katagiri, T.; Demura, H. Atrial and brain natriuretic peptides in cardiovascular diseases. *Hypertension* **1994**, *23*, I231–I234. [CrossRef]
88. Klinger, J.R.; Petit, R.D.; Curtin, L.A.; Warburton, R.R.; Wrenn, D.S.; Steinhelper, M.E.; Field, L.J.; Hill, N.S. Cardiopulmonary responses to chronic hypoxia in transgenic mice that overexpress ANP. *J. Appl. Physiol.* **1993**, *75*, 198–205. [CrossRef]
89. Hohne, C.; Drzimalla, M.; Krebs, M.O.; Boemke, W.; Kaczmarczyk, G. Atrial natriuretic peptide ameliorates hypoxic pulmonary vasoconstriction without influencing systemic circulation. *J. Physiol. Pharmacol.* **2003**, *54*, 497–510.
90. Klinger, J.R.; Warburton, R.R.; Pietras, L.; Oliver, P.; Fox, J.; Smithies, O.; Hill, N.S. Targeted disruption of the gene for natriuretic peptide receptor-A worsens hypoxia-induced cardiac hypertrophy. *Am. J. Physiol.* **2002**, *282*, H58–H65. [CrossRef]

91. Klinger, J.R.; Warburton, R.R.; Pietras, L.A.; Smithies, O.; Swift, R.; Hill, N.S. Genetic disruption of atrial natriuretic peptide causes pulmonary hypertension in normoxic and hypoxic mice. *Am. J. Physiol.* **1999**, *276*, L868–L874. [CrossRef]
92. Sun, J.Z.; Chen, S.J.; Li, G.; Chen, Y.F. Hypoxia reduces atrial natriuretic peptide clearance receptor gene expression in ANP knockout mice. *Am. J. Physiol. Lung Cell Mol. Physiol.* **2000**, *279*, L511–L519. [CrossRef]
93. Melo, L.G.; Sonnenberg, H. Requirement for prostaglandin synthesis in secretion of atrial natriuretic factor from isolated rat heart. *Regul. Pept.* **1995**, *60*, 79–87. [CrossRef]
94. Melo, L.G.; Veress, A.T.; Chong, C.K.; Pang, S.C.; Flynn, T.G.; Sonnenberg, H. Salt-sensitive hypertension in ANP knockout mice: Potential role of abnormal plasma renin activity. *Am. J. Physiol.* **1998**, *274*, R255–R261. [CrossRef]
95. Itoh, H.; Nakao, K.; Mukoyama, M.; Yamada, T.; Hosoda, K.; Shirakami, G.; Morii, N.; Sugawara, A.; Saito, Y.; Shiono, S.; et al. Chronic blockade of endogenous atrial natriuretic polypeptide (ANP) by monoclonal antibody against ANP accelerates the development of hypertension in spontaneously hypertensive and deoxycorticosterone acetate-salt-hypertensive rats. *J. Clin. Investig.* **1989**, *84*, 145–154. [CrossRef]
96. Felker, G.M.; Petersen, J.W.; Mark, D.B. Natriuretic peptides in the diagnosis and management of heart failure. *CMAJ* **2006**, *175*, 611–617. [CrossRef]
97. Reinhart, K.; Meisner, M.; Brunkhorst, F.M. Markers for sepsis diagnosis: What is useful? *Crit. Care Clin.* **2006**, *22*, 503–519. [CrossRef]
98. See, R.; de Lemos, J.A. Current status of risk stratification methods in acute coronary syndromes. *Curr. Cardiol. Rep.* **2006**, *8*, 282–288. [CrossRef]
99. Jaffe, A.S.; Babuin, L.; Apple, F.S. Biomarkers in acute cardiac disease: The present and the future. *J. Am. Coll. Cardiol.* **2006**, *48*, 1–11. [CrossRef]
100. Ellmers, L.J.; Scott, N.J.; Piuhola, J.; Maeda, N.; Smithies, O.; Frampton, C.M.; Richards, A.M.; Cameron, V.A. Npr1-regulated gene pathways contributing to cardiac hypertrophy and fibrosis. *J. Mol. Endocrinol.* **2007**, *38*, 245–257. [CrossRef]
101. Zhao, D.; Das, S.; Pandey, K.N. Interactive roles of Npr1 gene-dosage and salt diets on cardiac angiotensin II, aldosterone and pro-inflammatory cytokines levels in mutant mice. *J. Hypertens.* **2013**, *31*, 134–144. [CrossRef]
102. Chen, H.H.; Burnett, J.C., Jr. The natriuretic peptides in heart failure: Diagnostic and therapeutic potentials. *Proc. Assoc. Am. Physicians* **1999**, *111*, 406–416. [CrossRef]
103. Nakao, K.; Itoh, H.; Saito, Y.; Mukoyama, M.; Ogawa, Y. The natriuretic peptide family. *Curr. Opin. Nephrol. Hypertens.* **1996**, *5*, 4–11. [CrossRef]
104. Tsutamoto, T.; Kanamori, T.; Morigami, N.; Sugimoto, Y.; Yamaoka, O.; Kinoshita, M. Possibility of downregulation of atrial natriuretic peptide receptor coupled to guanylate cyclase in peripheral vascular beds of patients with chronic severe heart failure. *Circulation* **1993**, *87*, 70–75. [CrossRef]
105. Mukoyama, M.; Nakao, K.; Hosoda, K.; Suga, S.; Saito, Y.; Ogawa, Y.; Shirakami, G.; Jougasaki, M.; Obata, K.; Yasue, H.; et al. Brain natriuretic peptide as a novel cardiac hormone in humans. Evidence for an exquisite dual natriuretic peptide system, atrial natriuretic peptide and brain natriuretic peptide. *J. Clin. Investig.* **1991**, *87*, 1402–1412. [CrossRef]
106. Knowles, J.W.; Esposito, G.; Mao, L.; Hagaman, J.R.; Fox, J.E.; Smithies, O.; Rockman, H.A.; Maeda, N. Pressure-independent enhancement of cardiac hypertrophy in natriuretic peptide receptor A-deficient mice. *J. Clin. Investig.* **2001**, *107*, 975–984. [CrossRef]
107. Xue, H.; Wang, S.; Wang, H.; Sun, K.; Song, X.; Zhang, W.; Fu, C.; Han, Y.; Hui, R. Atrial natriuretic peptide gene promoter polymorphism is associated with left ventricular hypertrophy in hypertension. *Clin. Sci. (Lond.)* **2008**, *114*, 131–137. [CrossRef]
108. Doust, J.A.; Pietrzak, E.; Dobson, A.; Glasziou, P. How well does B-type natriuretic peptide predict death and cardiac events in patients with heart failure: Systematic review. *BWJ* **2005**, *330*, 625. [CrossRef]
109. Khan, I.A.; Fink, J.; Nass, C.; Chen, H.; Christenson, R.; de Filippi, C.R. N-terminal pro-B-type natriuretic peptide and B-type natriuretic peptide for identifying coronary artery disease and left ventricular hypertrophy in ambulatory chronic kidney disease patients. *Am. J. Cardiol.* **2006**, *97*, 1530–1534. [CrossRef]
110. Girsen, A.; Ala-Kopsala, M.; Makikallio, K.; Vuolteenaho, O.; Rasanen, J. Cardiovascular hemodynamics and umbilical artery N-terminal peptide of proB-type natriuretic peptide in human fetuses with growth restriction. *UOG: ISUOG* **2007**, *29*, 296–303. [CrossRef]

111. Kocylowski, R.D.; Dubiel, M.; Gudmundsson, S.; Sieg, I.; Fritzer, E.; Alkasi, O.; Breborowicz, G.H.; von Kaisenberg, C.S. Biochemical tissue-specific injury markers of the heart and brain in postpartum cord blood. *Am. J. Obstet. Gynecol.* **2009**, *200*, 273 e1–273 e25. [CrossRef]
112. Zahabi, A.; Picard, S.; Fortin, N.; Reudelhuber, T.L.; Deschepper, C.F. Expression of constitutively active guanylate cyclase in cardiomyocytes inhibits the hypertrophic effects of isoproterenol and aortic constriction on mouse hearts. *J. Biol. Chem.* **2003**, *278*, 47694–47699. [CrossRef]
113. Deschepper, C.F.; Masciotra, S.; Zahabi, A.; Boutin-Ganache, I.; Picard, S.; Reudelhuber, T.L. Function alterations of the Nppa promoter are linked to cardiac ventricular hypertrophy in WKY/WKHA rat crosses. *Circ. Res.* **2001**, *88*, 223–228. [CrossRef]
114. Lattion, A.L.; Michel, J.B.; Arnauld, E.; Corvol, P.; Soubrier, F. Myocardial recruitment during ANF mRNA increase with volume overload in the rat. *Am. J. Physiol.* **1986**, *251*, H890–H896. [CrossRef]
115. Hunter, J.J.; Chien, K.R. Signaling pathways for cardiac hypertrophy and failure. *NEJM* **1999**, *341*, 1276–1283. [CrossRef]
116. Hasegawa, K.; Fujiwara, H.; Doyama, K.; Miyamae, M.; Fujiwara, T.; Suga, S.; Mukoyama, M.; Nakao, K.; Imura, H.; Sasayama, S. Ventricular expression of brain natriuretic peptide in hypertrophic cardiomyopathy. *Circulation* **1993**, *88*, 372–380. [CrossRef]
117. Hasegawa, K.; Fujiwara, H.; Doyama, K.; Mukoyama, M.; Nakao, K.; Fujiwara, T.; Imura, H.; Kawai, C. Ventricular expression of atrial and brain natriuretic peptides in dilated cardiomyopathy. An immunohistocytochemical study of the endomyocardial biopsy specimens using specific monoclonal antibodies. *Am. J. Pathol.* **1993**, *142*, 107–116.
118. Knowlton, K.U.; Rockman, H.A.; Itani, M.; Vovan, A.; Seidman, C.E.; Chien, K.R. Divergent pathways mediate the induction of ANF transgenes in neonatal and hypertrophic ventricular myocardium. *J. Clin. Investig.* **1995**, *96*, 1311–1318. [CrossRef]
119. Calderone, A.; Thaik, C.M.; Takahashi, N.; Chang, D.L.; Colucci, W.S. Nitric oxide, atrial natriuretic peptide, and cyclic GMP inhibit the growth-promoting effects of norepinephrine in cardiac myocytes and fibroblasts. *J. Clin. Investig.* **1998**, *101*, 812–818. [CrossRef]
120. Horio, T.; Nishikimi, T.; Yoshihara, F.; Matsuo, H.; Takishita, S.; Kangawa, K. Inhibitory regulation of hypertrophy by endogenous atrial natriuretic peptide in cultured cardiac myocytes. *Hypertension* **2000**, *35*, 19–24. [CrossRef]
121. Wu, C.F.; Bishopric, N.H.; Pratt, R.E. Atrial natriuretic peptide induces apoptosis in neonatal rat cardiac myocytes. *J. Biol. Chem.* **1997**, *272*, 14860–14866. [CrossRef]
122. Silberbach, M.; Gorenc, T.; Hershberger, R.E.; Stork, P.J.; Steyger, P.S.; Roberts, C.T., Jr. Extracellular signal-regulated protein kinase activation is required for the anti-hypertrophic effect of atrial natriuretic factor in neonatal rat ventricular myocytes. *J. Biol. Chem.* **1999**, *274*, 24858–24864. [CrossRef]
123. Kishimoto, I.; Tokudome, T.; Horio, T.; Garbers, D.L.; Nakao, K.; Kangawa, K. Natriuretic Peptide Signaling via Guanylyl Cyclase (GC)-A: An Endogenous Protective Mechanism of the Heart. *Curr. Cardiol. Rev.* **2009**, *5*, 45–51. [CrossRef]
124. Ogawa, Y.; Itoh, H.; Tamura, N.; Suga, S.; Yoshimasa, T.; Uehira, M.; Matsuda, S.; Shiono, S.; Nishimoto, H.; Nakao, K. Molecular cloning of the complementary DNA and gene that encode mouse brain natriuretic peptide and generation of transgenic mice that overexpress the brain natriuretic peptide gene. *J. Clin. Investig.* **1994**, *93*, 1911–1921. [CrossRef]
125. Chusho, H.; Tamura, N.; Ogawa, Y.; Yasoda, A.; Suda, M.; Miyazawa, T.; Nakamura, K.; Nakao, K.; Kurihara, T.; Komatsu, Y.; et al. Dwarfism and early death in mice lacking C-type natriuretic peptide. *Proc. Natl. Acad. Sci. USA* **2001**, *98*, 4016–4021. [CrossRef]
126. Yasoda, A.; Komatsu, Y.; Chusho, H.; Miyazawa, T.; Ozasa, A.; Miura, M.; Kurihara, T.; Rogi, T.; Tanaka, S.; Suda, M.; et al. Overexpression of CNP in chondrocytes rescues achondroplasia through a MAPK-dependent pathway. *Nat. Med.* **2004**, *10*, 80–86. [CrossRef]
127. Wang, Y.; de Waard, M.C.; Sterner-Kock, A.; Stepan, H.; Schultheiss, H.P.; Duncker, D.J.; Walther, T. Cardiomyocyte-restricted over-expression of C-type natriuretic peptide prevents cardiac hypertrophy induced by myocardial infarction in mice. *Eur. J. Heart Fail.* **2007**, *9*, 548–557. [CrossRef]
128. Lopez, M.J.; Wong, S.K.; Kishimoto, I.; Dubois, S.; Mach, V.; Friesen, J.; Garbers, D.L.; Beuve, A. Salt-resistant hypertension in mice lacking the guanylyl cyclase-A receptor for atrial natriuretic peptide. *Nature* **1995**, *378*, 65–68. [CrossRef]

129. Tamura, N.; Doolittle, L.K.; Hammer, R.E.; Shelton, J.M.; Richardson, J.A.; Garbers, D.L. Critical roles of the guanylyl cyclase B receptor in endochondral ossification and development of female reproductive organs. *Proc. Natl. Acad. Sci. USA* **2004**, *101*, 17300–17305. [CrossRef]
130. Langenickel, T.H.; Buttgereit, J.; Pagel-Langenickel, I.; Lindner, M.; Monti, J.; Beuerlein, K.; Al-Saadi, N.; Plehm, R.; Popova, E.; Tank, J.; et al. Cardiac hypertrophy in transgenic rats expressing a dominant-negative mutant of the natriuretic peptide receptor B. *Proc. Natl. Acad. Sci. USA* **2006**, *103*, 4735–4740. [CrossRef]
131. Kishimoto, I.; Dubois, S.K.; Garbers, D.L. The heart communicates with the kidney exclusively through the guanylyl cyclase-A receptor: Acute handling of sodium and water in response to volume expansion. *Proc. Natl. Acad. Sci. USA* **1996**, *93*, 6215–6219. [CrossRef]
132. Kishimoto, I.; Garbers, D.L. Physiological regulation of blood pressure and kidney function by guanylyl cyclase isoforms. *Curr. Opin. Nephrol. Hypertens.* **1997**, *6*, 58–63. [CrossRef]
133. Nakayama, T.; Soma, M.; Takahashi, Y.; Rehemudula, D.; Kanmatsuse, K.; Furuya, K. Functional deletion mutation of the 5′-flanking region of type A human natriuretic peptide receptor gene and its association with essential hypertension and left ventricular hypertrophy in the Japanese. *Circ. Res.* **2000**, *86*, 841–845. [CrossRef]
134. Pandey, K.N. Ligand-mediated endocytosis and intracellular sequestration of guanylyl cyclase/natriuretic peptide receptors: Role of GDAY motif. *Mol. Cell. Biochem.* **2010**, *334*, 81–98. [CrossRef]
135. Nakanishi, M.; Saito, Y.; Kishimoto, I.; Harada, M.; Kuwahara, K.; Takahashi, N.; Kawakami, R.; Nakagawa, Y.; Tanimoto, K.; Yasuno, S.; et al. Role of natriuretic peptide receptor guanylyl cyclase-A in myocardial infarction evaluated using genetically engineered mice. *Hypertension* **2005**, *46*, 441–447. [CrossRef]
136. Ellmers, L.J.; Knowles, J.W.; Kim, H.S.; Smithies, O.; Maeda, N.; Cameron, V.A. Ventricular expression of natriuretic peptides in Npr1(-/-) mice with cardiac hypertrophy and fibrosis. *Am. J. Physiol.* **2002**, *283*, H707–H714.
137. Scott, N.J.; Ellmers, L.J.; Lainchbury, J.G.; Maeda, N.; Smithies, O.; Richards, A.M.; Cameron, V.A. Influence of natriuretic peptide receptor-1 on survival and cardiac hypertrophy during development. *Biochim. Biophys. Acta* **2009**, *1792*, 1175–1184. [CrossRef]
138. Pandey, K.N.; Vellaichamy, E. Regulation of cardiac angiotensin-converting enzyme and angiotensin AT1 receptor gene expression in Npr1 gene-disrupted mice. *Clin. Exp. Pharmacol. Physiol.* **2010**, *37*, e70–e77. [CrossRef]
139. Parthasarathy, A.; Gopi, V.; Umadevi, S.; Simna, A.; Sheik, M.J.; Divya, H.; Vellaichamy, E. Suppression of atrial natriuretic peptide/natriuretic peptide receptor-A-mediated signaling upregulates angiotensin-II-induced collagen synthesis in adult cardiac fibroblasts. *Mol. Cell. Biochem.* **2013**, *378*, 217–228. [CrossRef]
140. Li, Y.; Kishimoto, I.; Saito, Y.; Harada, M.; Kuwahara, K.; Izumi, T.; Takahashi, N.; Kawakami, R.; Tanimoto, K.; Nakagawa, Y.; et al. Guanylyl cyclase-A inhibits angiotensin II type 1A receptor-mediated cardiac remodeling, an endogenous protective mechanism in the heart. *Circulation* **2002**, *106*, 1722–1728. [CrossRef]
141. Kilic, A.; Bubikat, A.; Gassner, B.; Baba, H.A.; Kuhn, M. Local actions of atrial natriuretic peptide counteract angiotensin II stimulated cardiac remodeling. *Endocrinology* **2007**, *148*, 4162–4169. [CrossRef]
142. Holtwick, R.; van Eickels, M.; Skryabin, B.V.; Baba, H.A.; Bubikat, A.; Begrow, F.; Schneider, M.D.; Garbers, D.L.; Kuhn, M. Pressure-independent cardiac hypertrophy in mice with cardiomyocyte-restricted inactivation of the atrial natriuretic peptide receptor guanylyl cyclase-A. *J. Clin. Investig.* **2003**, *111*, 1399–1407. [CrossRef]
143. Morita, E.; Yasue, H.; Yoshimura, M.; Ogawa, H.; Jougasaki, M.; Matsumura, T.; Mukoyama, M.; Nakao, K. Increased plasma levels of brain natriuretic peptide in patients with acute myocardial infarction. *Circulation* **1993**, *88*, 82–91. [CrossRef]
144. Phillips, P.A.; Sasadeus, J.; Hodsman, G.P.; Horowitz, J.; Saltups, A.; Johnston, C.I. Plasma atrial natriuretic peptide in patients with acute myocardial infarction: Effects of streptokinase. *Br. Heart J.* **1989**, *61*, 139–143. [CrossRef]
145. Tomoda, H. Atrial natriuretic peptide in acute myocardial infarction. *Am. J. Cardiol.* **1988**, *62*, 1122–1123. [CrossRef]
146. Newton-Cheh, C.; Larson, M.G.; Vasan, R.S.; Levy, D.; Bloch, K.D.; Surti, A.; Guiducci, C.; Kathiresan, S.; Benjamin, E.J.; Struck, J.; et al. Association of common variants in NPPA and NPPB with circulating natriuretic peptides and blood pressure. *Nat. Genet.* **2009**, *41*, 348–353. [CrossRef]

147. Rubattu, S.; Bigatti, G.; Evangelista, A.; Lanzani, C.; Stanzione, R.; Zagato, L.; Manunta, P.; Marchitti, S.; Venturelli, V.; Bianchi, G.; et al. Association of atrial natriuretic peptide and type a natriuretic peptide receptor gene polymorphisms with left ventricular mass in human essential hypertension. *J. Am. Coll. Cardiol.* **2006**, *48*, 499–505. [CrossRef]
148. Webber, M.A.; Marder, S.R. Better pharmacotherapy for schizophrenia: What does the future hold? *Curr. Psychiatry Rep.* **2008**, *10*, 352–358. [CrossRef]
149. Pitzalis, M.V.; Sarzani, R.; Dessi-Fulgheri, P.; Iacoviello, M.; Forleo, C.; Lucarelli, K.; Pietrucci, F.; Salvi, F.; Sorrentino, S.; Romito, R.; et al. Allelic variants of natriuretic peptide receptor genes are associated with family history of hypertension and cardiovascular phenotype. *J. Hypertens.* **2003**, *21*, 1491–1496. [CrossRef]
150. Usami, S.; Kishimoto, I.; Saito, Y.; Harada, M.; Kuwahara, K.; Nakagawa, Y.; Nakanishi, M.; Yasuno, S.; Kangawa, K.; Nakao, K. Association of CT dinucleotide repeat polymorphism in the 5′-flanking region of the guanylyl cyclase (GC)-A gene with essential hypertension in the Japanese. *Hypertens. Res.* **2008**, *31*, 89–96. [CrossRef]
151. Wang, J.; Wang, Z.; Yu, C. Association of Polymorphisms in the Atrial Natriuretic Factor Gene with the Risk of Essential Hypertension: A Systematic Review and Meta-Analysis. *Int. J. Environ. Res. Public Health* **2016**, *13*, 458. [CrossRef]
152. Ji, W.; Foo, J.N.; O'Roak, B.J.; Zhao, H.; Larson, M.G.; Simon, D.B.; Newton-Cheh, C.; State, M.W.; Levy, D.; Lifton, R.P. Rare independent mutations in renal salt handling genes contribute to blood pressure variation. *Nat. Genet.* **2008**, *40*, 592–599. [CrossRef]
153. Lifton, R.P.; Gharavi, A.G.; Geller, D.S. Molecular mechanisms of human hypertension. *Cell* **2001**, *104*, 545–556. [CrossRef]
154. Cannone, V.; Cefalu, A.B.; Noto, D.; Scott, C.G.; Bailey, K.R.; Cavera, G.; Pagano, M.; Sapienza, M.; Averna, M.R.; Burnett, J.C., Jr. The atrial natriuretic peptide genetic variant rs5068 is associated with a favorable cardiometabolic phenotype in a Mediterranean population. *Diabetes Care* **2013**, *36*, 2850–2856. [CrossRef]
155. Cannone, V.; Scott, C.G.; Decker, P.A.; Larson, N.B.; Palmas, W.; Taylor, K.D.; Wang, T.J.; Gupta, D.K.; Bielinski, S.J.; Burnett, J.C., Jr. A favorable cardiometabolic profile is associated with the G allele of the genetic variant rs5068 in African Americans: The Multi-Ethnic Study of Atherosclerosis (MESA). *PLoS ONE* **2017**, *12*, e0189858. [CrossRef]
156. Knowles, J.W.; Erickson, L.M.; Guy, V.K.; Sigel, C.S.; Wilder, J.C.; Maeda, N. Common variations in noncoding regions of the human natriuretic peptide receptor A gene have quantitative effects. *Hum. Genet.* **2003**, *112*, 62–70. [CrossRef]
157. Levy, D.; DeStefano, A.L.; Larson, M.G.; O'Donnell, C.J.; Lifton, R.P.; Gavras, H.; Cupples, L.A.; Myers, R.H. Evidence for a gene influencing blood pressure on chromosome 17. Genome scan linkage results for longitudinal blood pressure phenotypes in subjects from the framingham heart study. *Hypertension* **2000**, *36*, 477–483. [CrossRef]
158. Suwa, M.; Seino, Y.; Nomachi, Y.; Matsuki, S.; Funahashi, K. Multicenter prospective investigation on efficacy and safety of carperitide for acute heart failure in the 'real world' of therapy. *Circ. J.* **2005**, *69*, 283–290. [CrossRef]
159. Park, K.; Itoh, H.; Yamahara, K.; Sone, M.; Miyashita, K.; Oyamada, N.; Sawada, N.; Taura, D.; Inuzuka, M.; Sonoyama, T.; et al. Therapeutic potential of atrial natriuretic peptide administration on peripheral arterial diseases. *Endocrinology* **2008**, *149*, 483–491. [CrossRef]
160. Colucci, W.S.; Elkayam, U.; Horton, D.P.; Abraham, W.T.; Bourge, R.C.; Johnson, A.D.; Wagoner, L.E.; Givertz, M.M.; Liang, C.S.; Neibaur, M.; et al. Intravenous nesiritide, a natriuretic peptide, in the treatment of decompensated congestive heart failure. Nesiritide Study Group. *NEJM* **2000**, *343*, 246–253. [CrossRef]
161. Nishikimi, T.; Kuwahara, K.; Nakagawa, Y.; Kangawa, K.; Minamino, N.; Nakao, K. Complexity of molecular forms of B-type natriuretic peptide in heart failure. *Heart* **2013**, *99*, 677–679. [CrossRef]
162. Tokudome, T.; Kishimoto, I.; Yamahara, K.; Osaki, T.; Minamino, N.; Horio, T.; Sawai, K.; Kawano, Y.; Miyazato, M.; Sata, M.; et al. Impaired recovery of blood flow after hind-limb ischemia in mice lacking guanylyl cyclase-A, a receptor for atrial and brain natriuretic peptides. *Arterioscler. Thromb. Vasc. Biol.* **2009**, *29*, 1516–1521. [CrossRef]

163. Kuhn, M.; Volker, K.; Schwarz, K.; Carbajo-Lozoya, J.; Flogel, U.; Jacoby, C.; Stypmann, J.; van Eickels, M.; Gambaryan, S.; Hartmann, M.; et al. The natriuretic peptide/guanylyl cyclase—A system functions as a stress-responsive regulator of angiogenesis in mice. *J. Clin. Investig.* **2009**, *119*, 2019–2030. [CrossRef]
164. Dohi, K.; Ito, M. Novel diuretic strategies for the treatment of heart failure in Japan. *Circ. J.* **2014**, *78*, 1816–1823. [CrossRef]
165. Hayek, S.; Nemer, M. Cardiac natriuretic peptides: From basic discovery to clinical practice. *Cardiovasc. Ther.* **2011**, *29*, 362–376. [CrossRef]
166. Saito, Y. Roles of atrial natriuretic peptide and its therapeutic use. *J. Cardiol.* **2010**, *56*, 262–270. [CrossRef]
167. McMurray, J.J.; Packer, M.; Desai, A.S.; Gong, J.; Lefkowitz, M.P.; Rizkala, A.R.; Rouleau, J.L.; Shi, V.C.; Solomon, S.D.; Swedberg, K.; et al. Investigators, Committees, Angiotensin-neprilysin inhibition versus enalapril in heart failure. *NEJM* **2014**, *371*, 993–1004. [CrossRef]

© 2019 by the author. Licensee MDPI, Basel, Switzerland. This article is an open access article distributed under the terms and conditions of the Creative Commons Attribution (CC BY) license (http://creativecommons.org/licenses/by/4.0/).

Article

PCSK9 is Expressed in Human Visceral Adipose Tissue and Regulated by Insulin and Cardiac Natriuretic Peptides

Marica Bordicchia [1], Francesco Spannella [1,2], Gianna Ferretti [3], Tiziana Bacchetti [3], Arianna Vignini [3], Chiara Di Pentima [1,2], Laura Mazzanti [3] and Riccardo Sarzani [1,2,*]

[1] Internal Medicine and Geriatrics, Department of Clinical and Molecular Sciences, University "Politecnica delle Marche", 60126 Ancona, Italy; marica.bordicchia@gmail.com (M.B.); fspannella@gmail.com (F.S.); chiara.dipentima@live.it (C.D.P.)

[2] Internal Medicine and Geriatrics, "Hypertension Excellence Centre" of the European Society of Hypertension, IRCCS-INRCA, 60127 Ancona, Italy

[3] Department of Clinical Sciences, Section of Biochemistry, Biology and Physics, School of Nutrition, University "Politecnica delle Marche", 60126 Ancona, Italy; g.ferretti@univpm.it (G.F.); t.bacchetti@univpm.it (T.B.); a.vignini@univpm.it (A.V.); l.mazzanti@univpm.it (L.M.)

* Correspondence: r.sarzani@univpm.it; Tel.: +39-071-596-4595

Received: 9 November 2018; Accepted: 4 January 2019; Published: 9 January 2019

Abstract: Proprotein convertase subtilisin/kexin type 9 (PCSK9) binds to and degrades the low-density lipoprotein receptor (LDLR), contributing to hypercholesterolemia. Adipose tissue plays a role in lipoprotein metabolism, but there are almost no data about PCSK9 and LDLR regulation in human adipocytes. We studied PCSK9 and LDLR regulation by insulin, atrial natriuretic peptide (ANP, a potent lipolytic agonist that antagonizes insulin), and LDL in visceral adipose tissue (VAT) and in human cultured adipocytes. *PCSK9* was expressed in VAT and its expression was positively correlated with body mass index (BMI). Both intracellular mature and secreted PCSK9 were abundant in cultured human adipocytes. Insulin induced PCSK9, LDLR, and sterol-regulatory element-binding protein-1c (SREBP-1c) and -2 expression (SREBP-2). ANP reduced insulin-induced PCSK9, especially in the context of a medium simulating hyperglycemia. Human LDL induced both mature and secreted PCSK9 and reduced LDLR. ANP indirectly blocked the LDLR degradation, reducing the positive effect of LDL on PCSK9. In conclusion, *PCSK9* is expressed in human adipocytes. When the expression of PCSK9 is induced, LDLR is reduced through the PCSK9-mediated degradation. On the contrary, when the induction of PCSK9 by insulin and LDL is partially blocked by ANP, the LDLR degradation is reduced. This suggests that NPs could be able to control LDLR levels, preventing PCSK9 overexpression.

Keywords: PCSK9; natriuretic peptides; adipose tissue; lipid metabolism; LDL receptor; insulin

---

1. Introduction

Maintenance of optimal blood lipid levels is central to vascular health. Liver plays a key role in lipoprotein metabolism, but much less is known about the role of adipose tissue. The adipose tissue is involved in energy balance and energy storage. It has endocrine functions and plays a fundamental role in the metabolism of triglyceride-rich lipoproteins. Recent studies suggest that thermogenic "brown" adipocytes are also involved in lipoprotein metabolism. Brown adipose tissue not only takes up triglycerides derived from plasma triglyceride-rich lipoproteins, but is also actively involved in the metabolic flux of high-density lipoprotein (HDL)-cholesterol to the liver [1].

Besides its role in triglyceride storage, adipose tissue contains a very large pool of free cholesterol, and adipocytes are known to support cholesterol efflux to HDL and apoA-I in vitro [2,3]. In fact, the main functional receptors for HDL, such as the ATP-binding cassette subfamily A member 1 (ABCA1) and the scavenger receptor class B type I (SR-BI), are expressed in mature adipocytes. Zhang et al. demonstrated that these receptors control cholesterol efflux. Furthermore, they suggested that adipose dysfunction caused by "inflammation", as seen in the insulin-resistance conditions, may impair HDL lipidation in the adipocytes, reducing circulating HDL-C levels [4].

On the contrary, very little is known about the adipocyte role in low-density lipoprotein (LDL) handling and about the involvement of proprotein convertase subtilisin kexin type 9 (PCSK9) on LDL receptor (LDLR) regulation. About forty years ago, some landmarking studies were carried on LDLR and adipocytes. In 1979, Angel et al. demonstrated that isolated human adipose cells contain a high-affinity receptor which can bind, internalize, and degrade LDL, suggesting that adipose tissue is an important site of LDL and HDL interactions [5]. From that time, the role of adipose tissue in lipoprotein metabolism has been largely forgotten or not widely studied.

PCSK9, a member of the proprotein convertase family, behaves mainly as a chaperon and it is highly expressed in human liver [6]. A PCSK9 gain-of-function mutation was identified as a cause of autosomal dominant familial hypercholesterolemia [7]. Indeed, PCSK9 plays a critical role in the regulation of cholesterol homeostasis. Many studies have demonstrated its involvement in the regulation of LDL cholesterol (LDL-C) levels by controlling the recyclable LDLR on hepatocyte [8,9]. When PCSK9 is released from the Golgi apparatus into the circulation, it is able to induce the degradation of LDLR after its binding with LDLR-LDL-C hepatocyte surface complex [7,10,11]. The LDLR, the hydroxy-methyl-glutaryl CoA (HMG-CoA) reductase, and the PCSK9 are co-regulated by the sterol regulatory element binding protein-2 (SREBP-2), to prevent excessive cholesterol uptake and preserve cholesterol homeostasis [12]. Therefore, pharmacological activation of the *SREBP* pathway by HMG-CoA reductase inhibitors (statins) induce *PCSK9* expression in experimental and clinical settings [13,14]. Furthermore, *SREBP-1c* also appeared to be involved in the induction of *PCSK9* by insulin [12,15,16]. The role of SREBP-1c in the regulation of PCSK9 levels has also been observed in humans, where PCSK9 is positively correlated with insulin resistance, liver steatosis, and very low-density lipoprotein-triglyceride (VLDL-TG) levels [17]. This evidence suggests that PCSK9 may be also implicated in the metabolism of TG-rich lipoproteins, such as VLDL and intermediate-density lipoprotein (IDL), that can be uptaken by the LDLR via apoB and apoE binding.

Taken together, we suggest a role of adipose tissue in PCSK9-mediated lipoprotein metabolism, although there are almost no data about PCSK9 in human adipocytes. Recently, *PCSK9* was detected in mice perigonadal fat [18,19]. Mice lacking *PCSK9 (PCSK9−/−)* exhibit normal weight but increased visceral adipose tissue (VAT) due to adipocyte hypertrophy, independently from the LDLR. *PCSK9−/−* mice increased both fatty acid uptake and triglyceride synthesis in the adipose tissue, together with an increased surface density of VLDL receptor [18,19]. Insulin resistance, a condition found in PCSK9−/− mice, is linked with an impaired expression of natriuretic peptides (NPs) receptors in human VAT. Moreover, opposite effects of insulin and NPs have been documented regarding lipid storage in adipocytes, with NPs antagonizing insulin-stimulating TG storage [20,21].

Cardiac NPs, including type-A (ANP) and type-B (BNP), play a crucial role in maintaining cardiovascular homeostasis, given their impact not only on blood pressure regulation, but also on glucose and lipid metabolism [22,23]. They exert several actions on adipocytes through the activation of cGMP-dependent pathway, including activation of lipolysis [24,25] and lipid oxidation [26], together with thermogenic program [27]. An important inverse association has been found between plasma LDL-C and circulating NPs in subjects with a wide range of NT-proBNP levels [28].

In our study, the first step was to verify *PCSK9* expression and secretion in human VAT and human adipocytes. Once confirmed, we studied the reciprocal role of insulin and ANP in the regulation of *PCSK9* and *LDLR* expression, as well as their respective regulatory genes in human adipocyte cell model. We hypothesized that NP activity might influence adipose tissue lipoprotein metabolism, by regulating those proteins (PCSK9 and LDLR) that play a significant role in atherogenic dyslipidemia.

2. Results

2.1. PCSK9 Expression in Human VAT

General characteristics of the studied population are summarized in Table 1. It is known that *PCSK9* is abundantly expressed in liver, small intestine, and kidney. Our present data show that *PCSK9* is abundantly expressed in human adipose tissue as well. Figure 1A shows *PCSK9* gene expression in VAT, even if highly variable among patients. The protein analysis of PCSK9 in human adipose tissue and liver revealed that the pre-form of *PCSK9* is easily detectable (72 kDa; Figure 1B). In fact, human PCSK9 is synthesized as a precursor that undergoes autocatalytic cleavage of its N-terminal prosegment in the endoplasmic reticulum (ER) necessary for its activation and function [6]. The mature form (63 kDa) is also detectable, but it is clearly weaker than pre-form, as expected. As shown in Figure 1C, *PCSK9* expression levels are significantly and positively correlated with the body mass index (BMI) of the 26 patients studied (p = 0.024), even after adjustment for gender and age (β = 0.429; 95% CI 0.023–1.224; p = 0.016). No significant correlation emerges between *PCSK9* in VAT and LDL cholesterol, according to linear regression model adjusted for gender and age (p = 0.654).

Table 1. General characteristics of studied population.

Variables	N	Mean ± SE
Gender (male/female)	26	15/11
Age (y)	26	66.9 ± 1.4
BMI (kg/m^2)	26	25.6 ± 0.8
Waist (cm)	26	96.9 ± 1.9
SBP (mmHg)	24	139.1 ± 3.1
DBP (mmHg)	24	78.9 ± 2.0
Triglycerides (mg/dL)	17	125.4 ± 9.7
Total Cholesterol (mg/dL)	17	176.1 ± 11.7
HDL Cholesterol (mg/dL)	17	39.5 ± 2.7
LDL Cholesterol (mg/dL)	17	118.2 ± 10.1
Non-HDL Cholesterol (mg/dL)	17	139.7 ± 9.9

BMI: body mass index; SBP: systolic blood pressure; DBP: diastolic blood pressure; HDL: high-density lipoprotein; LDL: low-density lipoprotein.

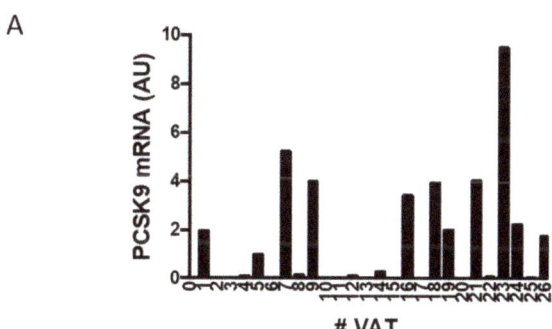

Figure 1. *Cont.*

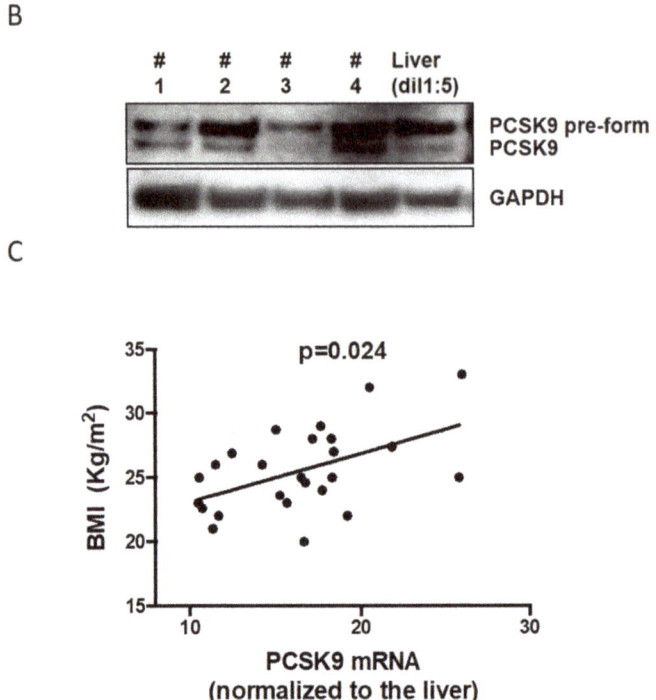

Figure 1. (**A**) *PCSK9* gene expression in VAT. Cycle threshold of *PCSK9* during real-time gene expression analysis for adipose tissue were between 21 and 24. (**B**) *PCSK9* levels in differentiated human adipocytes. (**C**) *PCSK9* in VAT according to BMI. # number of sample.

2.2. PCSK9 and LDLR Regulation in Human Adipocytes by Insulin

Differentiated adipocytes in multiple wells were treated for 1, 2, or 4 h with 10 nM of insulin. As shown in Figure 2A, *PCSK9* gene expression is 20-fold increased by insulin after 4 h, but a significant increase is already seen after 2 h. Similarly, *LDLR* (Figure 2B), *SREBP-1c* (Figure 2C), and *SREBP-2* (Figure 2D) are significantly increased by insulin with an earlier time course for the regulatory protein *SREBP-1c*. Protein analysis of PCSK9 confirms the induction of PCSK9 after insulin treatment. Interestingly, we analyzed the cell lysate as well as supernatant, and we observed that PCSK9 after 4 h treatment is increased, especially in the mature form secreted by adipocytes into the media (Figure 2E). LDLR protein is also induced by insulin in human adipocytes (Figure 2E).

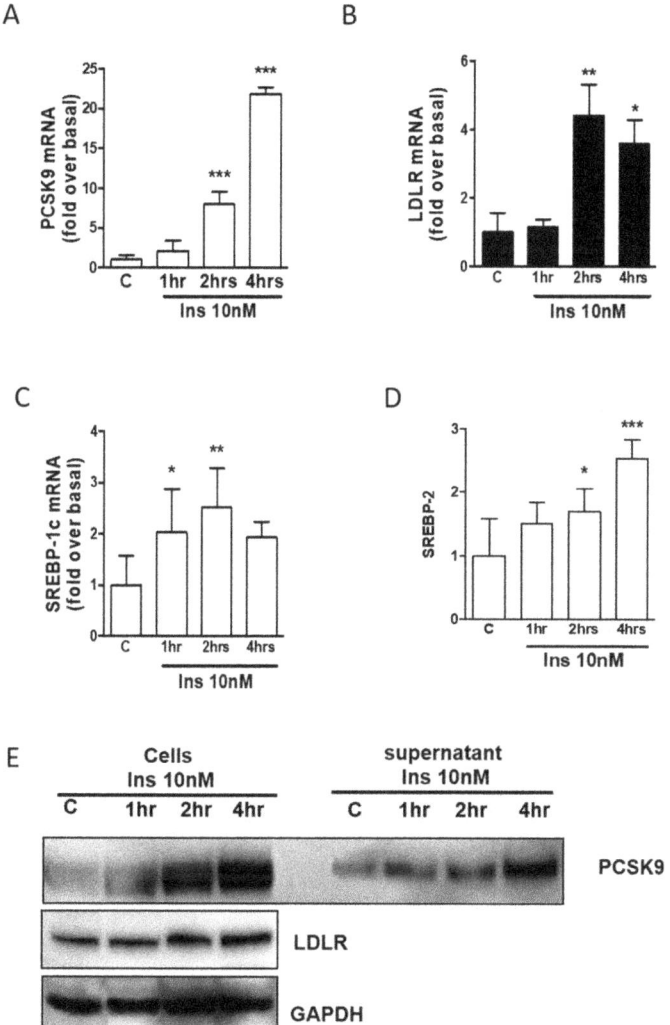

Figure 2. (**A**) *PCSK9* induction by insulin in differentiated human adipocytes. (**B**) *LDLR* induction by insulin in differentiated human adipocytes. (**C**) *SREBP-1c* induction by insulin in differentiated human adipocytes. (**D**) *SREBP-2* induction by insulin in differentiated human adipocytes. (**E**) PCSK9, LDLR, and GAPDH levels after treatment with insulin. * $p < 0.05$, ** $p < 0.01$ and *** $p < 0.001$ vs control.

2.3. PCSK9/LDLR Modulation by Cardiac Natriuretic Peptides

To understand whether PCSK9 and LDLR respond not only to insulin but also to NPs, that are physiologic antagonists of insulin effects on lipid metabolism in adipose tissue, adipocytes were treated for 4 h with insulin (10 nM), or ANP (100 nM), or insulin together with ANP. Gene expression analysis shows that *PCSK9* and *LDLR* are significantly induced by insulin, as described above, but also that ANP is able to partially block the insulin effect (Figure 3A,B). Considering the main genes involved in the regulation of cholesterologenesis, we found that *SREBP-2* is induced by insulin and ANP is able to block the insulin effect (Figure 3C). On the contrary, ANP is not able to reduce the induction of

SREBP-1c by insulin (Figure 3D). Similar results for PCSK9 and LDLR were obtained using Western blot analysis (Figure 3E).

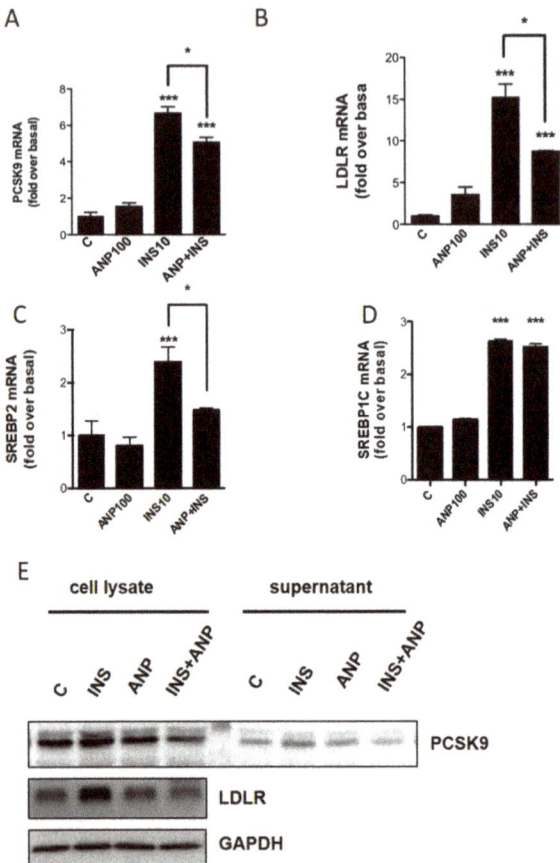

Figure 3. One-way ANOVA. (**A**) Effects of insulin and ANP on *PCSK9* in differentiated human adipocytes. (**B**) Effects of insulin and ANP on *LDLR* in differentiated human adipocytes. (**C**) Effects of insulin and ANP on *SREBP-2* in differentiated human adipocytes. (**D**) Effects of insulin and ANP on *SREBP-1c* in differentiated human adipocytes. (**E**) PCSK9, LDLR, and GAPDH levels after treatment with insulin and ANP. *** $p < 0.001$ vs control; * $p < 0.05$ insulin 10nM vs insulin+ANP treatment.

2.4. The Influence of LDL from Human Plasma

To study the physiological mechanism that links adipocyte PCSK9 with circulating LDL, human adipocytes were treated with increasing concentrations of isolated LDL from human plasma (from 25 to 100 ng/mL) for 4 and 18 h. Gene expression analysis shows that at the first time point (4 h, Figure 4A–D) the treatment with LDL induces a significant increase of all the target genes: *PCSK9*, *LDLR*, *SREBP-1c*, and *SREBP-2*. Interestingly, at the second time point (18 h) it is evident that human LDL decreases *LDLR* but, at the same time, significantly increases *PCSK9* together with *SREBP-1c* and *SREBP-2* (Figure 4F–I). Protein analysis clearly shows that, during the incubation times, PCSK9 initially increases as a mature form into the cells (Figure 4E) and then, after 18 h, increases as the mature form secreted into the media (Figure 4J). At the same time, LDLR increases after 4 h treatment (Figure 4E) and, thereafter, is significantly reduced after 18 h (Figure 4J). These time-dependent responses suggest that, at the beginning, the presence of LDL induces all the physiological pathways involved in cholesterol uptake, but after 18 h, the strong induction of PCSK9, especially for the secreted mature form, may induce LDLR reduced function by degradation.

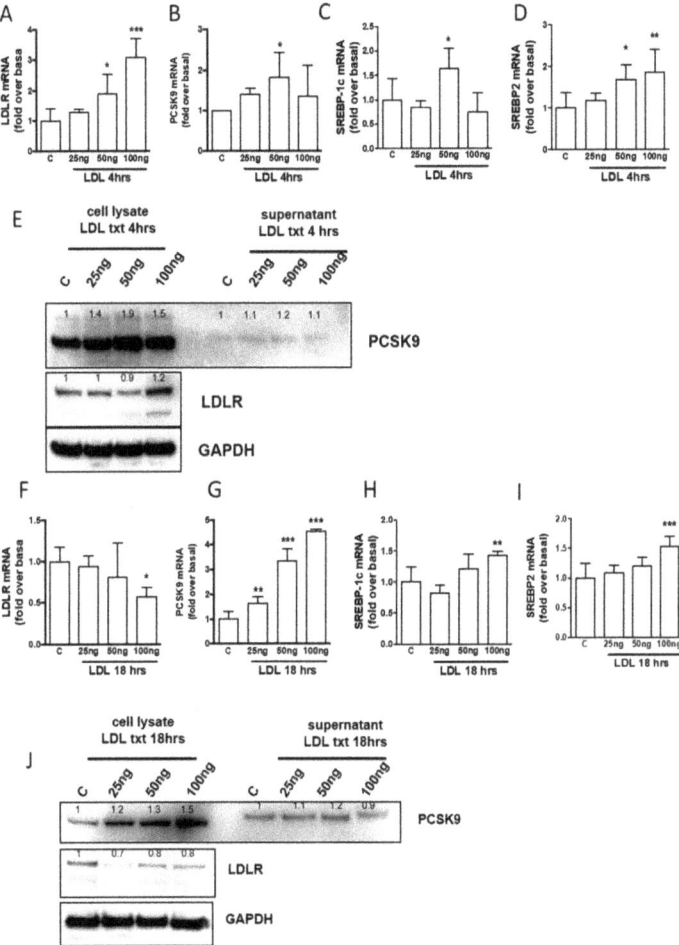

Figure 4. One-way ANOVA. (**A**) Effects of LDL treatment for 4 h on *LDLR* in differentiated human adipocytes. (**B**) Effects of LDL treatment for 4 h on *PCSK9* in differentiated human adipocytes. (**C**) Effects of LDL treatment for 4 h on *SREBP-1c* in differentiated human adipocytes. (**D**) Effects of LDL treatment for 4 h on *SREBP-2* in differentiated human adipocytes. (**E**) PCSK9, LDLR, and GAPDH proteins levels after LDL treatment for 4 h. (**F**) Effects of LDL treatment for 18 h on *LDLR* in differentiated human adipocytes. (**G**) Effects of LDL treatment for 18 h on *PCSK9* in differentiated human adipocytes. (**H**) Effects of LDL treatment for 18 h on *SREBP-1c* in differentiated human adipocytes. (**I**) Effects of LDL treatment for 18 h on SREBP-2 in differentiated human adipocytes. (**J**) PCSK9, LDLR, and GAPDH proteins levels after LDL treatment for 18 h. * $p < 0.05$, ** $p < 0.01$ and *** $p < 0.001$ vs. control.

2.5. *LDL and ANP Effects on Human Adipocytes*

PCSK9 and *LDLR* expression were also analyzed in presence of isolated LDL (50 ng/mL) and ANP (100 nM). As shown in Figure 5, protein analysis reveals that 4 h of treatment with LDL significantly induces PCSK9, as well as LDLR, and that this effect is blocked by ANP. After 18 h, it is still evident that, while PCSK9 is clearly induced also in the secreted form, LDLR is reduced, as described in Figure 4. Interestingly, ANP is still able to block the effect of LDL on target genes after 4 and 18 h treatments (Figure 5).

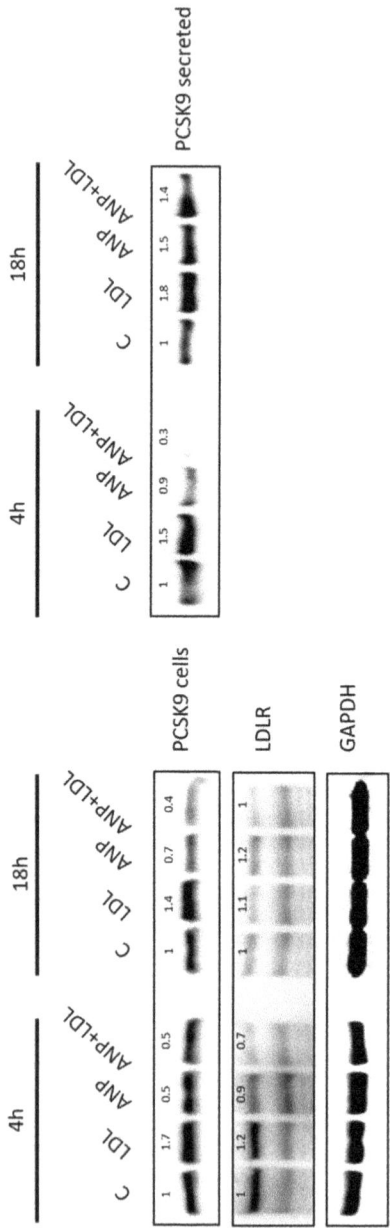

Figure 5. PCSK9, LDLR, and GAPDH levels after treatment with LDL and ANP for 4 and 18 h.

3. Discussion

The focus of this study was to investigate the expression and regulation of PCSK9 in human adipose tissue and adipocytes, using insulin and ANP, two opposing hormones on lipid metabolism [20]. It is known that the active form of PCSK9 is derived from three different steps: the cleavage of the signal sequences, the intramolecular proteolysis of proPCSK9, and trafficking through the *trans*-Golgi network before the secretion into extracellular space, where it reaches its target, LDLR [6,29,30]. Gene expression and protein analysis revealed that the precursor form of *PCSK9* is expressed and easily detectable in human visceral perirenal fat, even if there are wide differences among subjects. The concentrations of *PCSK9* in adipose tissue positively correlated with BMI values, the most used clinical index of adiposity. Interestingly, PCSK9 plasma concentrations were also related to carotid artery intima-media thickness (cIMT) in overweight and obese individuals, in comparison with the normal weight group, suggesting that PCSK9 could be an indicator as well as a player of cardiometabolic and vascular changes induced by excessive adiposity [31].

Therefore, we investigated the regulation of PCSK9 using the Simpson–Golabi–Behmel syndrome (SGBS) human adipocyte cell line. Differentiated SGBS adipocytes were used to verify the ability of insulin, ANP, and LDL to regulate PCSK9. First, we showed that 10 nm of insulin strongly induced PCSK9 and LDLR after 2 and 4 h of treatment. It has been reported that LDLR and PCSK9 share SREBP-2 as a common regulatory pathway, and that *PCSK9* expression is also regulated by insulin via the *SREBP-1c* in primary mouse and rat hepatocytes, as well as in vivo, during hyperinsulinemic-euglycemic clamps. [32]. These *SREBPs* seem to be active also in adipocytes. Indeed, *SREBP-1c* is activated by insulin earlier than *SREBP-2*, being already significantly increased after 1 h. This suggests that *SREBP-1c* is involved in *PCSK9* regulation in human adipocytes.

We recently described that higher insulin levels, together with higher glucose concentration, simulate insulin resistance found in obese patients, and reduced the ability of NPs to induce lipolysis and the thermogenic pathway [20]. Some interactions between the PCSK family and NPs have been recently discovered, such as for PCSK6, that activates the corin, a key enzyme in the activation of ANP precursors [33,34]. Here, we show that ANP is able to reduce the insulin-mediated induction of PCSK9 and LDLR in human adipocytes. Our data also show that ANP is able to reduce the regulation of SREBP-2, but not of SREBP-1c. SREBP-1c preferentially activates genes involved in fatty acid biosynthesis or carbohydrate metabolism, including fatty acid synthetase, acetyl-CoA carboxylase, or glucokinase (GK) [32]. GK converts glucose into glucose 6-phosphate and, therefore, SREBP-1c is thought to have a permissive action on glucose-dependent gene regulation [35]. Moreover, *SREBP-1c* is induced by *LXRalpha*, that is stimulated by insulin independently by glucose concentration, as we previously described [20]. In the liver, SREBP-2 processing is also enhanced by peroxisome proliferator-activated receptor gamma (PPAR-gamma) activation, affecting both PCSK9 and LDLR expression [36]. It is known that PPAR-gamma has an important role in both cell differentiation and energy metabolism in adipocytes [37,38]. The role of PPAR-gamma activation on these proteins in human adipocytes could be an interesting aspect to be explored in future studies.

Given the close relation between circulating PCSK9 and LDL levels in human plasma [39–41], we tested the role of different concentrations of LDL on PCSK9 and LDLR expression in our human adipocyte cell model. We used LDL isolated from human plasma to test the functionality of the LDLR/PCSK9 system, and our data indicate a fully functional system. The object of the present study was only to evaluate the expression and regulation of PCSK9 mediated by insulin and ANP. Future research, focused on the direct role of PCSK9 and LDLR in VAT and the relationship between PCSK9 and the morphology of adipocytes (i.e., lipid droplet formation or differentiation status), is warranted.

Published data reported that expression of PCSK9 in cultured cells has variable effects on LDLR. In some cell types, such as human hepatoma cells (HepG2 and HuH7) or human embryonic kidney cells (HEK-293 cells), PCSK9 expression dramatically reduces LDLR levels [29,42,43]. In other cells types, including fibroblasts, Chinese hamster ovarian (CHO-K1), monkey kidney cells (COS7) and rat liver cells (McArdle RH7777), PCSK9 is not likely to affect LDLR expression [42–44]. Here, we found that human LDL, added to the cells culture media, initially—after 4 h treatment—induces LDLR and pre-form PCSK9. After 18 h of treatment, the high levels of mature PCSK9 are secreted into the media, and are able to induce the degradation of LDLR. In fact, after 18 h, LDLR protein levels are reduced compared with LDLR after 4 h LDL treatment.

Moreover, when we tested the effect of LDL combined with ANP, we observed that LDL induced LDLR and the non-secreted form of PCSK9 after 4 h. At the same time, ANP is able to partially block the LDL-induced regulation. After 18 h, we confirmed that the LDL-mediated induction of PCSK9 was still present, and the PCSK9 secreted form was also significantly induced.

The elevated secretion of PCSK9 could represent the mechanism for the reduction of LDLR. Indeed, in the cells treated with ANP, where PCSK9 is reduced, the levels of LDLR are stable, suggesting again that ANP, by blocking the induction of PCSK9, indirectly reduced the degradation of LDLR.

Although in our research the regulatory effect of ANP appears to be modest, to the best of our knowledge, this is the first study demonstrating that *PCSK9* is expressed in adipose tissue, and ANP is able to interact with PCSK9 pathways. Several studies showed the role of NPs on lipid metabolism and adipose tissue, but the interaction of ANP with cholesterol metabolism through PCSK9 and LDLR regulation is absolutely new. ANP partially blocked the effect of insulin and LDL in adipocytes, but it was more effective in culture conditions simulating hyperglycemia. Overall, NPs are likely to reduce triglycerides and to increase cholesterol in human adipocytes, opposing insulin effects. Such activities may also have systemic consequences on blood lipid levels (Figure 6). We believe that this study could be the first of future studies on human adipose tissue, in order to better explain which mechanisms are involved in lipid metabolism.

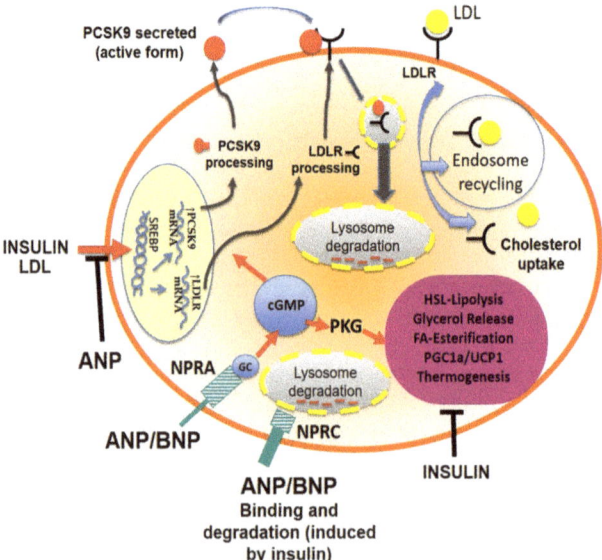

Figure 6. Interactions between insulin and cardiac natriuretic peptides (NPs) in lipid metabolism and PCSK9-LDL receptor handling. The potent lipolytic activity of NPs (ANP and BNP secreted from the heart), mediated by NPRA, is strongly reduced by insulin through the induction of the clearance receptor (NPRC), which binds and degrades NPs. Insulin also induces PCSK9, but this effect is opposed by NPs, especially in conditions simulating hyperglycemia. Overall, the NPs are likely to reduce the triglycerides and to increase cholesterol in human adipocytes, opposing insulin effects. Such activities may also have systemic consequences on blood lipid levels. T-bar arrow indicates inhibitory activity.

In conclusion, our data show a potentially relevant role for adipose tissue and adipocytes in the regulation of PCSK9-LDLR in humans, similar to what is known for human liver. NPs appear to play a key role in this pathway, with potentially important implications, especially in patients with obesity and hypertension, in which NPs could be able to regulate not only the blood pressure, but also the LDLR levels, preventing PCSK9 overexpression.

4. Materials and Methods

4.1. Reagents and Antibodies

Insulin, dexamethasone, isobutylmethylxanthine, tri-iodothyronine, transferrin, and wortmannin were obtained from Sigma-Aldrich (St. Louis, Missouri, USA). Antisera against PCSK9 and LDLR were from Abcam (Abcam, Cambridge, MA, USA). GAPDH antibody (cat #SC25778) and secondary antibody anti-rabbit (cat #SC2054) were from Santa Cruz Biotech; CA, USA. SuperSignal West Femto Maximum Sensitivity Substrate was from Thermo Scientific, Rockford, IL, USA.

4.2. Human Adipose Tissue

A set of human VAT samples ($n = 26$) were obtained from patients undergoing radical nephrectomy for localized clear cell renal carcinoma (without any evidence of local or metastatic cancer spread: T1/T2, N0, M0) at the "Ospedali Riuniti" University Hospital of Ancona, Italy. All women were in menopause. Patients with diabetes were excluded from the study, therefore, no patients took insulin or any other medications. The study was conducted in accordance with the guidelines proposed in The Declaration of Helsinki and the local Ethics Committee approved the study protocol (ID: 206964, date: 28 September 2006). All patients gave written informed consent for the collection of clinical data and tissue samples.

4.3. Adipocyte Cells Culture

Cells from Simpson–Golabi–Behmel syndrome cell line (SGBS) were grown in growth medium (DMEM/F12 with 10% fetal calf serum). When adipocytes reached 85%–90% confluence, they were differentiated in differentiation medium that included insulin, as previously described. To test the acute effect of insulin on NP receptors, we removed insulin from differentiation media at day 7 instead of day 11/12, as previously described [20]. Thus, on the seventh day, the cells were washed and deprived of insulin at least for three days, and were then treated with insulin and/or atrial natriuretic peptide (ANP). To assess whether PCSK9 expression and related genes are modulated by insulin, differentiated adipocytes were treated with 10 nM of insulin for 1, 2, and 4 h. PCSK9 regulation was also evaluated in response to cardiac NPs. Adipocytes were treated with 100 nM of ANP, as used in previous work [20]. ANP was also used together with insulin to evaluate the relative power of these two physiological counteracting (lipolytic vs. lipogenic) hormones on PCSK9/LDLR. At least 6 different wells in 3 different experiments were performed for each treatment.

4.4. Isolation of Human Plasma LDL

The role of LDL on PCK9 and LDLR expression was studied using LDL separated from human plasma. Fresh pooled human plasma from fasting healthy young volunteers (coauthors of these manuscript) was used for the preparation of LDL. LDL (density between 1.025 and 1.063 g/mL) were isolated by single vertical spin gradient ultracentrifugation, as described by Chung et al. [45]. After dialysis at 4 °C for 24 h against 10 mmol/L PBS (pH 7.4), LDLs were concentrated in a SpeedVac Concentrator and LDL protein concentration was determined by the method of Bradford [46].

The first set of experiments was performed to verify the effect of different concentrations of LDLs, from 25 to 100 ng/mL, after 4 and 18 h of incubations. Subsequently, PCSK9 and LDLR were analyzed after 4 and 18 h of treatment with human LDL (50 ng/mL), ANP, or together (ANP + LDL).

4.5. RNA Isolation and Gene Expression Analysis

Total RNA was extracted using TRIzol (Invitrogen, Carlsbad, CA, USA) and RNA reverse transcription of 2 µg was performed with High-Capacity cDNA Reverse Transcription Kit with RNase Inhibitor (Applied Biosystems, Warrington, UK). All gene expression experiments in SGBS and primary VAT adipocytes cultures were analyzed with SYBR Select Master Mix (Applied Biosystems Darmstadt, Germany). Each single gene expression experiment was performed in triplicate. Differences in total RNA or different efficiency of cDNA synthesis among samples were normalized using human GAPDH expression.

4.6. Western Blotting

Treated cells were lysed and sonicated in an appropriate buffer, as previously described [27]. Protein concentrations were determined using the Bradford Assay (Biorad, Hercules, CA, USA) and 50 µg of total proteins was resolved in 12% sodium dodecyl sulfate-polyacrylamide gel electrophoresis (SDS-PAGE), transferred to a PVDF membrane (Immobilon P, Millipore, Burlington, MA, USA), and probed overnight at 4 °C with specific PCSK9 or LDLR primary antibodies. Secondary antisera against rabbit IgG conjugated with peroxidase was used for specific protein detection. Target proteins were visualized using an enhanced chemiluminescent substrate (SuperSignal West Femto Maximum Sensitivity Substrate, Pierce) and were measured in comparison with GAPDH (Santa Cruz Biotech, Dallas, TX, USA). Image acquisition was performed on an ALLIANCE MINI HD9 (UVITEC, Cambridge UK). All lines were quantified with UVI-TEC NineAlliance analysis software and each line sample of Western blot shows the relative number of quantification, compared with control, that was considered as 1 (100%). In some cases, membranes were 'stripped' by incubation in a buffer (0.76 g Tris, 2 g SDS, 700 µL β-mercaptoethanol in 100 mL) at 37 °C for 45 min, in order to be subsequently probed with additional antibodies.

4.7. Statistical Analysis

Results are presented as mean ± SEM, unless otherwise indicated. Data were analyzed using two-tailed Student's *t*-test, one-way ANOVA, followed by post hoc Newman–Keuls tests when F was significant. A non-parametric test for two related samples (Wilcoxon's signed ranks test) was used to identify differences between each treated group and controls, differences between more than 2 groups were analyzed by analysis of variance and *post hoc* Holm-Bonferroni test. Pearson's correlation coefficient was used to assess the association between PCSK9 gene expression and BMI. Multiple linear regression was used to create adjusted models. SPSS 11.0 software was used for statistical analysis (SPSS Inc., Chicago, IL, USA) and a $p < 0.05$ was considered significant.

Author Contributions: Conceptualization, R.S. and M.B.; methodology, L.M. and G.F.; formal analysis, M.B. and T.B.; investigation, M.B. and A.V.; resources, F.S., G.F. and C.D.P.; writing—original draft preparation, M.B., A.V. and F.S.; writing—review and editing, L.M. and R.S.; visualization, M.B.; supervision, R.S.; project administration, R.S. and L.M.; funding acquisition, R.S.

Funding: This research was funded by University "Politecnica delle Marche" (Ricerca di Ateneo to R. Sarzani).

Acknowledgments: We thank Saverio Cinti for the use of microscopes and Martin Wabitsch for the SGBS cell line.

Conflicts of Interest: The authors declare no conflict of interest.

References

1. Khedoe, P.P.; Hoeke, G.; Kooijman, S.; Dijk, W.; Buijs, J.T.; Kersten, S.; Havekes, L.M.; Hiemstra, P.S.; Berbée, J.F.; Boon, M.R.; et al. Brown adipose tissue takes up plasma triglycerides mostly after lipolysis. *J. Lipid Res.* **2015**, *56*, 51–59. [CrossRef] [PubMed]
2. Krause, B.R.; Hartman, A.D. Adipose tissue and cholesterol metabolism. *J. Lipid Res.* **1984**, *25*, 97–110. [PubMed]

3. Prattes, S.; Hörl, G.; Hammer, A.; Blaschitz, A.; Graier, W.F.; Sattler, W.; Zechner, R.; Steyrer, E. Intracellular distribution and mobilization of unesterified cholesterol in adipocytes: Triglyceride droplets are surrounded by cholesterol-rich ER-like surface layer structures. *J. Cell Sci.* **2000**, *113*, 2977–2989. [PubMed]
4. Zhang, Y.; McGillicuddy, F.C.; Hinkle, C.C.; O'Neill, S.; Glick, J.M.; Rothblat, G.H.; Reilly, M.P. Adipocyte modulation of high-density lipoprotein cholesterol. *Circulation* **2010**, *121*, 1347–1355. [CrossRef] [PubMed]
5. Angel, A.; D'Costa, M.A.; Yuen, R. Low density lipoprotein binding, internalization, and degradation in human adipose cells. *Can. J. Biochem.* **1979**, *57*, 578–587. [CrossRef] [PubMed]
6. Seidah, N.G.; Prat, A. The biology and therapeutic targeting of the proprotein convertases. *Nat. Rev. Drug Discov.* **2012**, *11*, 367–383. [CrossRef] [PubMed]
7. Abifadel, M.; Varret, M.; Rabès, J.P.; Allard, D.; Ouguerram, K.; Devillers, M.; Cruaud, C.; Benjannet, S.; Wickham, L.; Erlich, D.; et al. Mutations in PCSK9 cause autosomal dominant hypercholesterolemia. *Nat. Genet.* **2003**, *34*, 154–156. [CrossRef]
8. Shimada, Y.J.; Cannon, C.P. PCSK9 (Proprotein convertase subtilisin/kexin type 9) inhibitors: Past, present, and the future. *Eur. Heart J.* **2015**, *36*, 2415–2424. [CrossRef]
9. Chapman, M.J.; Stock, J.K.; Ginsberg, H.N.; PCSK9 Forum. PCSK9 inhibitors and cardiovascular disease: Heralding a new therapeutic era. *Curr. Opin. Lipidol.* **2015**, *26*, 511–520. [CrossRef]
10. Ferri, N.; Corsini, A.; Macchi, C.; Magni, P.; Ruscica, M. Proprotein convertase subtilisin kexin type 9 and high-density lipoprotein metabolism: Experimental animal models and clinical evidence. *Transl. Res.* **2016**, *173*, 19–29. [CrossRef]
11. McNutt, M.C.; Lagace, T.A.; Horton, J.D. Catalytic activity is not required for secreted PCSK9 to reduce low density lipoprotein receptors in HepG2 cells. *J. Biol. Chem.* **2007**, *282*, 20799–20803. [CrossRef] [PubMed]
12. Maxwell, K.N.; Soccio, R.E.; Duncan, E.M.; Sehayek, E.; Breslow, J.L. Novel putative SREBP and LXR target genes identified by microarray analysis in liver of cholesterol-fed mice. *J. Lipid Res.* **2003**, *44*, 2109–2119. [CrossRef]
13. Guo, Y.L.; Liu, J.; Xu, R.X.; Zhu, C.G.; Wu, N.Q.; Jiang, L.X.; Li, J.J. Short-term impact of low-dose atorvastatin on serum proprotein convertase subtilisin/kexin type 9. *Clin. Drug Investig.* **2013**, *33*, 877–883. [CrossRef] [PubMed]
14. Careskey, H.E.; Davis, R.A.; Alborn, W.E.; Troutt, J.S.; Cao, G.; Konrad, R.J. Atorvastatin increases human serum levels of proprotein convertase subtilisin/kexin type 9. *J. Lipid Res.* **2008**, *49*, 394–398. [CrossRef] [PubMed]
15. Horton, J.D.; Shah, N.A.; Warrington, J.A.; Anderson, N.N.; Park, S.W.; Brown, M.S.; Goldstein, J.L. Combined analysis of oligonucleotide microarray data from transgenic and knockout mice identifies direct SREBP target genes. *Proc. Natl. Acad. Sci. USA* **2003**, *14*, 12027–12032. [CrossRef] [PubMed]
16. Dubuc, G.; Tremblay, M.; Paré, G.; Jacques, H.; Hamelin, J.; Benjannet, S.; Boulet, L.; Genest, J.; Bernier, L.; Seidah, N.G.; et al. A new method for measurement of total plasma PCSK9: Clinical applications. *J. Lipid Res.* **2010**, *51*, 140–149. [CrossRef]
17. Cariou, B.; Langhi, C.; Le Bras, M.; Bortolotti, M.; Lê, K.A.; Theytaz, F.; Le May, C.; Guyomarc'h-Delasalle, B.; Zaïr, Y.; Kreis, R.; et al. Plasma PCSK9 concentrations during an oral fat load and after short term high-fat, high-fat high-protein and high-fructose diets. *Nutr. Metab.* **2013**, *10*, 4. [CrossRef]
18. Roubtsova, A.; Munkonda, M.N.; Awan, Z.; Marcinkiewicz, J.; Chamberland, A.; Lazure, C.; Cianflone, K.; Seidah, N.G.; Prat, A. Circulating proprotein convertase subtilisin/kexin 9 (PCSK9) regulates VLDLR protein and triglyceride accumulation in visceral adipose tissue. *Arterioscler. Thromb. Vasc. Biol.* **2011**, *31*, 785–791. [CrossRef]
19. Roubtsova, A.; Chamberland, A.; Marcinkiewicz, J.; Essalmani, R.; Fazel, A.; Bergeron, J.J.; Seidah, N.G.; Prat, A. PCSK9 deficiency unmasks a sex- and tissue-specific subcellular distribution of the LDL and VLDL receptors in mice. *J. Lipid Res.* **2015**, *56*, 2133–2142. [CrossRef]
20. Bordicchia, M.; Ceresiani, M.; Pavani, M.; Minardi, D.; Polito, M.; Wabitsch, M.; Cannone, V.; Burnett, J.C., Jr.; Dessì-Fulgheri, P.; Sarzani, R. Insulin/glucose induces natriuretic peptide clearance receptor in human adipocytes: A metabolic link with the cardiac natriuretic pathway. *Am. J. Physiol. Regul. Integr. Comp. Physiol.* **2016**, *311*, R104–R114. [CrossRef]
21. Sarzani, R.; Spannella, F.; Giulietti, F.; Balietti, P.; Cocci, G.; Bordicchia, M. Cardiac Natriuretic Peptides, Hypertension and Cardiovascular Risk. *High Blood Press Cardiovasc. Prev.* **2017**, *24*, 115–126. [CrossRef] [PubMed]

22. Coué, M.; Barquissau, V.; Morigny, P.; Louche, K.; Lefort, C.; Mairal, A.; Carpéné, C.; Viguerie, N.; Arner, P.; Langin, D.; et al. Natriuretic peptides promote glucose uptake in a cGMP-dependent manner in human adipocytes. *Sci. Rep.* **2018**, *8*, 1097. [CrossRef] [PubMed]
23. Volpe, M.; Rubattu, S.; Burnett, J., Jr. Natriuretic peptides in cardiovascular diseases: Current use and perspectives. *Eur. Heart J.* **2014**, *35*, 419–425. [CrossRef] [PubMed]
24. Sengenès, C.; Berlan, M.; De Glisezinski, I.; Lafontan, M.; Galitzky, J. Natriuretic peptides: A new lipolytic pathway in human adipocytes. *FASEB J.* **2000**, *14*, 1345–1351. [CrossRef] [PubMed]
25. Moro, C.; Galitzky, J.; Sengenes, C.; Crampes, F.; Lafontan, M.; Berlan, M. Functional and pharmacological characterization of the natriuretic peptide-dependent lipolytic pathway in human fat cells. *J. Pharmacol. Exp. Ther.* **2004**, *308*, 984–992. [CrossRef]
26. Schlueter, N.; de Sterke, A.; Willmes, D.M.; Spranger, J.; Jordan, J.; Birkenfeld, A.L. Metabolic actions of natriuretic peptides and therapeutic potential in the metabolic syndrome. *Pharmacol. Ther.* **2014**, *144*, 12–27. [CrossRef] [PubMed]
27. Bordicchia, M.; Liu, D.; Amri, E.Z.; Ailhaud, G.; Dessì-Fulgheri, P.; Zhang, C.; Takahashi, N.; Sarzani, R.; Collins, S. Cardiac natriuretic peptides act via p38 MAPK to induce the brown fat thermogenic program in mouse and human adipocytes. *J. Clin. Investig.* **2012**, *122*, 1022–1036. [CrossRef]
28. Spannella, F.; Giulietti, F.; Cocci, G.; Landi, L.; Borioni, E.; Lombardi, F.E.; Rosettani, G.; Bernardi, B.; Bordoni, V.; Giordano, P.; et al. N-terminal pro B-Type natriuretic peptide is inversely correlated with low density lipoprotein cholesterol in the very elderly. *Nutr. Metab. Cardiovasc. Dis.* **2018**, *28*, 629–635. [CrossRef] [PubMed]
29. Benjannet, S.; Rhainds, D.; Essalmani, R.; Mayne, J.; Wickham, L.; Jin, W.; Asselin, M.C.; Hamelin, J.; Varret, M.; Allard, D.; et al. NARC-1/PCSK9 and its natural mutants: Zymogen cleavage and effects on the low density lipoprotein (LDL) receptor and LDL cholesterol. *J. Biol. Chem.* **2004**, *279*, 48865–48875. [CrossRef]
30. Naureckiene, S.; Ma, L.; Sreekumar, K.; Purandare, U.; Lo, C.F.; Huang, Y.; Chiang, L.W.; Grenier, J.M.; Ozenberger, B.A.; Jacobsen, J.S.; et al. Functional characterization of Narc 1, a novel proteinase related to proteinase K. *Arch. Biochem. Biophys.* **2003**, *420*, 55–67. [CrossRef]
31. Tóth, Š.; Fedačko, J.; Pekárová, T.; Hertelyová, Z.; Katz, M.; Mughees, A.; Kuzma, J.; Štefanič, P.; Kopolovets, I.; Pella, D. Elevated Circulating PCSK9 Concentrations Predict Subclinical Atherosclerotic Changes in Low Risk Obese and Non-Obese Patients. *Cardiol. Ther.* **2017**, *6*, 281–289. [CrossRef] [PubMed]
32. Costet, P.; Cariou, B.; Lambert, G.; Lalanne, F.; Lardeux, B.; Jarnoux, A.L.; Grefhorst, A.; Staels, B.; Krempf, M. Hepatic PCSK9 expression is regulated by nutritional status via insulin and sterol regulatory element-binding protein 1c. *J. Biol. Chem.* **2006**, *281*, 6211–6218. [CrossRef] [PubMed]
33. Chen, S.; Cao, P.; Dong, N.; Peng, J.; Zhang, C.; Wang, H.; Zhou, T.; Yang, J.; Zhang, Y.; Martelli, E.E.; et al. PCSK6-mediated corin activation is essential for normal blood pressure. *Nat. Med.* **2015**, *21*, 1048–1053. [CrossRef] [PubMed]
34. Volpe, M.; Rubattu, S. Novel insights into the mechanisms regulating pro-atrial natriuretic peptide cleavage in the heart and blood pressure regulation: Proprotein convertase subtilisin/kexin 6 is the corin activating enzyme. *Circ. Res.* **2016**, *118*, 196–198. [CrossRef] [PubMed]
35. Dentin, R.; Pégorier, J.P.; Benhamed, F.; Foufelle, F.; Ferré, P.; Fauveau, V.; Magnuson, M.A.; Girard, J.; Postic, C. Hepatic glucokinase is required for the synergistic action of ChREBP and SREBP-1c on glycolytic and lipogenic gene expression. *J. Biol. Chem.* **2004**, *279*, 20314–20326. [CrossRef] [PubMed]
36. Duan, Y.; Chen, Y.; Hu, W.; Li, X.; Yang, X.; Zhou, X.; Yin, Z.; Kong, D.; Yao, Z.; Hajjar, D.P.; et al. Peroxisome Proliferator-activated receptor γ activation by ligands and dephosphorylation induces proprotein convertase subtilisin kexin type 9 and low density lipoprotein receptor expression. *J. Biol. Chem.* **2012**, *287*, 23667–23677. [CrossRef] [PubMed]
37. Imai, T.; Takakuwa, R.; Marchand, S.; Dentz, E.; Bornert, J.M.; Messaddeq, N.; Wendling, O.; Mark, M.; Desvergne, B.; Wahli, W.; et al. Peroxisome proliferator-activated receptor gamma is required in mature white and brown adipocytes for their survival in the mouse. *Proc. Natl. Acad. Sci. USA* **2004**, *101*, 4543–4547. [CrossRef]
38. Botta, M.; Audano, M.; Sahebkar, A.; Sirtori, C.R.; Mitro, N.; Ruscica, M. PPAR Agonists and Metabolic Syndrome: An Established Role? *Int. J. Mol. Sci.* **2018**, *19*. [CrossRef]

39. Huijgen, R.; Fouchier, S.W.; Denoun, M.; Hutten, B.A.; Vissers, M.N.; Lambert, G.; Kastelein, J.J. Plasma levels of PCSK9 and phenotypic variability in familial hypercholesterolemia. *J. Lipid Res.* **2012**, *53*, 979–983. [CrossRef]
40. Lambert, G.; Ancellin, N.; Charlton, F.; Comas, D.; Pilot, J.; Keech, A.; Patel, S.; Sullivan, D.R.; Cohn, J.S.; Rye, K.A.; et al. Plasma PCSK9 concentrations correlate with LDL and total cholesterol in diabetic patients and are decreased by fenofibrate treatment. *Clin. Chem.* **2008**, *54*, 1038–1045. [CrossRef]
41. Pisciotta, L.; Priore Oliva, C.; Cefalù, A.B.; Noto, D.; Bellocchio, A.; Fresa, R.; Cantafora, A.; Patel, D.; Averna, M.; Tarugi, P.; et al. Additive effect of mutations in LDLR and PCSK9 genes on the phenotype of familial hypercholesterolemia. *Atherosclerosis* **2006**, *186*, 433–440. [CrossRef] [PubMed]
42. Park, S.W.; Moon, Y.A.; Horton, J.D. Post-transcriptional regulation of low density lipoprotein receptor protein by proprotein convertase subtilisin/kexin type 9a in mouse liver. *J. Biol. Chem.* **2004**, *279*, 50630–50638. [CrossRef] [PubMed]
43. Lagace, T.A.; Curtis, D.E.; Garuti, R.; McNutt, M.C.; Park, S.W.; Prather, H.B.; Anderson, N.N.; Ho, Y.K.; Hammer, R.E.; Horton, J.D. Secreted PCSK9 decreases the number of LDL receptors in hepatocytes and in livers of parabiotic mice. *J. Clin. Investig.* **2006**, *116*, 2995–3005. [CrossRef] [PubMed]
44. Sun, X.M.; Eden, E.R.; Tosi, I.; Neuwirth, C.K.; Wile, D.; Naoumova, R.P.; Soutar, A.K. Evidence for effect of mutant PCSK9 on apolipoprotein B secretion as the cause of unusually severe dominant hypercholesterolaemia. *Hum. Mol. Genet.* **2005**, *14*, 1161–1169. [CrossRef] [PubMed]
45. Chung, B.H.; Segrest, J.P.; Ray, M.J.; Brunzell, J.D.; Hokanson, J.E.; Krauss, R.M.; Beaudrie, K.; Cone, J.T. Single vertical spin density gradient ultracentrifugation. *Methods Enzymol.* **1986**, *128*, 181–209.
46. Bradford, M. A rapid and sensitive method for the quantitation of microgram quantities of protein utilizing the principle of protein-dye binding. *Anal. Biochem.* **1976**, *72*, 248–254. [CrossRef]

© 2019 by the authors. Licensee MDPI, Basel, Switzerland. This article is an open access article distributed under the terms and conditions of the Creative Commons Attribution (CC BY) license (http://creativecommons.org/licenses/by/4.0/).

Review

Molecular Implications of Natriuretic Peptides in the Protection from Hypertension and Target Organ Damage Development

Speranza Rubattu [1,2,*], Maurizio Forte [2], Simona Marchitti [2] and Massimo Volpe [1,2]

1. Department of Clinical and Molecular Medicine, School of Medicine and Psychology, Sapienza University of Rome, 00189 Rome, Italy; massimo.volpe@uniroma1.it
2. IRCCS Neuromed, 86077 Pozzilli, Italy; maurizio.forte@neuromed.it (M.F.); simona.marchitti@neuromed.it (S.M.)
* Correspondence: rubattu.speranza@neuromed.it

Received: 10 January 2019; Accepted: 8 February 2019; Published: 13 February 2019

Abstract: The pathogenesis of hypertension, as a multifactorial trait, is complex. High blood pressure levels, in turn, concur with the development of cardiovascular damage. Abnormalities of several neurohormonal mechanisms controlling blood pressure homeostasis and cardiovascular remodeling can contribute to these pathological conditions. The natriuretic peptide (NP) family (including ANP (atrial natriuretic peptide), BNP (brain natriuretic peptide), and CNP (C-type natriuretic peptide)), the NP receptors (NPRA, NPRB, and NPRC), and the related protease convertases (furin, corin, and PCSK6) constitute the NP system and represent relevant protective mechanisms toward the development of hypertension and associated conditions, such as atherosclerosis, stroke, myocardial infarction, heart failure, and renal injury. Initially, several experimental studies performed in different animal models demonstrated a key role of the NP system in the development of hypertension. Importantly, these studies provided relevant insights for a better comprehension of the pathogenesis of hypertension and related cardiovascular phenotypes in humans. Thus, investigation of the role of NPs in hypertension offers an excellent example in translational medicine. In this review article, we will summarize the most compelling evidence regarding the molecular mechanisms underlying the physiological and pathological impact of NPs on blood pressure regulation and on hypertension development. We will also discuss the protective effect of NPs toward the increased susceptibility to hypertensive target organ damage.

Keywords: natriuretic peptides; hypertension; stroke; cardiac hypertrophy; linkage analysis; genetic variants; animal models

1. Introduction

Hypertension is a complex trait that results from both environmental and genetic factors [1]. Several mechanisms have been highlighted as potential contributors to the etiopathogenesis of essential hypertension, mostly based on common knowledge of blood pressure regulation. In this regard, the biological properties of the cardiac natriuretic peptides (NPs)—that is, natriuresis, diuresis, and vasorelaxation [2], implying a direct modulatory role on blood pressure homeostasis—make all components of the NP family, including the natriuretic peptide receptors (NPRA (type A natriuretic peptide receptor), NPRB (type B natriuretic peptide receptor), and NPRC (type C natriuretic peptide receptor)), and the related protease convertases (furin, corin, PCSK6), as major etiopathogenetic candidates for hypertension. Moreover, the functional properties of these hormones involve anti-hypertrophic, anti-proliferative, and anti-inflammatory effects [3] that may significantly impact on the process of cardiovascular remodeling, such as that observed in hypertension.

Herein, we present evidence supporting the contribution of NPs to hypertension development obtained by both experimental and clinical investigations. As will be outlined below, the most compelling evidence derives from the genetic approach. The impact of NPs on the pathogenesis of hypertensive target organ damage will be also discussed.

2. Role of NP Circulating Levels in the Development of High Blood Pressure

Several physiological experiments have shown that αANP, the product of the cleavage of proANP by corin, along with the N-terminal residues 1–98, is a key regulator of systemic blood pressure through its natriuretic, diuretic, and vasorelaxant properties, all mediated by the type A natriuretic peptide receptor (NPRA) [3]. As a consequence, a reduction of ANP favors high blood pressure, whereas its increase may serve as a pharmacological tool to lower blood pressure levels. In fact, evidence that ANP levels contribute to hypertension has been initially shown by testing the effects of its infusion in hypertensive patients [4,5]. Subsequently, the anti-hypertensive effect of NPs has been documented in different pathological contexts, such as primary aldosteronism [6], pheochromocytoma [7], and hyperthyroidism [8]. In each of these conditions, the increase of NPs is able to counteract the pro-hypertensive effect of the specific endocrine disorder. Moreover, since diuresis represents a major action of NPs, the diuretic therapy reduces NP levels [9].

The genetic/molecular approach, undertaken over the last three decades, has the merit to provide further support to the significant association between NP levels and hypertension. In fact, it was shown that the knockout of *Nppa* (the gene encoding ANP) led to salt-sensitive hypertension in mice [10]. Consistently, the overexpression of *Nppa* led to hypotension [11]. Similarly, lack of NPRA caused salt-sensitive hypertension in mice [12].

On the other hand, the biologically active carboxy-terminal peptide (BNP 1–22), derived from the cleavage of the proBNP precursor by both furin and corin [13], appears to have a weaker impact on the pathogenesis of hypertension compared to ANP at the experimental level. In fact, deletion of *Nppb* in mice led to cardiac fibrosis rather than to hypertension development [14]. A hypertensive effect due to lack of *Nppb* was documented only in a rat model [15].

The results of the genetic manipulations in rodents stimulated several studies aimed at identifying the contribution of NPs to human hypertension. In this regard, the association of human *NPPA/NPPB* gene variations with circulating ANP and BNP levels was investigated for selected single nucleotide polymorphisms (SNPs) with the achievement of some remarkable results. Of note, we reported an association of the −664C > G minor allele located within the NPPA promoter and associated with lower plasma ANP levels, early onset of blood pressure increase, and the predisposition to develop hypertension in a general population from Southern Italy [16]. Contrasting evidence was obtained for the same SNP in a Japanese cohort of hypertensive patients [17].

Another NPPA variant, rs5063 (−664G > A), falling within the exon 1 of the gene and responsible of a Val-to-Met transition, was associated with blood pressure progression in the Women's Genome Health Study and with reduced blood pressure levels in a Chinese population [18,19]. A SNP in linkage disequilibrium with rs5063 (1837G > A, detected by a *Hpa*II restriction fragment length polymorphism) was associated with higher occurrence of hypertension in whites and African–American subjects [20]. Interestingly, two studies associated variants near the NPPA/NPPB locus with proANP, BNP, and NT-proBNP levels [21,22]. Among others, the minor allele at rs5068 NPPA variant turned out to be associated with higher ANP levels and with hypertension risk [21]. In this regard it has been shown that this variant falling within the 3'-UTR of NPPA, affects the quantity of the ANP transcript through an interference with the microRNA-425 [23]. In particular, in the presence of rs5068 minor allele, the microRNA-425 cannot bind to NPPA to inhibit the gene transcription. As a consequence, the ANP levels increase [23]. Moreover, lower blood pressure levels and reduced risk of hypertension were observed [21,24–26].

Subsequently four genome-wide association studies (GWAS) analyzed circulating BNP or NT-proBNP levels in association with *trans* loci near LOXL2, SLC39A8, KLKB1, and GALNT4 [27–30].

As a result of the abovementioned studies, genetic variants at the MTHFR-NPPB locus (mapping on human chromosome 1 and containing both NPPA and NPPB) appeared to act through increased ANP/BNP production to lower blood pressure levels and, consequently, to influence susceptibility to hypertension development. However, there was a need to more precisely identify the variants truly associated with a change in NP levels within the MTHFR-NPPB locus and, therefore, responsible for the hypertensive effects, which prompted subsequent investigations. In fact, a recent study testing eight independent genetic variants in two known loci (NPPA-NPPB and POC1B-GALNT4) and one novel locus (PPP3CC) found that only those variants correlated with midregional proANP levels had a statistically significant, albeit weak, impact on blood pressure, whereas variants affecting BNP levels did not [31].

Although the latter evidence appeared to further support the experimental findings in favor of a major role of ANP, rather than BNP, on blood pressure regulation and hypertension development, NPPB cannot be completely ruled out as a "hypertensive" gene. A single SNP, the rs198389 functional variant in the NPPB promoter region, is associated with NT-proBNP levels in several populations [32]. In a large biracial prospective cohort study, the rs198389 NPPB promoter variant was found to be highly associated with large differences in NT-proBNP levels in both black and white populations. Patients with the AG and GG genotypes had progressively higher NT-proBNP levels compared to those with AA genotype. Patients with the GG genotype had reduced systolic blood pressure and diastolic blood pressure levels and were 15% less likely to take anti-hypertensive medications and 19% less likely to have a diagnosis of hypertension [33].

3. Role of Other Components of the NP Family

The involvement of other members of the NP family in protection from the development of hypertension has been mainly discovered through the genetic approach, starting with the experimental evidence in mice and then moving to the human disease. Thus, both gene deletions in mice and functional variants of the corresponding human genes encoding corin, furin, NPRA, and NPRC receptors have been associated with hypertension.

Corin is the physiological proANP convertase that activates proANP in a sequence-specific manner [34]. Blocking corin expression inhibits proANP processing in cardiomyocytes [35]. The fundamental relevance of corin for the maintenance of normal blood pressure levels was revealed by the corin knockout mice model. This model carries undetectable levels of mature ANP and it develops salt-sensitive hypertension with cardiac hypertrophy [36].

A recent study revealed that proprotein convertase subtilisin/kexin-6 (PCSK6), a protease belonging to the PCSK family, is the long sought-after specific corin-activating enzyme and, as such, is a critical regulator of the whole cascade leading to proANP processing and to αANP release into circulation, ultimately controlling water–electrolyte and blood pressure homeostasis [37,38]. PCSK6 cleaves corin at the conserved activation site, converting zymogen corin to an active enzyme. PCSK6 knockout mice have no detectable corin activity in the heart and develop salt-sensitive hypertension [37], indicating a key role of PCSK6 in regulating corin activity and blood pressure levels. Interestingly, PCSK6-mediated processing of corin is reduced in the presence of corin variants (T555I and Q568P) previously associated to hypertension and to heart disease in black people [39,40], as well as in the presence of corin variants (K317E, S472G, and R539C) previously identified in patients with preeclampsia and hypertension [41]. More recently, nine corin variants were identified in a Chinese population, of which eight were characterized for the first time. Among them, the p.Arg530Ser and p.Thr924Met variants had reduced proANP processing activity, due to endoplasmic reticulum retention and impaired PCSK6-mediated zymogen activation, respectively [42].

The proprotein processing enzyme furin is the mammalian prototype of the subtilisin-like serine endoprotease family. These enzymes possess cleavage specificity for sites involving multiple basic amino acid residues and process several precursor proteins belonging to a variety of regulatory peptides and proteins [43]. In fact, furin is the enzyme responsible for the cleavage of proBNP and proCNP. It also cleaves pro-renin receptor, epithelial sodium channel, and TGF-β [43]. Based on its functional properties, the gene encoding furin can be considered a suitable candidate for hypertension development. However, despite its involvement in several pathways, only weak evidence in favor of a role of furin gene in hypertension has been provided, so far, by both a GWA study and a few case–control association studies [44,45].

The deletion of *NPRA* gene leads to salt-sensitive hypertension and left ventricular hypertrophy (LVH) in mice [12]. An insertion–deletion mutation in the 5′-UTR region of the human *NPRA* gene has been described [46]. The mutant allele lacks eight nucleotides and alters binding sites for the AP2 and *zeste* transcriptional factors. The transcriptional activity of the deletion allele is shown to be 30% lower than that of the wild-type allele. Subjects carrying the deletion allele were significantly more common in a Japanese population of hypertensive patients compared to the normotensive group [46]. These findings suggested that the deletion of the *NPRA* reduces receptor activity in Japanese individuals and may confer increased susceptibility to developing essential hypertension or left ventricular hypertrophy.

More interesting findings were obtained for the *NPRC* gene encoding the NP clearance receptor. In fact, in two large studies, a gene-centric array and a GWA study identified few blood pressure loci, including one containing NPRC [44,47]. A small case-control association study, utilizing SNPs belonging to NPRC, detected an association with hypertension and even with family history of hypertension [48]. A recent study had the merit to confirm, first of all, using a genome-wide approach, the existence of two independent blood pressure-related signals within NPRC. Moreover, the biological relevance of one variant (rs1173771) was demonstrated by showing its association with lower NPRC gene and protein expression in vascular smooth muscle cells [49].

In contrast to the significant results obtained with both NPRA and NPRC, no relevant variants were identified in the *NPRB* gene in association with essential hypertension.

4. Impact of NPs on Predisposition to Develop Hypertensive Target Organ Damage

The anti-remodeling properties of NPs may certainly explain their implication in the protection from the development of micro- and macrovascular damage, as well as of cardiac damage, that is associated with hypertension.

One of the most dramatic results of the hypertensive vascular damage is the occurrence of stroke [50]. The first evidence that *Nppa* was involved in the pathogenesis of stroke associated with hypertension was obtained in the animal model of the stroke-prone spontaneously hypertensive rat (SHRSP) through a genetic linkage analysis approach [51]. The latter was performed in a F2-hybrid cohort obtained from the SHRSP/stroke resistant SHR (SHRSR) intercross, considering the latency to stroke occurrence upon a high-salt dietary regimen as the stroke phenotype [51]. In fact, *Nppa* turned out to map, together with *Nppb*, at the peak of linkage of a quantitative trait locus (QTL) for stroke on rat chromosome 5 in this rat model [51]. The sequence analysis of both genes revealed the presence of two functional variants within *Nppa* (but none within *Nppb*), one located within the promoter region, the second one located within the coding part (exon 2). The *Nppa* variants were shown to cause an altered regulation, pro-peptide processing, and peptide function [52,53], further supporting their contribution to the development of cerebrovascular damage in SHRSP. Furthermore, the altered regulation of *Nppa* expression in the brain was found to co-segregate with increased susceptibility to stroke in the SHRSP/SHRSR F2-hybrid cohort [54].

Of note, the *Nppa* was excluded as a candidate gene for stroke in the SHRSP/Wistar Kyoto (WKY) genetic linkage analysis approach [55,56]. The different genetic background and the different stroke phenotype used in the latter study (the brain injury area produced by the middle cerebral artery occlusion, MCAO, as opposed to the stroke latency upon high salt diet used in the previous model) may certainly explain, at least in part, the different result. Despite the negative result of the genetic analysis in the study by Jeffs et al. [55], more recent evidence demonstrates that cerebral overexpression of ANP is vital to protection from brain damage in rats exposed to MCAO [57]. In fact, increased levels of ANP, achieved through an intraventricular injection, caused a significant reduction of the brain injury area [57]. Consistently it is noteworthy that a pharmacological treatment of high-salt fed SHRSP with a novel class of drugs (the type 1 angiotensin II receptor and neprilysin (NEP) inhibitor, ARNi) led to a dramatic reduction of stroke occurrence along with a significant increase of ANP levels in both circulation and brain tissue [58]. The latter evidence reinforces the beneficial cerebrovascular properties of ANP.

Moreover, when translating the experimental findings to human disease, NPPA was confirmed as a risk factor for stroke in the presence of specific variants responsible for either altered gene expression or an altered protein function [26]. In particular, the T2238C variant, falling within exon 3 of the gene and causing the synthesis of a peptide carrying two extra arginines, is the most common NPPA variant associated with stroke [59,60]. The mutant peptide favors stroke through the production of an altered endothelial function, reduced endothelium-dependent vasodilation, vascular smooth muscle cells constriction, increased platelet aggregation through a NPRC deregulated activation [61–67]. Interestingly, subjects carrying this variant are also at increased risk of ischemic heart disease [66]. Finally, a gene-by-treatment interaction was observed for the T2238C NPPA variant and the use of diuretic drug (chlortalidone) in a large cohort of hypertensive patients from the ALLHAT trial [68]. In this study, patients carrying the C variant allele had a better response to the diuretic therapy [68], suggesting a defective diuretic action of the mutant peptide.

The anti-hypertrophic and anti-fibrotic effects of ANP underlie its involvement in the pathogenesis of LVH in hypertension. The modulation of cardiac hypertrophy and fibrosis by ANP is achieved through the activation of several signaling pathways, such as the calcineurin/NFATt, sodium exchanger NHE-1, and TGFβ1/Smad pathways [69]. Thus, ANP behaves as a friend within the heart to protect it from stress stimuli [70]. At the experimental level the knockout of *Nppa* leads to cardiac hypertrophy in mice [10]. The same result is achieved when the *NPRA* gene is lacking [12] and also in the absence of corin [36]. In these models, the degree of LVH is unrelated to the levels of high blood pressure. In humans, we demonstrated that a promoter variant of *NPPA* causing reduced expression was associated with LVH in a cohort of hypertensive patients independently from the blood pressure levels [71]. Similarly, patients affected by the metabolic syndrome (that includes the diagnosis of hypertension) have plasma ANP levels inversely related to the degree of LVH such that the lower the ANP levels, the higher the degree of cardiac hypertrophy [72].

All components of the NP family are expressed in the kidney where they preserve renal function [73]. As a consequence, alteration of the NP system may contribute to renal damage occurrence in hypertension. In this regard, it has been reported that targeted disruption of NPRA provokes renal tubular damage and interstitial fibrosis with inflammation [74]. On the other hand, chronic ANP treatment ameliorates hypertension and end-organ damage in the kidney by reducing oxidative stress, increasing NO system activity, and diminishing collagen content and apoptosis in the SHR animal model [75]. Consistently, a significant protection from renal damage was also observed in the SHRSP receiving ARNi compared to SHR receiving valsartan alone [58]. As pointed out above, only ARNi treatment is able to increase the ANP levels.

NT-proBNP/BNP serves as a useful diagnostic and prognostic tool in heart failure, that is a frequent complication of the hypertensive cardiac damage. Of note, this component of the NP family has revealed a significant role into the prognosis of target organ damage in hypertensive patients [76,77]. However, in contrast to ANP, a direct involvement of BNP in the susceptibility to develop hypertensive target organ damage has not been proven either in animal models or in humans.

NPRC plays a fundamental role in NP clearance [2]. Specifically, it recognizes an 8 aa linear fragment of ANP molecule to perform peptide clearance. This process requires the ANP-NPRC internalization and is followed by ANP hydrolysis by lysosomes. In addition to its well-known function in NP clearance, NPRC can produce biological effects through the inhibition of adenylate cyclase and the involvement of Gi inhibitory proteins. Of note, its activation mediates the deleterious effects of the abovementioned T2238C ANP variant [62–64]. Evidence of a role of NPRC variants in the pathogenesis of hypertensive target organ damage, such as stroke, has been suggested by a GWA study [78]. We demonstrated that NPRC –55C > A transition contributes to the risk of early-onset ischemic stroke in an Italian cohort [79].

5. NP-Based Therapies for the Treatment of Hypertension and Related Organ Damage

This issue will be afforded by other articles from eminent scientists in this special issue of the Journal. As underscored by the experimental and human evidence discussed above, it is clear that within the NP family, ANP is the most suitable and efficacious anti-hypertensive agent. However, ANP cannot be administered orally. After many years of efforts, failures, and discoveries one later solution has been proposed in the form of a combination of a NEP inhibitor and an angiotensin II AT1 receptor blocker. NEP is responsible for the degradation of NPs, particularly ANP, and of angiotensin II [80]. This pharmacological combination, called ARNi, favors the protective effects of increased ANP levels along with inhibition of the negative effects of increased angiotensin II. ARNi is currently only indicated for the treatment of heart failure with reduced ejection fraction, but it shows potential to become a great solution for the treatment of hypertension, as documented by a few available experimental and human studies [58,81–86].

In summary, as represented in Figures 1 and 2, the molecular implications of the NP system, in the pathogenesis of hypertension and related vascular damage, are of key relevance. This system offers an excellent example in the field of translational medicine.

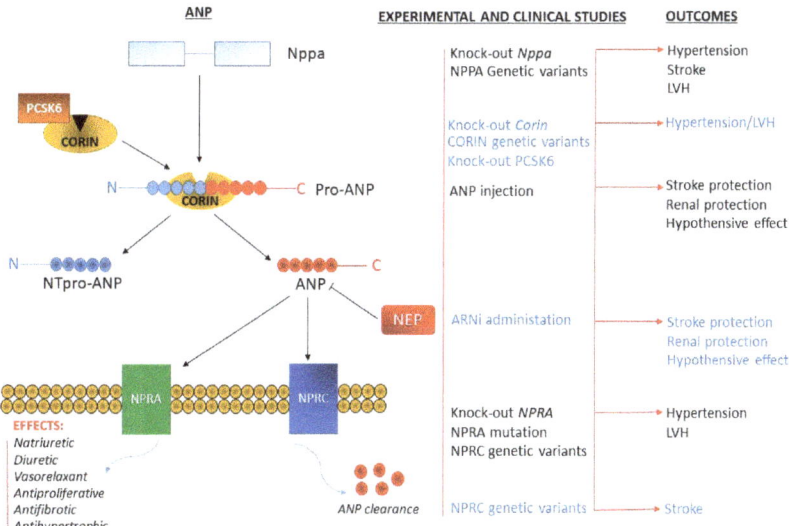

Figure 1. **Left side**: Schematic representation of the ANP processing, receptor interaction, and physiological functions. The ANP clearance by both NPRC and NEP is also shown. **Right side**: Summary of the experimental and clinical evidence showing the pathological consequences of the ANP, NPRA, corin gene deletions and gene mutations. ANP, atrial natriuretic peptide; ARNi, angiotensin type 1 receptor neprilysin inhibitor; LVH, left ventricular hypertrophy; NEP, neprilysin; NPPA, ANP gene; NPRA and NPRC, type A and type C natriuretic peptide receptors; NTpro-ANP, amino terminal pro-ANP; PCSK6, proprotein convertase subtilisin/kexin-6.

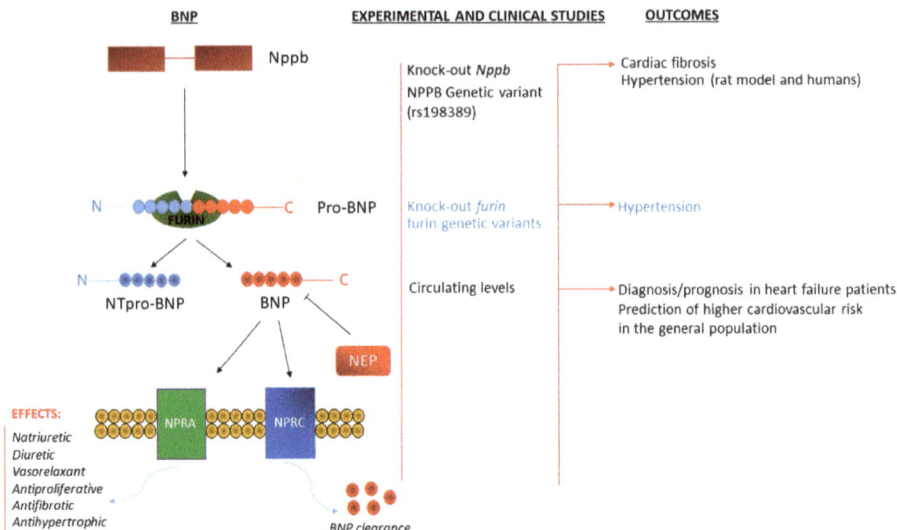

Figure 2. Left side: Schematic representation of the BNP processing, receptor interaction, clearance, and physiological functions. **Right side**: Summary of the experimental and clinical evidence showing the pathological consequences of the BNP and furin gene deletions; hBNP gene mutation and altered circulating BNP levels. BNP, brain natriuretic peptide; NEP, neprilysin; NPPB, BNP gene; NPRA and NPRC, type A and type C natriuretic peptide receptors; NT-proBNP, amino terminal pro-BNP.

Funding: This work was supported by a grant from the Italian Ministry of Health and the "5 per mille" grant.

Conflicts of Interest: The authors declare no conflict of interest.

References

1. Kunes, J.; Zicha, J. The interaction of genetic and environmental factors in the etiology of hypertension. *Physiol. Res.* **2009**, *58* (Suppl. 2), S33–S41.
2. Levin, E.R.; Gardner, D.G.; Samson, W.K. Natriuretic peptides. *NEJM* **1998**, *339*, 321–328. [PubMed]
3. Volpe, M.; Rubattu, S.; Burnett, J., Jr. Natriuretic peptides in cardiovascular diseases: Current use and perspectives. *Eur. Heart J.* **2014**, *35*, 419–425. [CrossRef] [PubMed]
4. Volpe, M.; De Luca, N.; Indolfi, C.; Mele, A.F.; Rivera, M.P.; Condorelli, M.; Trimarco, B. Acute hemodynamic effects of atrial natriuretic peptide in essential hypertension. *Kidney Int. Suppl.* **1998**, *25*, 583–585.
5. Ishii, M.; Sugimoto, T.; Matsuoka, H.; Hirata, Y.; Ishimitsu, T.; Fukul, K.; Sugimoto, T.; Kanagawa, K.; Matsuo, H. A comparative study on the hemodynamic, renal and endocrine effects of alpha-human atrial natriuretic polypeptide in normotensive persons and patients with essential hypertension. *Jpn. Circ. J.* **1986**, *50*, 1181–1184. [CrossRef] [PubMed]
6. Yamaji, T.; Ishibashi, M.; Sekihara, H.; Takaku, F.; Nakaoka, H.; Fujii, J. Plasma levels of atrial natriuretic peptide in primary aldosteronism and essential hypertension. *J. Clin. Endocrinol. Metab.* **1986**, *63*, 815–818. [CrossRef] [PubMed]
7. Stepniakowski, K.; Januszewicz, A.; Lapiński, M.; Feltynowski, T.; Chodakowska, J.; Ignatowska-Switalska, H.; Wocial, B.; Januszewicz, W. Plasma atrial natriuretic peptide (ANP) concentration in patients with pheochromocytoma. *Blood Press* **1992**, *1*, 157–161. [CrossRef] [PubMed]
8. Vesely, D.L. Atrial natriuretic peptide prohormone gene expression: Hormones and diseases that upregulate its expression. *IUBMB Life* **2002**, *53*, 153–159. [CrossRef] [PubMed]
9. Miyata, M.; Sasaki, T.; Ikeda, Y.; Shinsato, T.; Kubozono, T.; Furusho, Y.; Kusumoto, A.; Hamasaki, S.; Tei, C. COLD-CHF Investigators. Comparative study of therapeutic effects of short- and long-acting loop diuretics in outpatients with chronic heart failure (COLD-CHF). *J. Cardiol.* **2012**, *59*, 352–358. [CrossRef]

10. John, S.W.; Krege, J.H.; Oliver, P.M.; Hagaman, J.R.; Pang, S.C.; Flynn, T.G.; Smithies, O. Genetic decreases in atrial natriuretic peptide and salt-sensitive hypertension. *Science* **1995**, *267*, 679–681. [CrossRef]
11. Steinhelper, M.E.; Cochrane, K.L.; Filed, L.J. Hypotension in transgenic mice expressing atrial natriuretic factor fusion genes. *Hypertension* **1990**, *16*, 301–307. [CrossRef] [PubMed]
12. Oliver, P.M.; Fox, J.E.; Kim, R.; Rockman, H.A.; Kim, H.S.; Reddick, R.L.; Pandey, K.N.; Milgram, S.L.; Smithies, O.; Maeda, N. Hypertension, cardiac hypertrophy, and sudden death in mice lacking natriuretic peptide receptor A. *Proc. Natl. Acad. Sci. USA* **1997**, *94*, 14730–14735. [CrossRef] [PubMed]
13. Ichiki, T.; Huntley, B.K.; Burnett, J., Jr. BNP molecular forms and processing by the cardiac serine protease corin. *Adv. Clin. Chem.* **2013**, *61*, 1–31. [PubMed]
14. Tamura, N.; Ogawa, Y.; Chusho, H.; Nakamura, K.; Nakao, K.; Suda, M.; Kasahara, M.; Hashimoto, R.; Katsuura, G.; Mukoyama, M.; et al. Cardiac fibrosis in mice lacking brain natriuretic peptide. *Proc. Natl. Acad. Sci. USA* **2000**, *97*, 4239–4244. [CrossRef] [PubMed]
15. Holditch, S.J.; Schreiber, C.A.; Nini, R.; Tonne, J.M.; Peng, K.W.; Geurts, A.; Jacob, H.J.; Burnett, J.C.; Cataliotti, A.; Ikeda, Y. B-type natriuretic peptide deletion leads to progressive hypertension, associated organ damage, and reduced survival: Novel model for human hypertension. *Hypertension* **2015**, *66*, 199–210. [PubMed]
16. Rubattu, S.; Evangelista, A.; Barbato, D.; Barba, G.; Stanzione, R.; Iacone, R.; Volpe, M.; Strazzullo, P. Atrial natriuretic peptide (ANP) gene promoter variant and increased susceptibility to early development of hypertension in humans. *J. Hum. Hypertens.* **2007**, *21*, 822–824. [CrossRef] [PubMed]
17. Kato, N.; Sugiyama, T.; Morita, H.; Nabika, T.; Kurihara, H.; Yamori, Y.; Yazaki, Y. Genetic analysis of the atrial natriuretic peptide gene in essential hypertension. *Clin. Sci.* **2000**, *98*, 251–258. [CrossRef]
18. Conen, D.; Glynn, R.J.; Buring, J.E.; Ridker, P.M.; Zee, R.Y. Natriuretic peptide precursor A gene polymorphisms and risk of blood pressure progression and incident hypertension. *Hypertension* **2007**, *50*, 1114–1119. [CrossRef]
19. Zhang, S.; Mao, G.; Zhang, Y.; Tang, G.; Wen, Y.; Hong, X.; Jiang, S.; Yu, Y.; Xu, X. Association between human atrial natriuretic peptide Val7Met polymorphism and baseline blood pressure, plasma trough irbesartan concentrations, and the antihypertensive efficacy of irbesartan in rural Chinese patients with essential hypertension. *Clin. Ther.* **2005**, *27*, 1774–1784. [CrossRef]
20. Beige, J.; Ringel, J.; Hohenbleicher, H.; Rubattu, S.; Kreutz, R.; Sharma, A.M. HpaII-polymorphism of the atrial natriuretic peptide gene and essential hypertension in whites. *Am. J. Hypertens.* **1997**, *10*, 1316–1318.
21. Newton-Cheh, C.; Johnson, T.; Gateva, V.; Tobin, M.D.; Bochud, M.; Coin, L.; Najjar, S.S.; Zhao, J.H.; Heath, S.C.; Eyheramendy, S.; et al. Genome-wide association study identifies eight loci associated with blood pressure. *Nat. Genet.* **2009**, *41*, 666–676. [CrossRef]
22. Pereira, N.L.; Tosakulwong, N.; Scott, C.G.; Jenkins, G.D.; Prodduturi, N.; Chai, Y.; Olson, T.M.; Rodeheffer, R.J.; Redfield, M.M.; Weinshilboum, R.M.; et al. Circulating atrial natriuretic peptide genetic association study identifies a novel gene cluster associated with stroke in whites. *Circ. Cardiovasc. Genet.* **2015**, *8*, 141–149. [CrossRef] [PubMed]
23. Arora, P.; Wu, C.; Khan, A.M.; Bloch, D.B.; Davis-Dusenbery, B.N.; Ghorbani, A.; Spagnolli, E.; Martinez, A.; Ryan, A.; Tainsh, L.T.; et al. Atrial natriuretic peptide is negatively regulated by microRNA-425. *J. Clin. Investig.* **2013**, *123*, 3378–3382. [CrossRef] [PubMed]
24. Cannone, V.; Boerrigter, G.; Cataliotti, A.; Costello-Boerrigter, L.C.; Olson, T.M.; McKie, P.M.; Heublein, D.M.; Lahr, B.D.; Bailey, K.R.; Averna, M.; et al. A genetic variant of the atrial natriuretic peptide gene is associated with cardiometabolic protection in the general community. *J. Am. Coll. Cardiol.* **2011**, *58*, 629–636. [CrossRef]
25. Cannone, V.; Cefalu', A.B.; Noto, D.; Scott, C.G.; Bailey, K.R.; Cavera, G.; Pagano, M.; Sapienza, M.; Averna, M.R.; Burnett, J.C., Jr. The atrial natriuretic peptide genetic variant rs5068 is associated with a favorable cardiometabolic phenotype in a Mediterranean population. *Diabetes Care* **2013**, *36*, 2850–2856. [CrossRef] [PubMed]
26. Rubattu, S.; Sciarretta, S.; Volpe, M. Atrial natriuretic peptide gene variants and circulating levels: Implications in cardiovascular diseases. *Clin. Sci.* **2014**, *127*, 1–13. [CrossRef]
27. Musani, S.K.; Fox, E.R.; Kraja, A.; Bidulescu, A.; Lieb, W.; Lin, H.; Beecham, A.; Chen, M.H.; Felix, J.F.; Fox, C.S.; et al. Genome-wide association analysis of B-type natriuretic peptide in blacks: The Jacson Heart Study. *Circ. Cardiovasc. Genet.* **2015**, *8*, 122–130. [CrossRef]

28. Johansson, A.; Eriksson, N.; Lindholm, D.; Varenhorst, C.; James, S.; Syvanen, A.C.; Axelsson, T.; Siegbahn, A.; Barratt, B.J.; Becker, R.C.; et al. Genome-wide association and Mendelian randomization study of NT-proBNP in patients with acute coronary syndrome. *Hum. Mol. Genet.* **2016**, *25*, 1447–1456. [CrossRef]
29. Del Greco, M.F.; Pattaro, C.; Luchner, A.; Pichler, I.; Winkler, T.; Hicks, A.A.; Fuchsberger, C.; Franke, A.; Melville, S.A.; Peters, A.; et al. Genome-wide association analysis and fine mapping of NT-proBNP level provide novel insight into the role of the MTHFR-CLCN6-NPPA-NPPB gene cluster. *Hum. Mol. Genet.* **2011**, *20*, 1660–1671.
30. Folkersen, L.; Fauman, E.; Sabater-Lleal, M.; Strawbridge, R.J.; Franberg, M.; Sennblad, B.; Baldassarre, D.; Veglia, F.; Humphries, S.E.; Rauramaa, R.; et al. Mapping of 79 loci for 83 plasma protein biomarkers in cardiovascular disease. *PLoS Genet.* **2017**, *13*, e1006706. [CrossRef]
31. Salo, P.P.; Havulinna, A.S.; Tukiainen, T.; Raitakari, O.; Lehtimaki, T.; Kahonen, M.; Kettunen, J.; Männikkö, M.; Eriksson, J.G.; Jula, A.; et al. Genome-Wide Association Study implicates atrial natriuretic peptide rather than B-type natriuretic peptide in the regulation of blood pressure in the general population. *Circ. Cardiovasc. Genet.* **2017**, *10*, e001713. [CrossRef] [PubMed]
32. Lanfear, D.E.; Stolker, J.M.; Marsh, S.; Rich, M.W.; McLeod, H.L. Genetic variation in the B-type natriuretic peptide pathway affects BNP levels. *Cardiovasc. Drugs Ther.* **2007**, *21*, 55–62. [CrossRef]
33. Seidelmann, S.B.; Vardeny, O.; Claggett, B.; Yu, B.; Shah, A.M.; Ballantyne, C.M.; Selvin, E.; MacRae, C.A.; Boerwinkle, E.; Solomon, S.D. An NPPB promoter polymorphism associated with elevated N-terminal pro-B-type natriuretic peptide and lower blood pressure, hypertension, and mortality. *J. Am. Heart Assoc.* **2017**, *6*, e005257. [CrossRef] [PubMed]
34. Yan, W.; Wu, F.; Morser, J.; Wu, Q. Corin, a transmembrane cardiac serine protease, acts as a pro-atrial natriuretic peptide-converting enzyme. *Proc. Natl. Acad. Sci. USA* **2000**, *97*, 8525–8529. [CrossRef] [PubMed]
35. Wu, F.; Yan, W.; Pan, J.; Morser, J.; Wu, Q. Processing of pro-atrial natriuretic peptide by corin in cardiac myocytes. *J. Biol. Chem.* **2002**, *277*, 16900–16905. [CrossRef]
36. Chan, J.C.; Knudson, O.; Wu, F.; Morser, J.; Dole, W.P.; Wu, Q. Hypertension in mice lacking the proatrial natriuretic peptide convertase corin. *Proc. Natl. Acad. Sci. USA* **2005**, *102*, 785–790. [CrossRef]
37. Chen, S.; Cao, P.; Dong, N.; Peng, J.; Zhang, C.; Wang, H.; Zhou, T.; Yang, J.; Zhang, Y.; Martelli, E.E.; et al. PCSK6-mediated corin activation is essential for normal blood pressure. *Nat. Med.* **2015**, *21*, 1048–1053. [CrossRef]
38. Volpe, M.; Rubattu, S. Novel insights into the mechanisms regulating pro-atrial natriuretic peptide cleavage in the heart and blood pressure regulation. Proprotein Convertase Subtilisin/Kexin 6 is the corin activating enzyme. *Circ. Res.* **2016**, *118*, 196–198. [CrossRef]
39. Rame, J.E.; Drazner, M.H.; Post, W.; Peshock, R.; Lima, J.; Cooper, R.S.; Dries, D.L. Corin I555 (P568) is associated with enhanced cardiac hypertrophy response to increased systemic afterload. *Hypertension* **2007**, *49*, 857–864. [CrossRef]
40. Rame, J.E.; Tam, S.W.; McNamara, D.; Worcel, M.; Sabolinski, M.L.; Wu, A.H.; Dries, D.L. Dysfunctional corin I555(P568) allele is associated with impaired brain natriuretic peptide processing and advserse outcomes in blacks with systolic heart failure: Results from the genetic risk assessment in heart failure substudy. *Circ. Heart Fail.* **2009**, *2*, 541–548. [CrossRef]
41. Dong, N.; Zhou, T.; Zhang, Y.; Liu, M.; Li, H.; Huang, X.; Liu, Z.; Wu, Y.; Fukuda, K.; Qin, J.; et al. Corin mutations K317E and S472G from preeclamptic patients alter zymogen activation and cell surface targeting. *J. Biol. Chem.* **2014**, *289*, 17909–17916. [CrossRef] [PubMed]
42. Zhang, Y.; Zhou, T.; Niu, Y.; He, M.; Wang, C.; Liu, M.; Yang, J.; Zhang, Y.; Zhou, J.; Fukuda, K.; et al. Identification and functional analysis of CORIN variants in hypertensive patients. *Hum. Mutat.* **2017**, *38*, 1700–1710. [CrossRef] [PubMed]
43. Li, N.; Luo, W.; Juhong, Z.; Yang, J.; Wang, H.; Zhou, L.; Chang, J. Associations between genetic variations in the FURIN gene and hypertension. *BMC Med. Genet.* **2010**, *11*, 124. [CrossRef] [PubMed]
44. International Consortium for Blood Pressure Genome-Wide Association; Eehret, G.B.; Munroe, P.B.; Rice, K.M.; Bochud, M.; Johnson, A.D.; Chasman, D.I.; Smith, A.V.; Tobin, M.D.; Verwoert, G.C.; et al. Genetic variants in novel pathways influence blood pressure and cardiovascular disease risk. *Nature* **2011**, *478*, 103–109.
45. Zhang, H.; Mo, X.B.; Xu, T.; Bu, X.Q.; Lei, S.F.; Zhang, Y.H. Novel genes affecting blood pressure detected via gene-based association analysis. *G3* **2015**, *5*, 1035–1042. [CrossRef] [PubMed]

46. Nakayama, T.; Soma, M.; Takahashi, Y.; Rehemudula, D.; Kanmatsuse, K.; Furuya, K. Functional deletion mutation of the 5′-flanking region of type A human natriuretic peptide receptor gene and its association with essential hypertension and left ventricular hypertrophy in the Japanese. *Circ. Res.* **2000**, *86*, 841–845. [CrossRef]
47. Sun, H.; Yang, Z.Q.; Liu, S.Y.; Yu, L.; Huang, K.; Lin, K.Q. Correlation between natriuretic peptide receptor C (NPR3) gene polymorphisms and hypertension in the Dai people of China. *Genet. Mol. Res. GMR* **2015**, *14*, 8786–8795. [CrossRef]
48. Pitzalis, M.V.; Sarzani, R.; Dessi-Fulgheri, P.; Iacoviello, M.; Forleo, C.; Lucarelli, K.; Pietrucci, F.; Salvi, F.; Sorrentino, S.; Romito, R.; et al. Allelic variants of natriuretic peptide receptor genes are associated with family history of hypertension and cardiovascular phenotype. *J. Hypertens.* **2003**, *21*, 1491–1496. [CrossRef]
49. Ren, M.; Ng, F.L.; Warren, H.R.; Witkowska, K.; Baron, M.; Jia, Z.; Cabrera, C.; Zhang, R.; Mifsud, B.; Munroe, P.B.; et al. The biological impact of blood pressure-associated genetic variants in the natriuretic peptide receptor C gene on human vascular smooth muscle. *Hum. Mol. Genet.* **2018**, *27*, 199–210.
50. Pistola, F.; Sacco, S.; Degan, D.; Tiseo, C.; Ornello, R.; Carolei, A. Hypertension and stroke: Epidemiological aspects and clinical evaluation. *High Blood Press. Cardiovasc. Prev.* **2016**, *23*, 9–18. [CrossRef]
51. Rubattu, S.; Volpe, M.; Kreutz, R.; Ganten, U.; Ganten, D.; Lindpaintner, K. Chromosomal mapping of genetic loci contributing to stroke in an animal model of a complex human disease. *Nat. Genet.* **1996**, *13*, 429–434. [CrossRef] [PubMed]
52. Rubattu, S.; Lee, M.A.; De Paolis, P.; Giliberti, R.; Gigante, B.; Lombardi, A.; Volpe, M.; Lindpaintner, K. Altered structure, regulation and function of the gene encoding atrial natriuretic peptide in the stroke-prone spontaneously hypertensive rat. *Circ. Res.* **1999**, *85*, 900–905. [CrossRef] [PubMed]
53. Rubattu, S.; Giliberti, R.; De Paolis, P.; Stanzione, R.; Spinsanti, P.; Volpe, M. Effect of a regulatory mutation on the rat atrial natriuretic peptide gene transcription. *Peptides* **2002**, *23*, 555–560. [CrossRef]
54. Rubattu, S.; Giliberti, R.; Ganten, U.; Volpe, M. A differential brain ANP expression cosegregates with occurrence of early stroke in the stroke-prone phenotype of the spontaneously hypertensive rat. *J. Hypertens.* **1999**, *17*, 1849–1852. [CrossRef] [PubMed]
55. Jeffs, B.; Clark, J.S.; Anderson, N.H.; Gratton, J.; Brosnan, M.J.; Gauguier, D.; Reid, J.L.; Macrae, I.M.; Dominiczak, A.F. Sensitivity to cerebral ischaemic insult in a rat model of stroke is determined by a single genetic locus. *Nat. Genet.* **1997**, *16*, 364–367. [CrossRef] [PubMed]
56. Brosnan, M.J.; Clark, J.S.; Jeffs, B.; Negrin, C.D.; Van Vooren, P.; Arribas, S.M.; Carswell, H.; Aitman, T.J.; Szpirer, C.; Macrae, I.M.; et al. Genes encoding atrial and brain natriuretic peptides as candidates for sensitivity to brain ischemia in stroke-prone hypertensive rats. *Hypertension* **1999**, *33*, 290–297. [CrossRef] [PubMed]
57. Lopez-Morales, M.A.; Castello-Rruiz, M.; Burguete, M.C.; Jover-Mengual, T.; Aliena-Valero, A.; Centeno, J.M.; Alborch, E.; Salom, J.B.; Torregrosa, G.; Miranda, F.J. Molecular mechanisms underlying the neuroprotective role of atrial natriuretic peptide in experimental acute ischemic stroke. *Mol. Cell. Endocrinol.* **2018**, *472*, 1–9. [CrossRef]
58. Rubattu, S.; Cotugno, M.; Forte, M.; Stanzione, R.; Bianchi, F.; Madonna, M.; Marchitti, S.; Volpe, M. Effects of dual Angiotensin type 1 receptor/neprilysin inhibition versus Angiotensin type 1 receptor inhibition on target organ injury in the stroke-prone spontaneously hypertensive rat. *J. Hypertens.* **2018**, *36*, 1902–1914.
59. Rubattu, S.; Stanzione, R.; Di Angelantonio, E.; Zanda, B.; Evangelista, A.; Tarasi, D.; Gigante, B.; Pirisi, A.; Brunetti, E.; Volpe, M. Atrial natriuretic peptide gene polymorphisms and the risk of ischemic stroke in humans. *Stroke* **2004**, *35*, 814–818. [CrossRef]
60. Cannone, V.; Huntley, B.K.; Olson, T.M.; Heublein, D.M.; Scott, C.G.; Bailey, K.R.; Redfield, M.M.; Rodeheffer, R.J.; Burnett, J.C., Jr. Atrial natriuretic peptide genetic variant rs5065 and risk for cardiovascular disease in the general community: A 9-year follow-up study. *Hypertension* **2013**, *62*, 860–865. [CrossRef]
61. Scarpino, S.; Marchitti, S.; Stanzione, R.; Evangelista, A.; Di Castro, S.; Savoia, C.; Quarta, G.; Sciarretta, S.; Ruco, L.; Volpe, M.; et al. ROS-mediated differential effects of the human atrial natriuretic peptide T2238C genetic variant on endothelial cells in vitro. *J. Hypertens.* **2009**, *27*, 1804–1813. [CrossRef] [PubMed]
62. Sciarretta, S.; Marchitti, S.; Bianchi, F.; Moyes, A.; Barbato, E.; Di Castro, S.; Stanzione, R.; Cotugno, M.; Castello, L.; Calvieri, C.; et al. The C2238 Atrial Natriuretic Peptide Molecular Variant is Associated with Endothelial Damage and Dysfunction Through Natriuretic Peptide Receptor C Signaling. *Circ. Res.* **2013**, *112*, 1355–1364. [CrossRef] [PubMed]

63. Rubattu, S.; Marchitti, S.; Bianchi, F.; Di Castro, S.; Stanzione, R.; Cotugno, M.; Bozzao, C.; Sciarretta, S.; Volpe, M. The C2238/αANP variant is a negative modulator of both viability and function of coronary artery smooth muscle cells. *PLoS ONE* **2014**, *9*, e113108. [CrossRef] [PubMed]
64. Stanzione, R.; Sciarretta, S.; Marchitti, S.; Bianchi, F.; Di Castro, S.; Scarpino, S.; Cotugno, M.; Frati, G.; Volpe, M.; Rubattu, S. C2238/αANP modulates Apolipoprotein E through Egr-1/miR199a in vascular smooth muscle cells in vitro. *Cell Death Dis.* **2015**, *6*, e2033. [CrossRef] [PubMed]
65. Carnevale, R.; Pignatelli, P.; Frati, G.; Nocella, C.; Stanzione, R.; Pastori, D.; Marchitti, S.; Valenti, V.; Santulli, M.; Barbato, E.; et al. C2238 ANP gene variant promotes increased platelet aggregation through the activation of Nox2 and the reduction of cAMP. *Sci. Rep.* **2017**, *7*, 3797. [CrossRef] [PubMed]
66. Strisciuglio, T.; Barbato, E.; De Biase, C.; Di Gioia, G.; Cotugno, M.; Stanzione, R.; Trimarco, B.; Sciarretta, S.; Volpe, M.; Wijns, W.; et al. T2238C atrial natriuretic peptide gene variant and the response to antiplatelet therapy in stable ischemic heart disease patients. *J. Cardiovasc. Transl. Res.* **2018**, *11*, 36–41. [CrossRef]
67. Rubattu, S.; Sciarretta, S.; Marchitti, S.; Bianchi, F.; Forte, M.; Volpe, M. The T2238C atrial natriuretic peptide molecular variant and the risk of cardiovascular diseases. *Int. J. Mol. Sci.* **2018**, *19*, 540. [CrossRef]
68. Lynch, A.I.; Boerwinkle, E.; Davis, B.R.; Ford, C.E.; Eckfeldt, J.H.; Leiendecker-Foster, C.; Arnett, D.K. Pharmacogenetic association of the NPPA T2238C genetic variant with cardiovascular disease outcomes in patients with hypertension. *JAMA* **2008**, *299*, 296–307. [CrossRef]
69. Calvieri, C.; Rubattu, S.; Volpe, M. Molecular mechanisms underlying cardiac antihypertrophic and antifibrotic effects of natriuretic peptides. *J. Mol. Med.* **2012**, *90*, 5–13. [CrossRef]
70. Molkentin, J.D. A friend within the heart: Natriuretic peptide receptor signaling. *J. Clin. Investig.* **2003**, *111*, 1275–1277. [CrossRef]
71. Rubattu, S.; Bigatti, G.; Evangelista, A.; Lanzani, C.; Stanzione, R.; Zagato, L.; Manunta, P.; Marchitti, S.; Venturelli, V.; Bianchi, G.; et al. Association of atrial natriuretic and type-A natriuretic peptide receptor gene polymorphisms with left ventricular mass in human essential hypertension. *J. Am. Coll. Cardiol.* **2006**, *48*, 499–505. [CrossRef] [PubMed]
72. Rubattu, S.; Sciarretta, S.; Ciavarella, G.M.; Venturelli, V.; De Paolis, P.; Tocci, G.; De Biase, L.; Ferrucci, A.; Volpe, M. Reduced levels of pro-atrial natriuretic peptide in hypertensive patients with metabolic syndrome and their relationship with LVH. *J. Hypertens.* **2007**, *25*, 833–839. [CrossRef] [PubMed]
73. Dong, L.; Wang, H.; Dong, N.; Zhang, C.; Xue, B.; Wu, Q. Localization of corin and atrial natriuretic peptide expression in human renal segments. *Clin. Sci.* **2016**, *130*, 1655–1664. [CrossRef] [PubMed]
74. Yoshihara, F.; Tokudome, T.; Kishimoto, I.; Otani, K.; Kuwabara, A.; Horio, T.; Kawano, Y.; Kangawa, K. Aggravated renal tubular damage and interstitial fibrosis in mice lacking guanylyl cyclase A (GC-A), a receptor for atrial and B-type natriuretic peptides. *Clin. Exp. Nephrol.* **2015**, *19*, 197–207. [CrossRef] [PubMed]
75. Romero, M.; Caniffi, C.; Bouchet, G.; Costa, M.A.; Elesgaray, R.; Arranza, C.; Tomat, A.L. Chronic treatment with atrial natriuretic peptide in spontaneously hypertensive rats: Beneficial renal effects and sex differences. *PLoS ONE* **2015**, *10*, e0120362. [CrossRef] [PubMed]
76. Hildebrandt, P.; Boesen, M.; Olsen, M.; Wachtell, K.; Groenning, B. N-terminal pro brain natriuretic peptide in arterial hypertension—A marker for left ventricular dimensions and prognosis. *Eur. J. Heart Fail.* **2004**, *6*, 313–317. [CrossRef] [PubMed]
77. Olsen, M.H.; Wachtell, K.; Nielsen, O.W.; Hall, C.; Wergeland, R.; Ibsen, H.; Kjeldsen, S.E.; Devereux, R.B.; Dahlöf, B.; Hildebrandt, P.R. N-terminal brain natriuretic peptide predicted cardiovascular events stronger than high-sensitivity C-reactive protein in hypertension: A LIFE substudy. *J. Hypertens.* **2006**, *24*, 1531–1539. [CrossRef]
78. Ikram, M.A.; Seshadri, S.; Bis, J.C.; Fornage, M.; DeStefano, A.L.; Aulchenko, Y.S.; Debette, S.; Lumley, T.; Folsom, A.R.; van den Herik, E.G.; et al. Genomewide association studies of stroke. *NEJM* **2009**, *360*, 1718–1728. [CrossRef]
79. Rubattu, S.; Giusti, B.; Lotta, L.A.; Peyvandi, F.; Cotugno, M.; Stanzione, R.; Marchitti, S.; Palombella, A.M.; Di Castro, S.; Rasura, M.; et al. Association of a single nucleotide polymorphism of the NPR3 gene promoter with early onset ischemic stroke in an Italian cohort. *Eur. J. Intern. Med.* **2013**, *24*, 80–82. [CrossRef]
80. Volpe, M.; Tocci, G.; Battistoni, A.; Rubattu, S. Angiotensin II receptor blocker nephrilysin inhibitor (ARNI): New avenues in cardiovascular therapy. *High Blood Press. Cardiovasc. Prev.* **2015**, *22*, 241–246. [CrossRef]
81. Von Llueder, T.G.; Atar, D.; Krum, H. Current role of neprilysin inhibitors in hypertension and heart failure. *Pharmacol. Ther.* **2014**, *144*, 41–49. [CrossRef] [PubMed]

82. Kario, K.; Sun, N.; Chiang, F.T.; Supasyndh, O.; Baek, S.H.; Inubushi-Molessa, A.; Zhang, Y.; Gotou, H.; Lefkowitz, M.; Zhang, J. Efficacy and safety of LCZ696, a first-in-class angiotensin receptor neprilysin inhibitor, in Asian patients with hypertension: A randomized, double-blind, placebo-controlled study. *Hypertension* **2014**, *63*, 698–705. [CrossRef] [PubMed]
83. Ruilope, L.M.; Dukat, A.; Bohm, M.; Lacourciere, Y.; Gong, J.; Lefkowitz, M.P. Blood-pressure reduction with LCZ696, a novel dual-acting inhibitor of the angiotensin II receptor and neprilysin: A randomized, double-blind, placebo-controlled, active comparator study. *Lancet* **2010**, *375*, 1255–1266. [CrossRef]
84. Wang, T.D.; Tan, R.S.; Lee, H.Y.; Ihm, S.H.; Rhee, M.Y.; Tomlinson, B.; Pal, P.; Yang, F.; Hirschhorn, E.; Prescott, M.F.; et al. Effects of Sacubitril/Valsartan (LCZ696) on natriuresis, diuresis, blood pressures, and NT-proBNP in salt-sensitive hypertension. *Hypertension* **2017**, *69*, 32–41. [PubMed]
85. Kusaka, H.; Sueta, D.; Koibuchi, N.; Hasegawa, Y.; Nakagawa, T.; Lin, B.; Ogawa, H.; Kim-Mitsuyama, S. LCZ696, Angiotensin II receptor-neprilysin inhibitor, ameliorates high-salt-induced hypertension and cardiovascular injury more than valsartan alone. *Am. J. Hypertens.* **2015**, *28*, 1409–1417. [CrossRef] [PubMed]
86. Williams, B.; Cockcroft, J.R.; Kario, K.; Zappe, D.H.; Brunel, P.C.; Wang, Q.; Guo, W. Effects of sacubitril/valsartan versus olmesartan on central hemodynamics in the erderly with systolic hypertension: The PARAMETER study. *Hypertension* **2017**, *69*, 411–420. [PubMed]

 © 2019 by the authors. Licensee MDPI, Basel, Switzerland. This article is an open access article distributed under the terms and conditions of the Creative Commons Attribution (CC BY) license (http://creativecommons.org/licenses/by/4.0/).

Review

Natriuretic Peptides in Heart Failure with Preserved Left Ventricular Ejection Fraction: From Molecular Evidences to Clinical Implications

Daniela Maria Tanase [1,2], Smaranda Radu [1,3,*,†], Sinziana Al Shurbaji [1,4,†], Genoveva Livia Baroi [5,6,†], Claudia Florida Costea [7,8], Mihaela Dana Turliuc [9,10], Anca Ouatu [1,2] and Mariana Floria [1,2]

1. Department of Internal Medicine, "Grigore T. Popa" University of Medicine and Pharmacy, 700111 Iasi, Romania; tanasedm@gmail.com (D.M.T.); sanziana.alshurbaji@yahoo.com (S.A.S.); ank_mihailescu@yahoo.com (A.O.); floria_mariana@yahoo.com (M.F.)
2. Internal Medicine Clinic, "Sf. Spiridon" County Clinical Emergency Hospital Iasi, 700115 Iasi, Romania
3. Cardiology Clinic, "Prof. Dr. George I.M. Georgescu" Institute of Cardiovascular Diseases, 700503 Iasi, Romania
4. Institute of Gastroenterology and Hepatology, 700115 Iasi, Romania
5. Department of Surgery, "Grigore T. Popa" University of Medicine and Pharmacy, 700111 Iasi, Romania; hauliviagenoveva@hotmail.com
6. Vascular Surgery Clinic, "Sf. Spiridon" County Clinical Emergency Hospital Iasi, 700115 Iasi, Romania
7. Department of Ophthalmology, "Grigore T. Popa" University of Medicine and Pharmacy, 700115 Iasi, Romania; costea10@yahoo.com
8. 2nd Ophthalmology Clinic, "Prof. Dr. Nicolae Oblu" Emergency Clinical Hospital, 700115 Iași, Romania
9. Department of Neurosurgery, "Grigore T. Popa" University of Medicine and Pharmacy, 700115 Iași, Romania; turliuc_dana@yahoo.com
10. 2nd Neurosurgery Clinic, "Prof. Dr. Nicolae Oblu" Emergency Clinical Hospital, 700115 Iași, Romania

* Correspondence: radu.smaranda@gmail.com; Tel.: +40-232-240-822
† These authors contributed equally to this work.

Received: 7 April 2019; Accepted: 24 May 2019; Published: 28 May 2019

Abstract: The incidence of heart failure with preserved ejection fraction (HFpEF) is increasing and its challenging diagnosis and management combines clinical, imagistic and biological data. Natriuretic peptides (NPs) are hormones secreted in response to myocardial stretch that, by increasing cyclic guanosine monophosphate (cGMP), counteract myocardial fibrosis and hypertrophy, increase natriuresis and determine vasodilatation. While their role in HFpEF is controversial, most authors focused on b-type natriuretic peptides (BNPs) and agreed that patients may show lower levels. In this setting, newer molecules with an increased specificity, such as middle-region pro-atrial natriuretic peptide (MR-proANP), emerged as promising markers. Augmenting NP levels, either by NP analogs or breakdown inhibition, could offer a new therapeutic target in HFpEF (already approved in their reduced EF counterparts) by increasing the deficient cGMP levels found in patients. Importantly, these peptides also retain their prognostic value. This narrative review focuses on NPs' physiology, diagnosis, therapeutic and prognostic implication in HFpEF.

Keywords: heart failure; natriuretic peptides; preserved ejection fraction

1. Introduction

The incidence of heart failure (HF) is increasing. If for HF with reduced ejection fraction (HFrEF) there are well-established methods of diagnosis and treatment, this is far from true in HF with preserved ejection fraction (HFpEF) patients. This increasing incidence justifies the need for proper diagnostic, therapeutic and prognostic tools. In the era of cardiac biomarkers, natriuretic peptides (NPs) have a well-established role in HF pathophysiology and patient management. However, both the lack of consensus regarding NPs in HFpEF and the heterogeneous population makes diagnosis and management difficult and unstandardized, leading to a decrease in quality of life and increase in mortality and hospitalization.

2. Heart Failure with Preserved Left Ventricular Ejection Fraction

Heart failure represents a clinical syndrome affecting 2%–3% of the general population [1]. Its prevalence is increasing and it leads to an increased risk of death, increased hospitalization rates, a decrease in quality of life and higher costs through complex therapeutic strategies [2,3].

One of the classifications of HF is made by determining the left ventricular ejection fraction (LVEF). As such, HF can be either with preserved ejection fraction, HFpEF (LVEF > 50%), or reduced ejection fraction, HFrEF (LVEF < 40%), and patients with an LVEF between 40% and 50% are labeled as HF with mid-range EF-HFmrEF [3]. The European Society of Cardiology (ESC) highlights the fact that HFpEF diagnosis is challenging, emphasizing the current lack of consensus. According to ESC guidelines, there are four diagnostic criteria for HFpEF: the presence of HF symptoms and signs (dyspnea, orthopnea, cough), a LVEF of >50% (with the remark that patients with a LVEF 40%–49% may be classified as HFmrEF and included in clinical trials as HFpEF), increased levels of NPs (B-type natriuretic peptide: BNP > 35 pg/mL and/or N-terminal-Pro-BNP: NT-proBNP > 125 pg/mL) and imagistic evidence of structural heart disease, including LV hypertrophy, diastolic dysfunction and/or left atrial (LA) enlargement.

Biomarkers reflect myocardial damage and stress; systemic inflammation and fibrosis and their routine measurements mirror myocardial structural disease [2]. Natriuretic peptides are the most frequently used biomarkers in HF patients, as their elevated levels constitute a diagnostic criterion irrespective of LVEF [1–4]. However, the underlying mechanisms behind their increase differ in the two types of HF, with HFpEF patients showing lower levels of NPs [2–5]. In these patients, higher levels of circulating NP are given by the increased left ventricle (LV) diastolic filling pressure and end-diastolic wall stress [2,4]. As compared with HFrEF, the circulating levels of NPs are lower, especially in the context of lower myocardial stretch, LV end-diastolic wall stress and volume overload [4]. ESC guidelines recommend taking into consideration lower diagnostic thresholds for BNP and NT-proBNP when assessing a potential HFpEF patients [3].

Essential hypertension and myocardial ischemia are frequent causes of HFpEF and one third of these patients have concomitant atrial fibrillation (AF) [4]. The presence of AF impacts NPs circulating levels, with different studies showing that AF patients with HFpEF had higher mean NP levels compared to sinus rhythm patients [3–7]. As such, NP values must be interpreted differently in AF patients both when diagnosing HFpEF and assessing prognosis [2,4,7,8]. Several other factors affect NP levels [5–9]. Obese patients tend to have lower baseline NPs, while renal dysfunction, feminine gender and age are associated with increased levels [7].

3. Natriuretic Peptides: From Molecular Evidences to Clinical Implications

There are three endogenous NPs secreted as pre-prohormones: atrial natriuretic peptide (ANP), B-type natriuretic peptide (BNP) and C-type natriuretic peptide (CNP). Subsequently, there are three natriuretic peptide receptors (NPR): natriuretic peptide receptor A (NPR-A), natriuretic peptide receptor B (NPR-B) and natriuretic peptide receptor C (NPR-C or clearance receptor) [7–16]. These peptides act as hormones with pleiotropic effects by binding to the first two receptors, contributing to cardiovascular homeostasis and pressure and volume overload counter-regulatory mechanisms [12]. Moreover, all NPs have various molecular forms, which are different in healthy subjects from HF patients, such as different ANP forms and glycosylated proBNP [17].

Biologically active ANP is a 28 amino acid (aa) peptide. Initially secreted as pre-proANP (151 amino acids), this molecule is cleaved into proANP (126 amino-acids), which is deposited in granules inside atrial myocardium. A transmembrane protease cleaves the secreting proANP into its biologically active short-lived form (ANP- 28 amino acids) and its inactive form, NT-proANP (98 amino-acids) [7–11]. The latter has a much longer half-life (60–120 min); however, its numerous subsequent fragments make it a still unpractical biomarker in routine clinical practice.

The majority of ANP is secreted in the atria in response to myocardial stretch, atrial concentrations being 1000 higher as compared to ventricular ANP levels [17]. Ventricular myocardium produces small amounts of ANP; however, in HF patients, hypertrophied ventricular myocardium becomes able to secrete ANP [10]. Extracardiac sources include hypothalamus, lung and thyroid gland [5].

There are three forms of circulating ANP: αANP, βANP and pro-ANP [17]. Found only in the human atria, the biosynthesis pathways of the βANP are still unclear. Its structure is of an anti-parallel dimer of αANP. When compared to αANP, βANP has lower bioavailability (40%), slower onset of action and lower receptor affinity (Table 1).

Table 1. Various forms of atrial natriuretic peptide. ANP- atrial natriuretic peptide; LV- left ventricle.

ANP Form	Structure	Effects	Additional Remarks	Reference
αANP	Compact ring structure	Prolonged bioavailability as compared to other ANP forms	Healthy subjects: αANP>>proANP>>βANP Heart failure patients: Increased βANP and pro-ANP concentrations with decreased circulating corin levels Increased βANP/total ANP correlates with LV dysfunction	[17]
βANP	Flexible extended structure αANP antiparallel homodimer	40% of ANP effects		
proANP	precursor	10% of ANP effects (weak natriuretic)		

ANP has numerous biological effects, including blood pressure lowering effects and renin-angiotensin-aldosterone system (RAAS) inhibition [16–20]. It may affect apical Na channel and Na/K ATPasis basal activity, determining a decreased sodium reabsorption and increased Na excretion, leading to increased natriuresis [20]. Moreover, ANP inhibits RAAS through renin (at the juxtaglomerular cyclic guanosine monophosphate- cGMP dependent cells level) and aldosterone inhibition [21], leading to blood pressure reduction. Moreover, studies have shown that ANP acts at the level of adrenal glands, directly inhibiting aldosterone production [16]. Not only does ANP inhibit RAAS, it also possesses antifibrotic and antihypertrophic effects through cGMP dependent angiotensin II and endothelin inhibition [12]. A study revealed that ANP might counteract myocardial hypertrophy by inhibiting calcium mediated epinephrine response through cGMP [15]. A different mechanism through which ANP affects blood pressure may be through baroreflex modulation, stimulating vagal afferent fibers [16].

Recent studies focusing on AF patients found that NT-proANP levels correlate with AF type and LA dimensions [18]. Seewoster et al. revealed that NT-proANP correlated with LA dimension as determined by cardiac magnetic resonance [18].

The gene coding ANP contains three exons: the first exon codes the 5' region (not translated), a signal peptide formed of 25aa (16aa of proANP). The second exon is responsible for coding proANP while the third exon has a role in coding terminal 3' tyrosine [5]. Recent studies have shown that a chorionic transmembrane enzyme cleaves proANP into pre-preoANP and NT-proANP [8].

The biologically active form of human BNP is BNP32 [8], widely varying across different species. The gene coding BNP is also formed of 3 exons: the first exon codes a 26 aa signal peptide and the first 15aa of proBNP; the second exon codes the majority of proBNP and the third exon codes terminal tyrosine and 3' region [12]. mRNA BNP is translated into pre-proBNP with 134 aa after which the signal peptide is removed, resulting BNP-108 [1]. Different forms of BNP exist in the atria and the ventricles- BNP 32 and 108, respectively [12]. BNP32 is mostly found in the atria, while BNP 108 is found in the ventricular myocardial [21]. At the level of the Golgi apparatus, proBNP is cleaved into BNP and NT-proBNP, to be later released in plasma [21]. As opposed to ANP, which is mostly stored in vesicles, BNP is secreted in response to myocardial stretch.

Several studies identified higher BNP concentrations at the level of anterior interventricular vein and coronary sinus, which supports the fact that BNP is mainly secreted by ventricular myocardium [1]. The difference between BNP and ANP mRNA resides in a repetitive unit which determines mRNA BNP degradation in a fashion similar to oncogenes [4]. BNP gene expression differs from that of ANP, being more dynamic [9]. BNP possess vasodilator effects, promoting natriuresis and diuresis. At the myocardial level, BNP inhibits fibrosis and necrosis [16–19]. Also, it possesses anti-inflammatory effects through monocytes, B lymphocytes and natural killer cell regulation. Importantly, BNP interferes with post-skeletal muscles ischemia angiogenesis.

Importantly, it seems that HF patients have decreased amounts of BNP32; the latter is cleaved to either BNP3-32 or BNP8-32 by dipeptidyl peptidase IV (DPP IV). The two forms are less biologically active than BNP32, probably due to faster degradation and may account for the presumed resistance to NPs in HF patients [1,16].

Both ANP and BNP can undergo post-translational changes, such as phosphorylation and glycosylation. The physiological impact of a phosphorylated proANP form is still uncertain [17]. As one of the most frequent change, glycosylation stabilizes proteins, thus preventing further processing. The glycosylation pattern is strikingly different between ANP and BNP [17]. O-glycosylation of proBNP inside the Golgi apparatus is multi-sited (approximately seven sites) and its degree may vary among patients. Heavily glycosylated pro-BNP molecules are recognized as a cause of decreased conversion to its biologically active form, with a subsequent increase in pro-BNP/total BNP ratio. Interestingly, in acute HF the glycosylation level of proBNP decreases as there is a tendency towards forming more mature BNP. In acute HF, the increased proBNP production is accompanied by decreased glycosylation and increased furin activity, leading to elevated BNP and NT-proBNP levels.

C-type natriuretic peptide is secreted in the myocardium, endothelium, chondrocytes, brain and blood cells [12]. C-type natriuretic peptide gene also contains 3 exons, with the first exon containing 23 aa signal peptide and 7 aa proCNP. The second exon contains the proCNP sequence while the third the exon 3' terminal. After removal from the 126 aa pre-proCNP of the 23 aa signal sequence results 103 aa proCNP. An intracellular endopeptidase cleaves proCNP into 53 aa CNP [16]. CNP_{53} is cleaved in return to CNP_{22}. Both forms have the same functions; however, CNP_{53} is found in the myocardium while CNP_{22} is found in plasma and brain [21]. CNP also possess vasodilator effects, being secreted by endothelial cells in response to vascular lesions. Further, it inhibits fibrosis, platelet aggregation and tissue plasminogen activation. CNP levels tend to increase in advanced HF as compared to incipient HF.

NPR-A and NPR-B determine NP functions. ANP and BNP activate NPR-A while CNP activates NPR-B [10–16]. NPR-A is found in the lungs, kidneys, adrenal glands, while NPR-B is predominantly found inside fibroblasts [18–20].

NPR-C is found in the brain, atrium, lungs and aorta. These receptors have an extracellular region for the ligand attachment and a intracellular region with a cGMP dependent proteinkinase [6,22].

As a second messenger, cGMP is formed from two possible precursors, either soluble guanylyl cyclase- sGC (found in cytosol; it requires nitric oxide binding) and particulate guanylyl cyclase (pGC), found in the cellular membrane and activated via NPR. As such, this activation leads to an increase in cGMP, which in turn increases protein-kinase G (PKG) levels [22–26]. The latter phosphorylates several proteins, including myocardial cytoskeletal titin [26]. Moreover, decreased levels of cGMP and subsequently of PKG have been associated with myocardial remodeling through increased cardiomyocyte hypertrophy and resting tension [26] (Figure 1). HTN, AF, chronic kidney disease- frequently found comorbidities in HFPEF patients, determine a decrease in cGMP through a pro-inflammatory state and subsequent decrease of nitric oxide. Importantly, augmenting cGMP concentrations may constitute therapeutic targets in HFpEF.

NPs have a wide range of biological effects, including endocrine and paracrine. They promote diuresis and natriuresis, vasodilation and inhibit sympathetic nervous system and renin-angiotensin-angiotensinogen. In advanced HF there is a resistance to the effects of NPs, together with either an increased turnover, increased biologically inactive NP secretion or decreased NPR-A activation due to the dephosphorylation of secondary receptors [1–5].

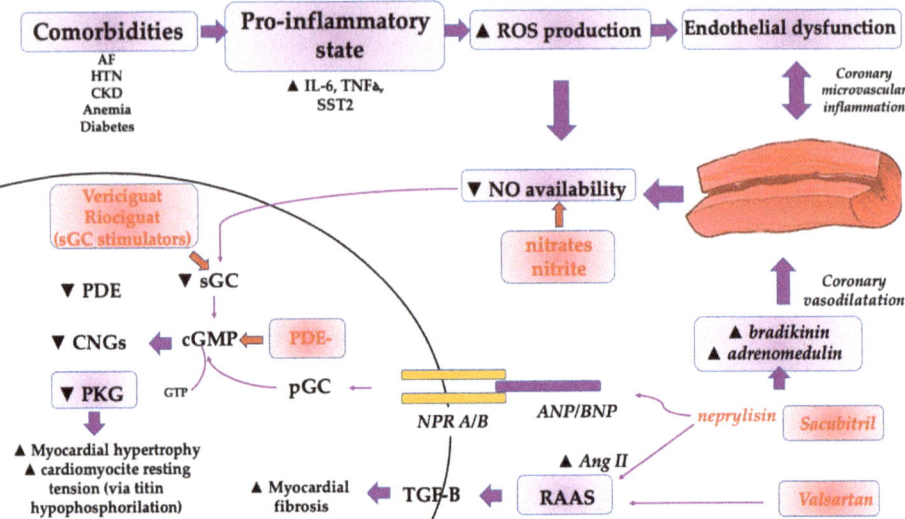

Figure 1. NPs, cGMP and RAAS in HFpEF patients and possible therapeutic targets. AF: atrial fibrillation; Ang II: angiotensin II; cGMP: cyclic guanosine monophosphate; CKD: chronic kidney disease; CNGs: cyclic nucleotide gated-ion channels; HTN: hypertension; IL: interleukin; NO: nitric oxide; NP: natriuretic peptide; NPR: natriuretic peptide receptor; pGC: particulate guanylyl cyclase; PKG: protein kinase G; PDE: phosphodiesterase; PDE-: phosphodiesterase inhibitors; RAAS: renin-angiotensin-aldosterone system; sGC: soluble guanylyl cyclase, TGF-B: transforming growth factor beta.

If ANP levels are influenced by atrial pressures, BNP concentrations are determined by ventricular stretch in response to underlying pressure and/or volume overload. ANP's short half-life (derived from its higher affinity for NPR-C) precludes its utilization in routine practice while BNP's stability makes it a desirable biomarker in HF diagnosis and prognosis (Table 2).

Table 2. The biological characteristics of natriuretic peptides. ANP- atria natriuretic peptide; BNP- B type natriuretic peptide; CNP- C-type natriuretic peptide; * values may vary slightly across studies.

Natriuretic Peptide	Mechanism	Time	Normal Levels	Reference
BNP	Ventricular wall stretch (pressure/volume overload)	20 min	3.5 pg/mL	[4–17]
NT-proBNP	Biologically inactive form of BNP	60–90 min	51 pg/mL	
ANP	Atrial wall stretch (pressure/volume overload)	2 min	20 pg/mL	
NT-proANP	Biologically inactive form of ANP	60–120 min	0.11–0.60 nmol/L	
CNP	Endothelial lesions	2.6 min	Nearly undetectable	

A novel NP, middle-range proANP (MR-proANP) has emerged as having a greater stability and both diagnostic and prognostic values, especially in HFpEF patients. It derives from an intermediate region of NT-proANP and exhibits increased stability. Moreover, it correlates with increased LA dimensions and it seems that its diagnostic [24] and prognostic [12] utility might be superior to that of NT-proBNP in HFpEF. Moreover, it correlates with NYHA class in several studies [24].

The degradation of NPs can be either receptor- mediated (NPR-C) or enzyme-mediated [5–10] and is recognized as a therapeutic target in both hypertension and HF [12,16]. While receptor mediated NP degradation is based on internalization (claritin-mediated) and hydrolysis [16], enzyme-mediated NP degradation occurs mainly through the neutral endopeptidase (NEP) zinc-dependent neprylisin [4,10]. Although it is expressed by varying tissues, it can mostly be found in the proximal renal tubules, myocardium, fibroblasts and endothelial cells [4–7]. This enzymatic degradation makes both ANP and BNP unstable in serum, their plasmatic levels being used in routine clinical practice. Contrarily, NT-proBNP shows an increased serum stability. Inhibiting NEP leads to an increase in NP levels, which benefits HF patients [21]. Insulin degrading enzyme has also proven to degrade NPs, specifically ANP in addition to insulin [5].

4. Implications of Natriuretic Peptides in Heart Failure with Preserved Left Ventricular Ejection Fraction Diagnosis

Diagnosing HFpEF remains difficult due to a lack of consensus, patients' heterogeneity and multiple concurrent pathologies that may mimic not only HF symptoms, but also lead to either increased (AF) or decreased (obesity) NP levels [1]. HFpEF diagnosis requires clinical and imagistic criteria, as well as elevated NPs levels. Given that one third of HFpEF have normal NP levels [1], relying solely on their values for diagnosis is not recommended and their value must be interpreted in the clinical context. As such, the diagnostic gold standard is cardiac catheterization showing increased LV filling pressures.

The longer plasma half-life of BNP and NT-proBNP (22 min and 70 min, respectively) as compared to ANP (2 min), makes the two the preferred NPs for guiding HF diagnosis and, probably, therapy. Moreover, despite having a longer half-life, NT-proANP failed to emerge as a diagnostic marker due to the increased number of cleaved fragments that limit its detection. European Society of Cardiology considers a BNP level of > 35 pg/mL and/or NT-proBNP > 125 pg/mL suggestive of chronic heart failure, with higher values being recommended in acute settings- over 100 pg/mL and > 300 pg/mL, respectively [1,3]. Moreover, it is agreed upon that acute HF patients exhibit increased NPs levels regardless of LVEF [3,27–35]. In a study, acute HFpEF patients showed NT-proBNP levels between 600–1000 pg/mL [36]. Although HFpEF patients/ patients previously treated with diuretics may exhibit lower levels, there is no consensus in what regards NPs diagnostic threshold. Given that NPs levels are affected by several factors, including the presence of AF and body mass index, their use rather resides in excluding the diagnosis of HF with a subsequent high-negative predictive value (0.94–0.98) [3]. As such, especially with respect to the diagnosis of HFpEF, NP levels must be corroborated with both the clinical context and other imagistic parameters. In the light of numerous affections that may mimic HFpEF symptoms, the ESC guidelines propose diagnostic thresholds of these NPs so as to limit over-diagnosing HFpEF.

Several studies agree that both BNP and NT-proBNP retain their diagnostic performance (although with a decrease in sensitivity and specificity) in acute heart failure, irrespective of LVEF [32]. However, factors that impact their levels should be taken into consideration when attempting HFpEF diagnosis. Both cardiac and non-cardiac causes may lead to a subsequent increase in NPs, including AF, recent cardioversion, myocarditis, acute coronary syndrome, age, severe renal impairment, pulmonary embolism, sepsis and critical illness [32,33]. It has been shown that the predictive values of NPs drop from 0.95 to 0.82 in patients > 75 years old [4]. As such, the expected NP values will be higher in the elderly, even in the absence of HF. Age-adjusted NT-proBNP have been proposed for acute HF diagnosis, considering cut-off values of >450 pg/mL, 900 pg/mL and >1800 pg/mL for patients <50 years, > 50 years and >75 years old, respectively [5]. In contrast, NPs levels tend to be lower in obese patients, irrespective of volume status [4]. The fact that NP concentrations vary among HFpEF patients is demonstrated by different medians across studies. The I-PRESERVE trial revealed a median NT-proBNP concentration of 341 pg/mL in HFpEF patients [34], while a different study highlighted that nearly a third of HFpEF patients had BNP levels below 100 pg/mL while displaying increased LV filling pressures as measured by cardiac catheterization [33].

Recent studies pointed out the need of different thresholds in HFpEF in regard to sinus rhythm patients. HFpEF and AF coexist in 30% of patients [37] while AF in itself is one of the most frequent causes of HFpEF. AF patients tend to have higher NT-proBNP levels and exercise intolerance in the absence of HF [1]. Moreover, they have increased LA dimensions and LV filling pressures, making echocardiographic findings of HFpEF difficult to interpret [1,37,38]. Although the use of higher levels is recommended, ESC has failed to provide a clear cut-off value for HFpEF diagnosis in AF patients. Several studies reported the use of different NPs cut-off values as inclusion criteria with respect to underlying cardiac rhythm (sinus versus atrial fibrillation): 600 pg/mL in the SOCRATES trial [39] and >900 pg/mL in PARAGON trial [40].

The use of different diagnostic methods in adjunction with increased NPs level is recommended by ESC guidelines. Structural and functional alterations as determined by transthoracic echocardiography include a left atrial indexed volume (LAVI) of >34 mL, increased LV mass index (115g/m^2 for men and 95 g/m^2 for females), E/e' >13 and mean septal and lateral wall e' of <9 cm/s [3]. This is supported by the correlation between BNP and structural and functional alterations in HFpEF. In a study conducted by Iwanaga et al., BNP levels correlated with both left ventricular end-diastolic pressure and end-diastolic wall stress, more significantly with the latter [41]. Moreover, it seems that a BNP of > 100 pg/mL or a NT-proBNP of > 600 pg/mL indicates a LV restrictive filling pattern [32]. NP levels also correlate with LA dimensions, this correlation being stronger in HFpEF patients [35]. Although LVEF is preserved, it seems that global systolic function is altered in HFpEF patients. In a study conducted by Kraigher-Krainer et al., the decreased LV systolic strain (both longitudinal and circumferential) noticed in HFpEF patients correlated with NT-proBNP levels [38]. The disposition of LV hypertrophy also influences BNP levels, being significantly elevated in patients with concentric as compared to eccentric LV hypertrophy [5].

In the setting of an acute coronary syndrome, ANP levels show an early elevation with a rapid decline, as opposed to BNP levels, who tend to exhibit a bimodal elevation [5]. The first peak has been reported in the first two days post-myocardial infarction, with the second occurring nearly one week after the event, reflecting the extent of LV remodeling.

Not only that the diagnostic performance remains constant, but the treatment is similar in decompensated HF with regard to LVEF [3,40–47]. In chronic HF however, the treatment differs significantly between HFpEF and HFrEF, being well-established for the latter.

Renal function is tightly related to NPs levels. NPs tend to be elevated in CKD and end-stage renal disease, up to a value of 200 pg/mL even in the absence of overt HF [5]. Haemodialysis but not peritoneal dialysis has been shown to lower BNP levels with nearly 40% [43]. Proposed cut-off values of NT-proBNP for diagnosing HF in CKD patients include a level of > 1200 pg/mL, with higher levels being suggested in older patients [5,44].

Recently, a new NP has emerged as a potential HFpEF diagnostic tool. Cui et al. have revealed increased MR-proANP levels in HFpEF patients, with a significantly higher AUC when compared to NT-proBNP (0.844 versus 0.518, p <0.001) [24]. Moreover, when comparing their levels based on NYHA class, MR-proANP concentrations differ in regard with NYHA class, as compared to NT-proBNP, which showed no variation. Taking into consideration the link with echocardiographic parameters, MR-proANP correlated with LAVI, as opposed to NT-proBNP. In another study, MR-proANP showed non-inferiority to NT-proBNP in acute HF diagnosis, being elevated even in patients who showed non-diagnostic NT-proBNP levels [27]. The authors of the BACH study found that a MR-proANP of >120 pmol/L was suggestive of HF; adding this parameter to BNP increased its diagnostic performance to 73.6% [27].

Not only have NPs proven their diagnostic utility, several studies have questioned their ability to identify patients at risk for HF development. The STOP-HF trial referred patients with BNP levels of >50 pg/mL to further echocardiographic investigations, leading to a decrease in LV dysfunction [42].

Given its difficulties, the diagnosis of HFpEF in its characteristically heterogeneous population usually requires more than one biomarker. As the separation of HF in reduced and preserved EF occurred rather recently, more studies focused on HFrEF. Although both European and American guidelines have yet to consider different thresholds for the two classes of HF, increasing evidence links their distinct pathophysiologies to unique therapeutic strategies. Moreover, taking into consideration the increasing HFpEF incidence and altered prognosis, proper identification of these patients become vital.

5. Therapeutic Implications of Natriuretic Peptides in Heart Failure with Preserved Left Ventricular Ejection Fraction

NPs have been shown to inhibit RAAS, suppressing angiotensin II mediated vasoconstriction, sodium reabsorption (proximal tubule) and aldosterone, endothelin and renin secretion [5,9]. Their use in HF therapy is bimodal, both as a therapeutic target per se and as an indicator evaluating therapy response. However, their use in HFpEF remains controversial.

The rationale behind using NPs as a therapeutic target in HF therapy [45–62] resides in the seemingly abnormal BNP processing with a subsequent deficiency in active forms and resistance to their biological effects in these patients [5]. It seems that HF patients are deficient in biologically active BNP32 with a subsequent increase in BNP1–108 [45]. Augmenting their effects can be done either by administering NPs or reducing their breakdown.

5.1. Natriuretic Peptides Analogs

The use of recombinant NPs in acute HF is recommended by ESC [3], but without any distinction regarding EF. Nesiritide is a recombinant form of BNP used as a vasodilator in acute HF which increases GMP levels by binding to NPR-A [3,5]. Apart from its vasodilator effects, is has shown to inhibit apoptosis and limit myocardial remodeling and subsequent hypertrophy [5,50]. ESC guidelines recommend its use especially in hypertensive acute HF, very frequently associated with a preserved LVEF. However, patients with acute HFpEF tend to respond differently to vasodilators than their counterparts with reduced EF. While BP reduction tends to be greater in response to vasodilators, improvement in stroke volume is minimal [47]. The ASCEND trial showed that nesiritide leads to minimal clinical improvement in HFpEF patients, with no impact on renal function, hospitalization rates or death [42]. A different study showed that nesiritide failed to impact both symptoms and clinical outcomes, with lower decreased sodium excretion and subsequent weight reduction and increased rates of therapy failure [49]. Despite not being formally contraindicated and while its current use as a vasodilator is a class IIaB indication, the role of nesiritide in acute HFpEF therapy remains controversial with further studies required to assess its benefits.

A different NP used in HF treatment is carperitide, a recombinant form of ANP. Although recommended in Japan for acute HF therapy, it has failed to enter either ESC or American guidelines, especially because of limited studies and decreased half-life [5,51]. Despite improving symptoms in the studied Japanese population, there was no impact reported on mortality. Furthermore, the impact on HFpEF patients is unknown. Urodilatin is a form of ANP which determined clinical improvement in acute HF patients, but at the cost of worsening renal function and hypotension [5]. Its routine use in acute HF therapy has yet to be approved due to the lack of impact on mortality, hospitalization rates and myocardial injury [52]. Its use in HFpEF has yet to been studied.

Attempts have been made to produce variants by the means of native NP gene alteration, resulting in NPR-A binding peptides resistant to degradation with subsequent increased half-life [1,5]. Several designer NPs (Table 3) have already been produced, seemingly retaining the anti-fibrotic, anti-hypertrophic and vasodilator properties of the native NPs: cenderitide-NP [53], CU-NP [56], M-ANP [58], AS-BNP [59]. Although they haven't been approved for HF therapy, some have entered phase II trials (CD-NP).

Table 3. Designer natriuretic peptides as potential therapies in heart failure.

Designer Natriuretic Peptide	Structure	Effects	Reference
CD-NP (Cenderitide)	Fusion between CNP-22 and 15 aa DNP C-terminal	Vasodilator Antifibrotic, antiproliferative ↑GFR, ↓LA pressure (via NPR-A, NPR-B) less hypotension than nesiritide (minimal changes in BP)	[53–55]
CU-NP (humanized version of cenderitide)	Fusion between 17 aa ring of CNP and urodilatin's N-terminal	+ cGMP => RAAS inhibition antihypertrophic (NHE-1/calcineurin pathway inhibition) renal function enhancement	[56,57]
M-ANP (Mutant-ANP)	12 aa extension to native ANP's C-terminal	Enhances natriuresis and diuresis RAAS and sympathetic nervous system inhibition Inhibits cellular proliferation Antifibrotic	[58]
ANX-042 (AS-BNP)	Fusion between AS-BNP's 16 aa of C terminal and 26 aa of native BNP	+ cGMP; RAAS inhibition natriuresis and diuresis stimulation	[57,59,60]
Nesiritide	Recombinant BNP	Vasodilator Hypotension Less renal function deterioration as compared to diuretic treated HFpEF patients	[61]
Carperitide	Recombinant ANP	Vasodilator Renoprotective Not widely recommended	[51]

ANP- atrial natriuretic peptide; AS-BNP- alternatively spliced BNP; aa-amino acids; BNP- b type natriuretic peptide; CNP- c type natriuretic peptide; cGMP- cyclic guanosine monophosphate; GFR- glomerular filtration rate; LA- left atrium; M-ANP: mutant ANP; NP- natriuretic peptide; NPR- natriuretic peptide receptor; NHE- natrium-proton exchanger; RAAS- renin-angiotensin-aldosterone system.

5.2. Natriuretic Peptide Breakdown Inhibitors

Another possible therapeutic strategy is inhibiting the breakdown of NPs, thus promoting their biological effects [63–81]. This can be done by inhibiting NEP, a zinc dependent enzyme that represents the final step in NPs cleavage. NEP has a large distribution, being found inside cardiomyocytes, fibroblasts, central nervous system (brain, cerebrospinal fluid), immune cells (neutrophils), lungs, endocrine glands (thyroid, adrenal glands), myocardium. Its substrate is as extended as its distribution, acting on angiotensin I, II, bradykinin, endothelin-1, glucagon and amyloid-beta [5,81].

There are several drugs that inhibit NEP (Table 4). Pure NEP inhibitors, such as candoxatril [62] and ecadotril [5,62] have been shown to raise NPs, but their ability to stimulate RAAS at the same time together with uncertain and contradictory effects on blood pressure, clinical improvement and survival precluded their entry in routine practice [81].

Table 4. Dual and triple endothelin converting enzyme, neutral endopeptidase and angiotensin converting enzyme inhibitors.

Class	Drug	Effects	References
Pure NEP inhibitors	Candoxatril	↑NPs and natriuresis ↑angiotensin II (RAAS stimulation) => vasoconstriction	[62]
	Ecadotril	No proven clinical benefit; may determine aplastic anemia	[63]

Table 4. Cont.

Class	Drug	Effects	References
Dual ECE/NEP inhibitors	Daglutril (SLV-306)	↑NPs, ET-1, ↓BP, LV hypertrophy and pressure	[67]
	SLV-338	↓ LV remodelling (independently of BP lowering effects)	[64]
Dual NEP/ACE inhibition	Sampatrilat	Despite clinical benefits, its short half-life precluded its clinical use	[68]
	Omapatrilat	Symptoms relief and improved survival; ↓BP, vascular resistance (more than candoxatril) LV remodelling, myocardial fibrosis hypotension and angioedema (bradykinin)	[68–70]
Triple ACE/ECE/NEP inhibitors	Benazepril (ACE) + CGS 26303 (dual ECE/NEP inhibitor)	↑NPs, bradykinin inhibits angiotensin II and ET-1 ↓ LV remodelling (including mass and end-diastolic pressure)	[71]

ACE- angiotensin converting enzyme; BP- blood pressure; ET-1- endothelin- 1; ECE- endothelin converting enzyme; NEP- neutral endopeptidase; NPs- natriuretic peptides, LV- left ventricle.

A different strategy that has been attempted targets dual endothelin and NP breakdown inhibition. It seems that endothelin-1 (ET-1) and angiotensin II share common pathophysiological mechanisms. Although ET-1 impacts HF progression through vasoconstrictor and fibrotic effects, antagonizing its receptors have failed to show prognostic effects in HF patients [5]. However, inhibiting endothelin-converting enzyme (ECE) and NP cleavage led to an improvement of symptoms through RAAS inhibition, with a subsequent decrease in ET-1 levels apart from an increase NPs (dual ECE/NEP inhibitors). Moreover, it seems that ECE also plays a role in NP degradation, as its structure chemically resembles that of NEP [65]. Different drugs have been created [64,67], however none implemented yet due to the lack of clinical trials. Drugs capable of inhibiting ECE, NEP and ACE have been produced; however there are serious concerns about their safety [71].

Sacubitril/valsartan (formerly known as LCZ 696) combines a NEP inhibitor (sacubiril) with an angiotensin receptor blocker (valsartan) in a 1:1 ratio, being the first in its class (ARNI). Its benefits come from combining neprilysin inhibition, which limits NPs breakdown and leads to a subsequent rise in both their levels, and cGMPs with RAAS inhibition given by valsartan [5,74]. Among the NPs, both ANP and BNP are substrates for neprilysin; however, its affinity is stronger for the former [1,9,74–86]. Subsequently, the increase in ANP levels (and subsequently in cGMP) will surpass that of BNP. However, due to its short plasma half-life, BNP is preferred over ANP for monitoring its effects. Of note, NT-proBNP is not a direct substrate for neprylisin inhibition; as such the decrease in its levels makes it the preferred biomarker for monitoring HF patients treated with sacubitril/valsartan, especially during the first 10 weeks, when the increase in BNP levels is maximum [74–76,87]. Compared with dual ACE/NEP inhibitors, the incidence of angioedema is lower due to a lower increase in bradykinin [74–77].

Several studies support the sacubitril/valsartan treatment class IB indication in HFrEF patients [3,73–76,88]. However, there is no clear indication for this drug in HFpEF and there are no clinical trials assessing outcomes in patients treated with the ARNI. Moreover, a Swedish study revealed an undiagnosed/lower NP level as the most frequent cause of HFrEF patients not receiving sacubitril/valsartan [72]. This could easily apply in HFpEF, given the controversy of NP levels in this set of patients [5,74].

Regarding cGMP and sacubitril/valsartan therapy, there is extensive focus on its beneficial effects on cGMP levels and, subsequently, PKG. cGMP deficiency is of paramount importance in HFpEF, as it lowers PKG levels and thus promotes myocardial remodeling (both hypertrophy and impaired relaxation through increased cardiomyocyte resting tension) [26]. Through oxidative stress [21] and a subsequent decrease in NO availability [26], numerous comorbidities contribute to the lower levels of cGMP found in HFpEF. It seems that the pro-inflammatory state induced by these comorbidities triggers coronary microvascular inflammation, which further decreases NO availability [26] and inhibits cGMP formation. Moreover, oxidative stress affects titin, a protein found in cardiomyocytes cytoskeleton in two ways: firstly by inhibiting PKG- dependent phosphorylation and secondly by determining the formation of disulfide bridges, which leads to a more stiffer titin molecule [26]. The authors agree that the cGMP deficiency in HFpEF can be ascribed to decreased production and not increased breakdown [25], which explains why in HFpEF patients, cGMP levels increase in response to neprylisin inhibitors and not phosphodiesterase inhibitors, which inhibit cGMP breakdown [21].

The PARAMOUNT trial assessed the benefits and safety profile of sacubitril/valsartan treatment in HFpEF patients [84]. NT-proBNP levels significantly decreased in sacubitril/valsartan group, especially in diabetic patients. After 36 weeks of treatment, there was an improvement in NYHA class and a reverse LA remodelling with a subsequent decrease in LA volume (especially in sinus rhythm patients), and no changes in LVEF or LV volumes.

Recently, the baseline characteristics of patients enrolled in PARAGON-HF trial, which assessed the impact of sacubitril/valsartan therapy in HFpEF on mortality, have been published [85]. 4822 patients with HF NYHA class II-IV, elevated NPs, LVEF >45% and increased LA dimensions/ LV hypertrophy were included. Regarding NPs levels, the median NT-proBNP was 885 pg/mL, with higher thresholds being used for AF and previously HF hospitalized patients' inclusion. In the absence of recent (<9 months) HF hospitalization, NT-proBNP inclusion concentration thresholds used were >300 pg/mL for sinus rhythm patients and >900 pg/mL for patients, the levels being lower in recent hospitalized patients (>200 pg/mL and >600 pg/mL, respectively). Again, the most common cause of exclusion was lower NPs levels. The final results will show whether combined ARB/NEP inhibition could be a therapeutic option in HF patients with an LVEF >45%.

An interesting possible side-effect of sacubitril/valsartan therapy may be an increase in amyloid-β concentration as NEP is responsible with enzymatic clearance of amyloid, thus augmenting the risk of developing Alzheimer's. An ongoing trial will assess the impact of the novel drug as compared to valsartan treated HFpEF patients on cognitive impairment, with results being scheduled for 2022 [79].

5.3. Mineralocorticoid Receptors Antagonists

Apart from novel therapies, several studies investigated the benefits of mineralocorticoid receptors antagonists (MRAs) therapy in HFpEF [6,89–92]. The sub analysis of the TOPCAT trial conducted by Anand et al. reveals that while spironolactone does not impact overall mortality and hospitalization rates, it proved beneficial in HFpEF patients with lower NPs levels and not in those with increased concentrations [88]. These results were similar to those of the I-PRESERVE trial [34], in which Irbesartan treatment was beneficial in HFpEF patients with lower NPs levels. The results of the two trials point out the fact that higher NPs levels in HFpEF translate an increased cardiovascular event rate, not necessarily increased treatment responsiveness. Moreover, it seems that higher NP levels in these patients relate to a more advanced degree of structural heart disease, including higher degree of myocardial fibrosis, less responsive to therapeutic intervention [88].

Patients with HFpEF present increased fibrosis biomarkers, as shown by Zile et al. [89] and Oikonomou et al. [6] In a study conducted by Cho-Kai Wu et al. [20], HFpEF patients with increased myocardial fibrosis degree as assessed by late-gadolinium enhancement magnetic resonance imaging scans had higher levels of NT-proBNP. Fibrosis biomarker concentrations (growth differentiation factor, galectin-3, tissue inhibitor of metalloproteinase, matrix metalloproteinase 2) correlated with the severity of fibrosis. However, the authors concluded that NPs rather reflect wall tension and volume overload with lower discriminative power in regard to the presence and/or degree of myocardial fibrosis.

It appears that HFpEF with lower matrix metalloproteinase 9 levels benefit most from eplerenone treatment [6]. Chen et al. [90] show that MRA therapy in HFpEF reduce both myocardial remodeling and amino terminal peptide of procollagen III levels, without however affecting NP levels. It may be that in the early stages of HFpEF, the progression of myocardial remodeling and subsequent fibrosis is preventable by therapeutic interventions, such as MRAs therapy, thus explaining the benefit that patients with lower fibrosis biomarkers and NPs levels have from this therapy.

6. Prognostic Implications of Natriuretic Peptides in Heart Failure with Preserved Left Ventricular Ejection Fraction

It is agreed upon that although NP values tend to be lower in HFpEF patients with no consensus regarding diagnostic thresholds, they retain their prognostic utility irrespective of LVEF [3,87].

In a different study, increased BNP levels were associated with increased mortality, irrespective of LVEF [93]. Also, it seems that higher BNP levels translate into increased risk for developing both AF and transient ischemic attacks [1].

6.1. Prognostic Value of BNP and NT-proBNP

The NPs concentrations retain their prognostic utility albeit their lower baseline levels in HFpEF patients [94]. Several studies reported similar death risks for a given NPs concentration irrespective of HF phenotype [93,95–99]. Levy and Anand compared patients from I-PRESERVE and VALHEFT trials, showing that a 1-log increase in NT-proBNP levels carries a mortality HR of 1.7 regardless of LVEF [96]. However, the same authors demonstrated that the mortality of HFrEF patients was two thirds higher than of those with HFpEF. Similarly, Salah et al. reported that hospitalized HFpEF patients have lower mortality rates when compared to HFrEF [98]; however, the difference is minor and the risk of death tends to equalize after discharge. More importantly, although patients with HFpEF display lower baseline NPs concentrations, for the same NT-proBNP level, prognosis is similar regardless of LVEF. This paradox can be firstly be explained by the different mechanisms involved in NPs secretion between the two phenotypes. It is known that at the same end-diastolic LV pressure, HFpEF patients display lower NT-proBNP levels. This may be due to the correlation of NPs with diastolic wall stress, which, according to the law of La Place, correlates with wall pressure and cavity diameter and is inversely related to wall thickness. As patients with HFpEF typically exhibit a concentric LV remodelling with increased wall thickness, this may explain in part their lower NPs levels for a given increased wedge pressure as compared to the HFrEF patients (who typically have an eccentrical LV remodelling). Secondly, this paradox can be explained by the distribution of comorbidities in regard with HF phenotype. As such, HFpEF patients tend to be older, with increased incidence of arterial hypertension, chronic kidney disease, AF and anemia, comorbidities that may account for similar prognosis between HF phenotypes despite lower NTpro-BNP of HFpEF [95,98]. This drives attention to the need of also addressing non-cardiovascular diseases in order to improve outcomes in HFpEF patients. The fact that while the same NP concentration may point out to the same relative risk of death and overall prognosis still differs between HF phenotypes underlines exactly the contribution of other risk factors and comorbidities to the mortality. That's why it is not advisable to regard NP levels as surrogate markers of mortality.

It appears that different NPs concentrations threshold might be necessary for prognosis in HFpEF patients as compared to HFrEF [94–106]. Accordingly, NPs prognostic values vary across different studies [34,87,88], several authors highlighting that NPs increase in addition to their baseline levels holds prognostic importance [34]. Moreover, relying on an NP prognosis threshold may not be necessary, as prognosis can and should be continuously assessed.

Anand et al. associated a baseline NT-proBNP of 339 pg/mL with a four year mortality of 21.1%, which translates into a 5% annual mortality for an NT-proBNP level between 300–500 pg/mL [34].

As NT-proBNP is also influenced by AF [34], increasing its concentration irrespective of HF presence [37], these levels should be judged accordingly when determining prognosis. Kristensen et al. showed that different NT-proBNP levels should be used in AF versus non-AF patients to determine prognosis. While an NT-proBNP level of < 400 pg/mL was associated with a better prognosis irrespective of rhythm; the presence of AF accounted for the different mortality rates in HFpEF patients with a NT-proBNP > 400 pg/mL [107]. However, higher NT-proBNP AF patients had increased risks of hospitalization as compared to patients with HFrEF and the same NPs level. In addition, it seems that patients with lower baseline NPs levels would benefit on long-term from supplementary explorations [87].

A recent study launched the possibility of increased NT-proBNP playing a causative role in AF [100]. This idea is controversial as several comorbidities that promote AF lead in turn to an increase in NT-proBNP, which remains a biologically inactive product. Moreover, an increase in both ANP and BNP in sacubitril/valsartan patients led to atrial reverse-remodeling, emphasizing their antifibrotic and antihypertrophic effects [87].

6.2. Prognostic Value of MR-proANP

Another emerging prognostic biomarker is MR-proANP [27]. It seems that MR-proANP accurately assesses mortality in both acute and chronic HF [27,30]. It seems that increased MR-proANP levels correlate with an increased four year mortality risk [106]. Moreover, in a Swedish study, the novel biomarker predicted both HF and AF, as opposed to NT-proBNP which failed to predict AF [97]. The study assessed the difference in AF versus non-AF HFpEF patients' prognostics as determined by their NT-proBNP level. Alehagen et al. also found MR-proANP to be prognostic of cardiovascular mortality at five years for HFpEF patients [108]. Similarly, Zabarovskaja and colleagues stated that a MR-proANP of >313 pmol/L predicted all-cause mortality (Table 5) [109]. The association between NP and AF has been studied [18,109–113]. Although ANP levels seem to decrease after electrical cardioversion [112] and are associated with AF progression [18], both BNP and NT-proBNP are preferred to assess AF recurrence rates post electrical cardioversion [114].

Table 5. Prognostic values of natriuretic peptides in heart failure with preserved ejection fraction patients.

Biomarker	Concentration	Utility	Reference	Additional Remarks
BNP	>540 pg/mL	Predicts in-hospital mortality	[93]	May be able to predict AF Levels affected by sacubitril/valsartan
NT-proBNP	>300–500 pg/mL (339 pg/mL)	5% 1 year mortality	[34]	AF patients hold better prognosis at the same NT-proBNP levels as their sinus rhythm counterparts
MR-proANP	>313 pmol/L	Predicts all-cause mortality	[109]	May be able to predict AF; correlates with LAVI

BNP- B-type natriuretic peptide; MR-proANP- middle region pro atrial natriuretic peptide; AF- atrial fibrillation; LAVI- indexed left atrial volume.

6.3. Prognostic Value of Natriuretic Peptides and Heart Failure with Preserved Ejection Fraction Therapy

There are controversies regarding the prognostic value of NPs in sacubitril/valsartan treated patients. As the formerly known LCZ696 interacts with NPs levels, there still are questions regarding their accuracy in determining HF patients' prognosis. Some studies agree that both BNP and NT-proBNP retain their prognostic utility in these patients, emphasizing the fact that during therapy initiation, NT-proBNP is the preferred biomarker (several months are required for BNP levels stabilization) [74]. However, the currently available trials did not include HFpEF patients treated with sacubitril/valsartan.

Another issue regarding NPs prognostic value in HFpEF is whether these patients would benefit from NPs guided therapy. So far, studies are controversial [3,46,115–127], especially taking into consideration that these patients respond differently to HF therapy than those with reduced LVEF. A study conducted by Khan et al. including patients with both HFrEF and HFpEF came to the conclusion that NP guided therapy was not beneficial [115]. A different study emphasized that regardless of LVEF, a constantly elevated NPs should not determine changes in patient management [117]. Maeder et al. revealed in TIME-CHF trial that NT-proBNP guided therapy was not beneficial in HFpEF patients, as compared to HFrEF [118]. Although it included HFrEF patients, the GUIDE-IT trial concluded that biomarker directed therapy may not benefit these patients [122]. The opposite was shown in a meta-analysis conducted by Troughton et al. [119]. The authors stated that BNP-guided therapy improved outcomes in patients < 75 years old and reduced HF hospitalization rates regardless of age and LVEF. Brunner la Roca et al. agreed that NP guided therapy either through BNP or NT-proBNP is safe and more importantly, cost-effective. [4]. A British team stated that BNP-guided therapy might be cost-effective for HFpEF patients < 75 years old [120].

Recent studies have focused on MR-proANP as a possible biomarker to guide HF therapy. So far, it seems that MR-proANP levels might be able to predict incident HF, as it shown in a study conducted by Sabatine et al. [124]. Interestingly, it seems that MR-proANP might predict cardiac resynchronization therapy responders which showed lower levels of the biomarker as compared to the non-responders [125].

Regardless of the biomarker used, NP guided medical therapy remains controversial, especially in HFpEF patients. In these patients, more studies are required to assess the benefits of optimizing medical therapy after serial NP measurements. The most studied NPs are BNP and its inactive form, NT-proBNP, especially due to their proven diagnostic capacities. The novel MR-proANP, with its increased half time as compared to ANP, shows promise both in prognosis and guiding medical therapy, but more studies are required to confirm its safety profile and utility.

7. Future Perspectives

Several on-going studies on NPs in HFpEF have been announced. The results of PARAGON-HF trial will clear the impact of sacubitril/valsartan therapy in HFpEF [85], while another on-going study conducted by Mayo clinic (ClinicalTrials.gov Identifier: NCT03506412) will determine how this therapy affects NPs levels and cGMP in HFpEF patients [127]. A different study finishing in December 2019 will assess the effects of sacubitril/valsartan as compared to either enalapril or valsartan in HFpEF patients (ClinicalTrials.gov Identifier: NCT03066804) [128].

Moreover, in the context of the pro-inflammatory state of HFpEF, there is increasing body of evidence linking non-NPs biomarkers to HFpEF as a better diagnostic tool [21]. Emerging markers reflecting myocardial fibrosis such as a soluble source of tumorigenicity 2 (sST2), growth differentiation factor-15 [21], galectin 3 [126] and interleukins (1 and 6) are showing promise, correlating in different degrees with transthoracic echocardiographic parameters of diastolic dysfunction [10,21]. Their utility in comparison to NPs remains to be evaluated.

8. Conclusions

The number of HFpEF patients is increasing; moreover, these patients show similar prognosis with HFrEF patients.

The role of NPs in HF is both diagnostic and prognostic. Guidelines regard a BNP > 35 pg/mL or an NT-proBNP > 125 pg/mL as being diagnostic in non-acute setting, while emphasizing the fact that although several factors alter NP concentrations, their diagnostic utility rather lies in their negative predictive ability. For decompensated HF, either a BNP > 100 pg/mL, an NT-proBNP > 300 pg/mL or a MR-proANP of > 120 pmol/L is diagnostic. Authors agree that lower levels may be considered for HFpEF, without any thresholds being agreed upon. The absence of a consensus regarding diagnostic thresholds for NPs in HFpEF has led to inhomogeneous inclusion criteria in large clinical trials, thus affecting the results.

The role of NPs extends beyond diagnosis, as they can be regarded as a therapeutic target per se through nesiritide and the novel ARNI, sacubitril/valsartan. Augmenting the concentrations of NPs leads to lower blood pressure levels as well as an improvement in symptoms and quality of life. A degree of reverse remodelling with a subsequent decrease in myocardial fibrosis and LA dimensions could be determined.

Higher NPs levels are associated with poorer prognosis, especially in acute settings, irrespective of LVEF. Moreover, BNP and MR-proANP predict not only incident HF, but also AF. As AF is one of the most frequent causes of HFpEF, this becomes of the utmost importance. For AF patients it must be remembered that NP levels tend to be higher, irrespective of the presence of HF. Also, for the same BNP concentration, they show a better prognostic as compared with their sinus-rhythm counterparts.

Finally, guiding medical therapy by serial NP measurements is controversial, especially in HFpEF. Authors recommend in the absence of consensus that no change in patient management should be taken if the NP levels remain constantly elevated throughout treatment.

The need for further studies assessing diagnostic NP thresholds and prognostic values in HFpEF patients is enormous. Furthermore, given the novel therapies that interact with NPs, such as sacubitril/valsartan, these levels might be altered.

Author Contributions: Conceptualization, S.A.S., M.F. and S.R.; methodology, S.A.S., M.F., G.L.B. and S.R.; formal analysis, M.D.T., C.F.C., G.L.B.; investigation, A.O., D.M.T., M.D.T., C.F.C., and G.L.B.; resources, S.A.S., M.F., S.R.; A.O., D.M.T., and G.L.B.; writing—S.A.S., S.R., M.F., G.L.B.; writing—review and editing, M.F., S.A.S., S.R., G.L.B.; visualization, A.O., D.M.T., C.F.C., M.D.T., and G.L.B.; supervision, M.F., A.O., D.M.T. and G.L.B.

Funding: This research received no external funding.

Conflicts of Interest: The authors declare no conflict of interest.

Abbreviations

The following abbreviations are used in this manuscript:

aa	amino-acid
ACE	angiotensin converting enzyme
Ang II	angiotensin II
AF	atrial fibrillation
ANP	atrial natriuretic peptide
ARB	angiotensin receptor blockers
ATP	adenylate triphosphates
BNP	B-type natriuretic peptide
CNP	C-type natriuretic peptide
CD-NP	Cenderitide
ECE	endothelin-converting enzyme
ET-1	endothelin- 1
ESC	European Society of Cardiology
GFR	glomerular filtration rate
GMP	guanylate mono phosphatase
HF	heart failure
HFrEF	heart failure with reduced ejection fraction
HFpEF	heart failure with preserved ejection fraction
HFmrEF	heart failure with mid-range ejection fraction
LA	left atrial
LAVI	left atrial indexed volume
LV	left ventricular
LVEF	left ventricular ejection fraction
mRNA	messenger ribonucleic acid
MR-proANP	middle range pro atrial natriuretic peptide
NEP	neutral endopeptidase
NHE	natrium-proton exchanger
NPs	natriuretic peptides
NPR-A	natriuretic peptide receptor A
NPR-B	natriuretic peptide receptor B
NPR-C	natriuretic peptide receptor C
MRAs	mineralocorticoid receptor antagonists
pGC	particulate guanylyl cyclase
RAAS	renin-angiotensin-aldosterone system

sGC   soluble guanylyl cyclase
sST2   soluble source of tumorigenicity 2
TGF ß   transforming growth factor beta

References

1. De Keulenaer, G.W.; Brutsaert, D.L. Systolic and diastolic heart failure are overlapping phenotypes within the heart failure spectrum. *Circulation* **2011**, *123*. [CrossRef]
2. Aronow, W.S. Epidemiology, pathophysiology, prognosis, and treatment of systolic and diastolic heart failure in elderly patients. *Heart Dis.* **2003**, *5*, 279–294. [CrossRef] [PubMed]
3. Ponikowski, P.; Voors, A.A.; Anker, S.D.; Bueno, H.; Cleland, J.G.; Coats, A.J.; Falk, V.; González-Juanatey, J.R.; Harjola, V.P.; Jankowska, E.A.; et al. 2016 ESC Guidelines for the diagnosis and treatment of acute and chronic heart failure: The Task Force for the diagnosis and treatment of acute and chronic heart failure of the European Society of Cardiology (ESC). Developed with the special contribution of the Heart Failure Association (HFA) of the ESC. *Eur. J. Heart Fail.* **2016**, *18*, 891–975. [CrossRef] [PubMed]
4. Brunner-La Rocca, H.P.; Sanders-van Wijk, S. Natriuretic Peptides in Chronic Heart Failure. *Card. Fail. Rev.* **2019**, *5*, 44–49. [CrossRef] [PubMed]
5. Fu, S.; Ping, P.; Wang, F.; Luo, L. Synthesis, secretion, function, metabolism and application of natriuretic peptides in heart failure. *J. Biol. Eng.* **2018**, *12*. [CrossRef] [PubMed]
6. Oikonomou, E.; Vogiatzi, G.; Tsalamandris, S.; Mourouzis, K.; Siasos, G.; Lazaros, G.; Skotsimara, G.; Marinos, G.; Vavuranakis, M.; Tousoulis, D. Non-natriuretic peptide biomarkers in heart failure with preserved and reduced ejection fraction. *Biomark. Med.* **2018**, *12*, 783–797. [CrossRef] [PubMed]
7. Meijers, W.C.; van der Velde, A.R.; de Boer, R.A. Biomarkers in heart failure with preserved ejection fraction. *Neth. Heart J.* **2016**, *24*, 252–258. [CrossRef]
8. D'Elia, E.; Vaduganathan, M.; Gori, M.; Gavazzi, A.; Butler, J.; Senni, M. Role of biomarkers in cardiac structure phenotyping in heart failure with preserved ejection fraction: Critical appraisal and practical use. *Eur. J. Heart Fail.* **2015**, *17*, 1231–1239. [CrossRef]
9. Nadar, S.K.; Shaikh, M.M. Biomarkers in Routine Heart Failure Clinical Care. *Card. Fail. Rev.* **2019**, *5*, 50–56. [CrossRef] [PubMed]
10. de Boer, R.A.; Nayor, M.; deFilippi, C.R.; Enserro, D.; Bhambhani, V.; Kizer, J.R.; Blaha, M.J.; Brouwers, F.P.; Cushman, M.; Lima, J.A.C.; et al. Association of Cardiovascular Biomarkers with Incident Heart Failure with Preserved and Reduced Ejection Fraction. *JAMA Cardiol.* **2018**, *3*, 215–224. [CrossRef]
11. Gruson, D.; Favvresse, J. Peptides natriurétiques : Dégradation, formes circulantes, dosages et nouvelles approches thérapeutiques. *Ann. Biol. Clin.* **2017**, *75*, 259–267. [CrossRef]
12. Maisel, A.S.; Duran, J.M.; Wettersten, N. Natriuretic Peptides in Heart Failure: Atrial and B-type Natriuretic Peptides. *Heart Fail. Clin.* **2018**, *14*, 13–25. [CrossRef] [PubMed]
13. Nishikimi, T.; Kuwahara, K.; Nakao, K. Current biochemistry, molecular biology, and clinical relevance of natriuretic peptides. *J. Cardiol.* **2011**, *57*, 131–140. [CrossRef]
14. Rahmutula, D.; Zhang, H.; Wilson, E.; Olgin, J.E. Absence of natriuretic peptide clearance receptor attenuates TGF-b1-induced selective atrial fibrosis and atrial fibrillation. *Cardiovasc. Res.* **2019**, *115*, 357–372. [CrossRef] [PubMed]
15. Calderone, A.; Thaik, C.M.; Takahashi, N.; Chang, D.L.; Colucci, W.S. Nitric oxide, atrial natriuretic peptide, and cyclic GMP inhibit the growth-promoting effects of norepinephrine in cardiac myocytes and fibroblasts. *J. Clin. Investig.* **1998**, *101*, 812–818. [CrossRef]
16. Volpe, M.; Carnovali, M.; Mastromarino, V. The natriuretic peptides system in the pathophysiology of heart failure: From molecular basis to treatment. *Clin. Sci.* **2016**, *130*, 57–77. [CrossRef]
17. Matsuo, A.; Nagai-Okatani, C.; Nishigori, M.; Kangawa, K.; Minamino, N. Natriuretic peptides in human heart: Novel insight into their molecular forms, functions, and diagnostic use. *Peptides* **2019**, *111*, 3–17. [CrossRef]
18. Seewöster, T.; Büttner, P.; Nedios, S.; Sommer, P.; Dagres, N.; Schumacher, K.; Bollmann, A.; Hilbert, S.; Jahnke, C.; Paetsch, I.; et al. Association Between Cardiovascular Magnetic Resonance-Derived Left Atrial Dimensions, Electroanatomical Substrate and NT-proANP Levels in Atrial Fibrillation. *J. Am. Heart Assoc.* **2018**, *7*, e009427. [CrossRef]

19. Nattel, S. Natriuretic peptide receptors and atrial-selective fibrosis: Potential role in atrial fibrillation. *Cardiovasc. Res.* **2019**, *115*, 258–260. [CrossRef]
20. Wu, C.K.; Su, M.M.; Wu, Y.F.; Hwang, J.J.; Lin, L.Y. Combination of Plasma Biomarkers and Clinical Data for the Detection of Myocardial Fibrosis or Aggravation of Heart Failure Symptoms in Heart Failure with Preserved Ejection Fraction Patients. *J. Clin. Med.* **2018**, *7*, 427. [CrossRef]
21. Zakeri, R.; Cowie, M.R. Heart failure with preserved ejection fraction: Controversies, challenges and future directions. *Heart* **2018**, *104*, 377–384. [CrossRef]
22. Reddy, Y.N.V.; Carter, R.E.; Obokata, M.; Redfield, M.M.; Borlaug, B.A. A Simple, Evidence-Based Approach to Help Guide Diagnosis of Heart Failure with Preserved Ejection Fraction. *Circulation* **2018**, *138*, 861–870. [CrossRef]
23. Michalska-Kasiczak, M.; Bielecka-Dabrowa, A.; von Haehling, S.; Anker, S.D.; Rysz, J.; Banach, M. Biomarkers, myocardial fibrosis and co-morbidities in heart failure with preserved ejection fraction: An overview. *Arch. Med. Sci.* **2018**, *14*, 890–909. [CrossRef] [PubMed]
24. Cui, K.; Huang, W.; Fan, J.; Lei, H. Midregional pro-atrial natriuretic peptide is a superior biomarker to N-terminal pro-B-type natriuretic peptide in the diagnosis of heart failure patients with preserved ejection fraction. *Medicine* **2018**, *97*, e12277. [CrossRef] [PubMed]
25. Paulus, W.J.; Tschöpe, C. A novel paradigm for heart failure with preserved ejection fraction: Comorbidities drive myocardial dysfunction and remodeling through coronary microvascular endothelial inflammation. *J. Am. Coll. Cardiol.* **2013**, *62*, 263–271. [CrossRef]
26. van Heerebeek, L.; Hamdani, N.; Falcão-Pires, I.; Leite-Moreira, A.F.; Begieneman, M.P.; Bronzwaer, J.G.; van der Velden, J.; Stienen, G.J.; Laarman, G.J.; Somsen, A.; et al. Low myocardial protein kinase G activity in heart failure with preserved ejection fraction. *Circulation* **2012**, *126*, 830–839. [CrossRef] [PubMed]
27. Maisel, A.; Mueller, C.; Nowak, R.M.; Peacock, W.F.; Ponikowski, P.; Mockel, M.; Hogan, C.; Wu, A.H.; Richards, M.; Clopton, P.; et al. Midregion prohormone adrenomedullin and prognosis in patients presenting with acute dyspnea: Results from the BACH (Biomarkers in Acute Heart Failure) trial. *J. Am. Coll. Cardiol.* **2011**, *58*, 1057–1067. [CrossRef] [PubMed]
28. Zheng, S.L.; Chan, F.T.; Nabeebaccus, A.A.; Shah, A.M.; McDonagh, T.; Okonko, D.O.; Ayis, S. Drug treatment effects on outcomes in heart failure with preserved ejection fraction: A systematic review and meta-analysis. *Heart* **2018**, *104*, 407–415. [CrossRef]
29. Salo, P.P.; Havulinna, A.S.; Tukiainen, T.; Raitakari, O.; Lehtimäki, T.; Kähönen, M.; Kettunen, J.; Männikkö, M.; Eriksson, J.G.; Jula, A.; et al. Genome-Wide Association Study Implicates Atrial Natriuretic Peptide Rather Than B-Type Natriuretic Peptide in the Regulation of Blood Pressure in the General Population. *Circ. Cardiovasc. Genet.* **2017**, *10*, e001713. [CrossRef]
30. Hausfater, P.; Claessens, Y.E.; Martinage, A.; Joly, L.M.; Lardeur, J.Y.; Der Sahakian, G.; Lemanski, C.; Ray, P.; Freund, Y.; Riou, B. Prognostic value of PCT, copeptin, MR-proADM, MR-proANP and CT-proET-1 for severe acute dyspnea in the emergency department: The BIODINER study. *Biomarkers* **2017**, *22*, 28–34. [CrossRef]
31. Kang, S.H.; Park, J.J.; Choi, D.J.; Yoon, C.H.; Oh, I.Y.; Kang, S.M.; Yoo, B.S.; Jeon, E.S.; Kim, J.J.; Cho, M.C.; et al. Prognostic value of NT-proBNP in heart failure with preserved versus reduced EF. *Heart* **2015**, *101*, 1881–1888. [CrossRef]
32. Richards, A.M.; Januzzi, J.L., Jr.; Troughton, R.W. Natriuretic peptides in heart failure with preserved ejection fraction. *Heart Fail. Clin.* **2014**, *10*, 453–470. [CrossRef]
33. Anjan, V.Y.; Loftus, T.M.; Burke, M.A.; Akhter, N.; Fonarow, G.C.; Gheorghiade, M.; Shah, S.J. Prevalence, clinical phenotype, and outcomes associated with normal B-type natriuretic peptide levels in heart failure with preserved ejection fraction. *Am. J. Cardiol.* **2012**, *110*, 870–876. [CrossRef]
34. Anand, I.S.; Rector, T.S.; Cleland, J.G.; Kuskowski, M.; McKelvie, R.S.; Persson, H.; McMurray, J.J.; Zile, M.R.; Komajda, M.; Massie, B.M.; et al. Prognostic value of baseline plasma amino-terminal pro-brain natriuretic peptide and its interactions with irbesartan treatment effects in patients with heart failure and preserved ejection fraction: Findings from the I-PRESERVE trial. *Circ. Heart Fail.* **2011**, *4*, 569–577. [CrossRef]
35. Jaubert, M.P.; Armero, S.; Bonello, L.; Nicoud, A.; Sbragia, P.; Paganelli, F.; Arques, S. Predictors of B-type natriuretic peptide and left atrial volume index in patients with preserved left ventricular systolic function: An echocardiographic-catheterization study. *Arch. Cardiovasc. Dis.* **2010**, *103*, 3–9. [CrossRef]
36. Maisel, A.S.; McCord, J.; Nowak, R.M.; Hollander, J.E.; Wu, A.H.; Duc, P.; Omland, T.; Storrow, A.B.; Krishnaswamy, P.; Abraham, W.T.; et al. Bedside B-Type natriuretic peptide in the emergency diagnosis

of heart failure with reduced or preserved ejection fraction. Results from the Breathing Not Properly Multinational Study. *J. Am. Coll. Cardiol.* **2003**, *41*, 2010–2017. [CrossRef]
37. Lam, C.S.; Rienstra, M.; Tay, W.T.; Liu, L.C.; Hummel, Y.M.; van der Meer, P.; de Boer, R.A.; Van Gelder, I.C.; van Veldhuisen, D.J.; Voors, A.A.; et al. Atrial Fibrillation in Heart Failure with Preserved Ejection Fraction: Association with Exercise Capacity, Left Ventricular Filling Pressures, Natriuretic Peptides, and Left Atrial Volume. *JACC Heart Fail.* **2017**, *5*, 92–98. [CrossRef] [PubMed]
38. Kraigher-Krainer, E.; Shah, A.M.; Gupta, D.K.; Santos, A.; Claggett, B.; Pieske, B.; Zile, M.R.; Voors, A.A.; Lefkowitz, M.P.; Packer, M.; et al. Impaired systolic function by strain imaging in heart failure with preserved ejection fraction. *J. Am. Coll. Cardiol.* **2014**, *63*, 447–456. [CrossRef] [PubMed]
39. Pieske, B.; Butler, J.; Filippatos, G.; Lam, C.; Maggioni, A.P.; Ponikowski, P.; Shah, S.; Solomon, S.; Kraigher-Krainer, E.; Samano, E.T.; et al. Rationale and design of the SOluble guanylate Cyclase stimulatoR in heArT failurE Studies (SOCRATES). *Eur. J. Heart Fail.* **2014**, *16*, 1026–1038. [CrossRef] [PubMed]
40. Solomon, S.D.; Rizkala, A.R.; Gong, J.; Wang, W.; Anand, I.S.; Ge, J.; Lam, C.S.P.; Maggioni, A.P.; Martinez, F.; Packer, M.; et al. Angiotensin Receptor Neprilysin Inhibition in Heart Failure with Preserved Ejection Fraction: Rationale and Design of the PARAGON-HF Trial. *JACC Heart Fail.* **2017**, *5*, 471–482. [CrossRef]
41. Iwanaga, Y.; Nishi, I.; Furuichi, S.; Noguchi, T.; Sase, K.; Kihara, Y.; Goto, Y.; Nonogi, H. B-type natriuretic peptide strongly reflects diastolic wall stress in patients with chronic heart failure comparison between systolic and diastolic heart failure. *J. Am. Coll. Cardiol.* **2006**, *47*, 742–748. [CrossRef] [PubMed]
42. Ledwidge, M.; Gallagher, J.; Conlon, C.; Tallon, E.; O'Connell, E.; Dawkins, I.; Watson, C.; O'Hanlon, R.; Bermingham, M.; Patle, A.; et al. Natriuretic peptide-based screening and collaborative care for heart failure: The STOP-HF randomized trial. *JAMA* **2013**, *310*, 66–74. [CrossRef]
43. Obineche, E.N.; Pathan, J.Y.; Fisher, S.; Prickett, T.C.; Yandle, T.G.; Frampton, C.M.; Cameron, V.A.; Nicholls, M.G. Natriuretic peptide and adrenomedullin levels in chronic renal failure and effects of peritoneal dialysis. *Kidney Int.* **2006**, *69*, 152–156. [CrossRef]
44. Santos-Araújo, C.; Leite-Moreira, A.; Pestana, M. Clinical value of natriuretic peptides in chronic kidney disease. *Nefrologia* **2015**, *35*, 227–233. [CrossRef]
45. Singh, J.S.S.; Burrell, L.M.; Cherif, M.; Squire, I.B.; Clark, A.L.; Lang, C.C. Sacubitril/valsartan: Beyond natriuretic peptides. *Heart* **2017**, *103*, 1569–1577. [CrossRef]
46. Brunner-La Rocca, H.P.; Eurlings, L.; Richards, A.M.; Januzzi, J.L.; Pfisterer, M.E.; Dahlström, U.; Pinto, Y.M.; Karlström, P.; Erntell, H.; Berger, R.; et al. Which heart failure patients profit from natriuretic peptide guided therapy? A meta-analysis from individual patient data of randomized trials. *Eur. J. Heart Fail.* **2015**, *17*, 1252–1261. [CrossRef] [PubMed]
47. Bishu, K.; Redfield, M.M. Acute heart failure with preserved ejection fraction: Unique patient characteristics and targets for therapy. *Curr. Heart Fail. Rep.* **2013**, *10*, 190–197. [CrossRef] [PubMed]
48. Dandamudi, S.; Chen, H.H. The ASCEND-HF trial: An acute study of clinical effectiveness of nesiritide and decompensated heart failure. *Expert Rev. Cardiovasc. Ther.* **2012**, *10*, 557–563. [CrossRef] [PubMed]
49. Wan, S.H.; Stevens, S.R.; Borlaug, B.A.; Anstrom, K.J.; Deswal, A.; Felker, G.M.; Givertz, M.M.; Bart, B.A.; Tang, W.H.; Redfield, M.M.; et al. Differential Response to Low-Dose Dopamine or Low-Dose Nesiritide in Acute Heart Failure with Reduced or Preserved Ejection Fraction: Results from the ROSE AHF Trial (Renal Optimization Strategies Evaluation in Acute Heart Failure). *Circ. Heart Fail.* **2016**, *9*, e002593. [CrossRef]
50. Chen, H.H.; Glockne, J.F.; Schirger, J.A.; Cataliotti, A.; Redfield, M.M.; Burnett, J.C., Jr. Novel protein therapeutics for systolic heart failure: Chronic subcutaneous B-type natriuretic peptide. *J. Am. Coll. Cardiol.* **2012**, *60*, 2305–2312. [CrossRef]
51. Travessa, A.M.; Meneze Falcão, L. Vasodilators in acute heart failure—Evidence based on new studies. *Eur. J. Intern. Med.* **2018**, *51*. [CrossRef] [PubMed]
52. Packer, M.; O'Connor, C.; McMurray, J.J.V.; Wittes, J.; Abraham, W.T.; Anker, S.D.; Dickstein, K.; Filippatos, G.; Holcomb, R.; Krum, H.; et al. Effect of Ularitide on Cardiovascular Mortality in Acute Heart Failure. *N. Engl. J. Med.* **2017**, *376*, 1956–1964. [CrossRef]
53. Lisy, O.; Huntley, B.K.; McCormick, D.J.; Kurlansky, P.A.; Burnett, J.C., Jr. Design, synthesis, and actions of a novel chimeric natriuretic peptide: CDNP. *J. Am. Coll. Cardiol.* **2008**, *52*, 60–68. [CrossRef]
54. McKie, P.M.; Sangaralingham, S.J.; Burnett, J.C., Jr. CD-NP: An innovative designer natriuretic peptide activator of particulate guanylyl cyclase receptors for cardiorenal disease. *Curr. Heart Fail. Rep.* **2010**, *7*, 93–99. [CrossRef]

55. Lee, C.Y.; Huntley, B.K.; McCormick, D.J.; Ichiki, T.; Sangaralingham, S.J.; Lisy, O.; Burnett, J.C., Jr. Cenderitide: Structural requirements for the creation of a novel dual particulate guanylyl cyclase receptor agonist with renal-enhancing in vivo and ex vivo actions. *Eur. Heart J. Cardiovasc. Pharmacother.* **2016**, *2*, 98–105. [CrossRef]
56. Martin, F.L.; Sangaralingham, S.J.; Huntley, B.K.; McKie, P.M.; Ichiki, T.; Chen, H.H.; Korinek, J.; Harders, G.E.; Burnett, J.C., Jr. CD-NP: A novel engineered dual guanylylcyclase activator with anti-fibrotic actions in the heart. *PLoS ONE* **2012**, *7*, e52422. [CrossRef]
57. von Lueder, T.G.; Sangaralingham, S.J.; Wang, B.H.; Kompa, A.R.; Atar, D.; Burnett, J.C.; Krum, H. Renin-angiotensin blockade combined with natriuretic peptide system augmentation: Novel therapeutic concepts to combat heart failure. *Circ. Heart Fail.* **2013**, *6*, 594–605. [CrossRef]
58. McKie, P.M.; Cataliotti, A.; Ichiki, T.; Sangaralingham, S.J.; Chen, H.H.; Burnett, J.C., Jr. M-atrial natriuretic peptide and nytroglycerin in a canine model of experimental acute hypertensive heart failure: A differential actions of 2 cGMP activating therapeutics. *J. Am. Heart Assoc.* **2014**, *3*, e000206. [CrossRef]
59. Pan, S.; Chen, H.H.; Dickey, D.M.; Boerrigter, G.; Lee, C.; Kleppe, L.S.; Hall, J.L.; Lerman, A.; Redfield, M.M.; Potter, L.R.; et al. Biodesign of a renal- protective peptide based on alternative splicing of B-type natriuretic peptide. *Proc. Natl. Acad. Sci. USA* **2009**, *106*, 11282–11287. [CrossRef]
60. Meems, L.M.G.; Burnett, J.C., Jr. Innovative Therapeutics: Designer Natriuretic Peptides. *JACC Basic Transl. Sci.* **2016**, *1*, 557–567. [CrossRef]
61. Kelesidis, I.; Mazurek, J.; Khullar, P.; Saeed, W.; Vittorio, T.; Zolty, R. The Effect of Nesiritide on Renal Function and Other Clinical Parameters in Patients with Decompensated Heart Failure and Preserved Ejection Fraction. *Congest. Heart Fail.* **2012**, *18*, 158–164. [CrossRef]
62. Bijkerk, R.; Aleksinskaya, M.A.; Duijs, J.M.G.J.; Veth, J.; Husen, B.; Reiche, D.; Prehn, C.; Adamski, J.; Rabelink, T.J.; De Mey, J.G.R.; et al. Neutral endopeptidase inhibitors blunt kidney fibrosis by reducing myofibroblast formation. *Clin. Sci.* **2019**, *133*, 239–252. [CrossRef] [PubMed]
63. Monteil, T.; Danvy, D.; Sihel, M.; Leroux, R.; Plaquevent, J.C. Strategies for access to enantiomerically pure ecadotril, dexecadotril and fasidotril: A review. *Mini-Rev. Med. Chem.* **2002**, *2*, 209–217. [CrossRef]
64. Kalk, P.; Sharkovska, Y.; Kashina, E.; von Websky, K.; Relle, K.; Pfab, T.; Alter, M.; Guillaume, P.; Provost, D.; Hoffmann, K.; et al. Endothelinconverting enzyme/neutral endopeptidase inhibitor SLV338 prevents hypertensive cardiac remodeling in a blood pressure-independent manner. *Hypertension* **2011**, *57*, 755–763. [CrossRef] [PubMed]
65. Nakayama, K.; Emoto, N.; Suzuki, Y.; Vignon-Zellweger, N.; Yagi, K.; Hirata, K. Physiological relevance of hydrolysis of atrial natriuretic peptide by endothelin-converting enzyme-1. *Kobe J. Med. Sci.* **2012**, *58*, E12–E18. [PubMed]
66. Emoto, N.; Raharjo, S.B.; Isaka, D.; Masuda, S.; Adiarto, S.; Jeng, A.Y.; Yokoyama, M.; Dual, E.C.E. NEP inhibition on cardiac and neurohumoral function during the transition from hypertrophy to heart failure in rats. *Hypertension* **2005**, *45*, 1145–1152. [CrossRef]
67. Seed, A.; Kuc, R.E.; Maguire, J.J.; Hillier, C.; Johnston, F.; Essers, H.; de Voogd, H.J.; McMurray, J.; Davenport, A.P. The dual endothelin converting enzyme/neutral endopeptidase inhibitor SLV-306 (daglutril), inhibits systemic conversion of big endothelin-1 in humans. *Life Sci.* **2012**, *91*, 743–748. [CrossRef]
68. Maki, T.; Nasa, Y.; Tanonaka, K.; Takahashi, M.; Takeo, S. Beneficial effects of sampatrilat, a novel vasopeptidase inhibitor, on cardiac remodeling and function of rats with chronic heart failure following left coronary artery ligation. *J. Pharmacol. Exp. Ther.* **2003**, *305*, 97–105. [CrossRef]
69. Kostis, J.B.; Packer, M.; Black, H.R.; Schmieder, R.; Henry, D.; Levy, E. Omapatrilat and enalapril in patients with hypertension: The Omapatrilat Cardiovascular Treatment vs. Enalapril (OCTAVE) trial. *Am. J. Hypertens.* **2004**, *17*, 103–111. [CrossRef]
70. Vesterqvist, O.; Reeves, R.A. Effects of omapatrilat on pharmacodynamic biomarkers of neutral endopeptidase and Angiotensin-converting enzyme activity in humans. *Curr. Hypertens. Rep.* **2001**, *3* (Suppl. 2), S22–S27. [CrossRef]
71. Mellin, V.; Jeng, A.Y.; Monteil, C.; Renet, S.; Henry, J.P.; Thuillez, C.; Mulder, P. Triple ACE-ECE-NEP inhibition in heart failure: A comparison with ACE and dual ECE-NEP inhibition. *J. Cardiovasc. Pharmacol.* **2005**, *46*, 390–397. [CrossRef]
72. Simpson, J.; Benson, L.; Jhund, P.S.; Dahlström, U.; McMurray, J.J.V.; Lund, L.H. "Real World" Eligibility for Sacubitril/Valsartan in Unselected Heart Failure Patients: Data from the Swedish Heart Failure Registry. *Cardiovasc. Drugs Ther.* **2019**. [CrossRef]

73. Martens, P.; Beliën, H.; Dupont, M.; Mullens, W. Insights into implementation of sacubitril/valsartan into clinical practice. *ESC Heart Fail.* **2018**, *5*, 275–283. [CrossRef]
74. Myhre, P.L.; Vaduganathan, M.; Claggett, B.; Packer, M.; Desai, A.S.; Rouleau, J.L.; Zile, M.R.; Swedberg, K.; Lefkowitz, M.; Shi, V.; et al. B-Type Natriuretic Peptide During Treatment with Sacubitril/Valsartan. The PARADIGM-HF Trial. *J. Am. Coll. Cardiol.* **2019**, *73*, 25886. [CrossRef]
75. McMurray, J.J.; Packer, M.; Desai, A.S.; Gong, J.; Lefkowitz, M.P.; Rizkala, A.R.; Rouleau, J.L.; Shi, V.C.; Solomon, S.D.; Swedberg, K.; et al. Angiotensin-neprilysin inhibition versus enalapril in heart failure. *N. Engl. J. Med.* **2014**, *371*, 993–1004. [CrossRef]
76. Desai, A.S.; McMurray, J.J.; Packer, M.; Swedberg, K.; Rouleau, J.L.; Chen, F.; Gong, J.; Rizkala, A.R.; Brahimi, A.; Claggett, B.; et al. Effect of the angiotensin-receptor-neprilysin inhibitor LCZ696 compared with enalapril on mode of death in heart failure patients. *Eur. Heart J.* **2015**, *36*, 1990–1997. [CrossRef]
77. Yandrapalli, S.; Aronow, W.S.; Mondal, P.; Chabbott, D.R. The evolution of natriuretic peptide augmentation in management of heart failure and the role of sacubitril/valsartan. *Arch. Med. Sci.* **2017**, *13*, 1207–1216. [CrossRef]
78. Cannon, J.A.; Shen, L.; Jhund, P.S.; Kristensen, S.L.; Køber, L.; Chen, F.; Gong, J.; Lefkowitz, M.P.; Rouleau, J.L.; Shi, V.C.; et al. Dementia-related adverse events in PARADIGM-HF and other trials in heart failure with reduced ejection fraction. *Eur. J. Heart Fail.* **2017**, *19*, 129–137. [CrossRef]
79. Efficacy and Safety of LCZ696 Compared to Valsartan on Cognitive Function in Patients with Chronic Heart Failure and Preserved Ejection Fraction (PERSPECTIVE). ClinicalTrials.gov Identifier: NCT02884206. Available online: https://clinicaltrials.gov/ct2/show/NCT02884206 (accessed on 29 March 2019).
80. Bramblett, T.; Teleb, M.; Albaghdadi, A.; Agrawal, H.; Mukherjee, D. Heart Failure with Preserved Ejection Fraction: Entresto a Possible Option. *Cardiovasc. Hematol. Disord. Drug Targets* **2017**, *17*, 80–85. [CrossRef]
81. Galli, A.; Lombardi, F. Neprilysin inhibition for heart failure. *N. Engl. J. Med.* **2014**, *371*, 2336–2337. [CrossRef]
82. Marques da Silva, P.; Aguiar, C. Sacubitril/valsartan: An important piece in the therapeutic puzzle of heart failure. *Rev. Port. Cardiol.* **2017**, *36*, 655–668. [CrossRef]
83. Gori, M.; D'Elia, E.; Senni, M. Sacubitril/valsartan therapeutic strategy in HFpEF: Clinical insights and perspectives. *Int. J. Cardiol.* **2019**, *281*, 158–165. [CrossRef]
84. Solomon, S.D.; Zile, M.; Pieske, B.; Voors, A.; Shah, A.; Kraigher-Krainer, E.; Shi, V.; Bransford, T.; Takeuchi, M.; Gong, J.; et al. The angiotensin receptor neprilysin inhibitor LCZ696 in heart failure with preserved ejection fraction: A phase 2 double-blind randomised controlled trial. *Lancet* **2012**, *380*, 1387–1395. [CrossRef]
85. Solomon, S.D.; Rizkala, A.R.; Lefkowitz, M.P.; Shi, V.C.; Gong, J.; Anavekar, N.; Anker, S.D.; Arango, J.L.; Arenas, J.L.; Atar, D.; et al. Baseline Characteristics of Patients with Heart Failure and Preserved Ejection Fraction in the PARAGON-HF Trial. *Circ. Heart Fail.* **2018**, *11*, e004962. [CrossRef]
86. Grodin, J.L.; Philips, S.; Mullens, W.; Nijst, P.; Martens, P.; Fang, J.C.; Drazner, M.H.; Tang, W.H.W.; Pandey, A. Prognostic implications of plasma volume status estimates in heart failure with preserved ejection fraction: Insights from TOPCAT. *Eur. J. Heart Fail.* **2019**. [CrossRef]
87. Myhre, P.L.; Vaduganathan, M.; Claggett, B.L.; Anand, I.S.; Sweitzer, N.K.; Fang, J.C.; O'Meara, E.; Shah, S.J.; Desai, A.S.; Lewis, E.F.; et al. Association of Natriuretic Peptides with Cardiovascular Prognosis in Heart Failure with Preserved Ejection Fraction: Secondary Analysis of the TOPCAT Randomized Clinical Trial. *JAMA Cardiol.* **2018**, *3*, 1000–1005. [CrossRef]
88. Anand, I.S.; Claggett, B.; Liu, J.; Shah, A.M.; Rector, T.S.; Shah, S.J.; Desai, A.S.; O'Meara, E.; Fleg, J.L.; Pfeffer, M.A.; et al. Interaction between spironolactone and natriuretic peptides in patients with heart failure and preserved ejection fraction: From the TOPCAT trial. *JACC Heart Fail.* **2017**, *5*, 241–252. [CrossRef]
89. Zile, M.R.; Baicu, C.F.; Ikonomidis, J.S.; Stroud, R.E.; Nietert, P.J.; Bradshaw, A.D.; Slater, R.; Palmer, B.M.; Van Buren, P.; Meyer, M.; et al. Myocardial stiffness in patients with heart failure and a preserved ejection fraction: Contributions of collagen and titin. *Circulation* **2015**, *131*, 1247–1259. [CrossRef] [PubMed]
90. Chen, Y.; Wang, H.; Lu, Y.; Huang, X.; Liao, Y.; Bin, J. Effects of mineralocorticoid receptor antagonists in patients with preserved ejection fraction: A meta-analysis of randomized clinical trials. *BMC Med.* **2015**, *13*, 10. [CrossRef]
91. Seidelmann, S.B.; Vardeny, O.; Claggett, B.; Yu, B.; Shah, A.M.; Ballantyne, C.M.; Selvin, E.; MacRae, C.A.; Boerwinkle, E.; Solomon, S.D. An NPPB promoter polymorphism associated with elevated n-terminal pro-B-type natriuretic peptide and lower blood pressure, hypertension, and mortality. *J. Am. Heart Assoc.* **2017**, *6*, e005257. [CrossRef] [PubMed]

92. Packer, M. Epicardial adipose tissue may mediate deleterious effects of obesity and inflammation on the myocardium. *J. Am. Coll. Cardiol.* **2018**, *71*, 2360–2372. [CrossRef]
93. Hsich, E.M.; Grau-Sepulveda, M.V.; Hernandez, A.F.; Eapen, Z.J.; Xian, Y.; Schwamm, L.H.; Bhatt, D.L.; Fonarow, G.C. Relationship between sex.; ejection fraction.; and B-type natriuretic peptide levels in patients hospitalized with heart failure and associations with inhospital outcomes: Findings from the Get with the Guideline-Heart Failure Registry. *Am. Heart J.* **2013**, *166*. [CrossRef]
94. Berry, C.; Doughty, R.N.; Granger, C.; Køber, L.; Massie, B.; McAlister, F.; McMurray, J.; Pocock, S.; Poppe, K.; Swedberg, K.; et al. The survival of patients with heart failure with preserved or reduced left ventricular ejection fraction: An individual patient data meta-analysis. *Eur. Heart J.* **2012**, *33*, 1750–1757. [CrossRef]
95. Lam, C.S.P.; Gamble, G.D.; Ling, L.H.; Sim, D.; Leong, K.T.G.; Yeo, P.S.D.; Ong, H.Y.; Jaufeerally, F.; Ng, T.P.; Cameron, V.A.; et al. Mortality associated with heart failure with preserved vs. reduced ejection fraction in a prospective international multi-ethnic cohort study. *Eur. Heart J.* **2018**, *39*, 1770–1780. [CrossRef]
96. Levy, W.C.; Anand, I.S. Heart failure risk prediction models: What have we learned? *JACC Heart Fail.* **2014**, *2*, 437–439. [CrossRef] [PubMed]
97. van Veldhuisen, D.J.; Linssen, G.C.; Jaarsma, T.; van Gilst, W.H.; Hoes, A.W.; Tijssen, J.G.; Paulus, W.J.; Voors, A.A.; Hillege, H.L. B-type natriuretic peptide and prognosis in heart failure patients with preserved and reduced ejection fraction. *J. Am. Coll. Cardiol.* **2013**, *61*, 1498–1506. [CrossRef]
98. Salah, K.; Stienen, S.; Pinto, Y.M.; Eurlings, L.W.; Metra, M.; Bayes-Genis, A.; Verdiani, V.; Tijssen, J.G.P.; Kok, W.E. Prognosis and NT-proBNP in heart failure patients with preserved versus reduced ejection fraction. *Heart* **2019**. [CrossRef] [PubMed]
99. Wang, T.J.; Larson, M.G.; Levy, D.; Benjamin, E.J.; Corey, D.; Leip, E.P.; Vasan, R.S. Heritability and genetic linkage of plasma natriuretic peptide levels. *Circulation* **2003**, *108*, 13–16. [CrossRef]
100. Richards, A.M. Do the Natriuretic Peptides Cause Atrial Fibrillation or is it Not So Black and White? *J. Am. Heart Assoc.* **2019**, *8*, e012242. [CrossRef] [PubMed]
101. Cheng, V.; Kazanagra, R.; Garcia, A.; Lenert, L.; Krishnaswamy, P.; Gardetto, N.; Clopton, P.; Maisel, A. A rapid bedside test for B-type peptide predicts treatment outcomes in patients admitted for decompensated heart failure: A pilot study. *J. Am. Coll. Cardio.l* **2001**, *37*, 386–391. [CrossRef]
102. Wang, T.J.; Larson, M.G.; Levy, D.; Benjamin, E.J.; Leip, E.P.; Omland, T.; Wolf, P.A.; Vasan, R.S. Plasma natriuretic peptide levels and the risk of cardiovascular events and death. *N. Engl. J. Med.* **2004**, *350*, 655–663. [CrossRef]
103. Cypen, J.; Ahmad, T.; Testani, J.M.; DeVore, A.D. Novel Biomarkers for the Risk Stratification of Heart Failure with Preserved Ejection Fraction. *Curr. Heart Fail. Rep.* **2017**, *14*, 434–443. [CrossRef]
104. Gegenhuber, A.; Struck, J.; Dieplinger, B.; Poelz, W.; Pacher, R.; Morgenthaler, N.G.; Bergmann, A.; Haltmayer, M.; Mueller, T. Comparative evaluation of B-type natriuretic peptide.; mid-regional pro-A-type natriuretic peptide.; mid-regional pro-adrenomedullin.; and copeptin to predict 1-year mortality in patients with acute destabilized heart failure. *J. Card. Fail.* **2007**, *13*, 42–49. [CrossRef] [PubMed]
105. Miller, W.L.; Hartman, K.A.; Grill, D.E.; Struck, J.; Bergmann, A.; Jaffe, A.S. Serial measurements of midregion proANP and copeptin in ambulatory patients with heart failure: Incremental prognostic value of novel biomarkers in heart failure. *Heart* **2012**, *98*, 389–394. [CrossRef] [PubMed]
106. Shah, R.V.; Truong, Q.A.; Gaggin, H.K.; Pfannkuche, J.; Hartmann, O.; Januzzi, J.L., Jr. Mid-regional pro-atrial natriuretic peptide and pro-adrenomedullin testing for the diagnostic and prognostic evaluation of patients with acute dyspnoea. *Eur. Heart J.* **2012**, *33*, 2197–2205. [CrossRef]
107. Kristensen, S.L.; Jhund, P.S.; Mogensen, U.M.; Rørth, R.; Abraham, W.T.; Desai, A.; Dickstein, K.; Rouleau, J.L.; Zile, M.R.; Swedberg, K.; et al. N-Terminal Pro-B-Type Natriuretic Peptide Levels for Risk Prediction in Patients with Heart Failure and Preserved Ejection Fraction According to Atrial Fibrillation Status. *Circ. Heart Fail.* **2019**, *12*, e005766. [CrossRef] [PubMed]
108. Alehagen, U.; Dahlström, U.; Rehfeld, J.F.; Goetze, J.P. Pro-A-type na- triuretic peptide.; proadrenomedullin.; and N-terminal pro-B-type natriuretic peptide used in a multimarker strategy in primary health care in risk assessment of patients with symptoms of heart failure. *J. Card. Fail.* **2013**, *19*, 31–39. [CrossRef]
109. Zabarovskaja, S.; Hage, C.; Linde, C.; Daubert, J.C.; Donal, E.; Gabrielsen, A.; Mellbin, L.; Lund, L.H. Adaptive cardiovascular hormones in a spectrum of heart failure phenotypes. *Int. J. Cardiol.* **2015**, *189*, 6–11. [CrossRef]

110. Smith, J.G.; Newton-Cheh, C.; Almgren, P.; Struck, J.; Morgenthaler, N.G.; Bergmann, A.; Platonov, P.G.; Hedblad, B.; Engström, G.; Wang, T.J.; et al. Assessment of conventional cardiovascular risk factors and multiple biomarkers for the prediction of incident heart failure and atrial fibrillation. *J. Am. Coll. Cardiol.* **2010**, *56*, 1712–1719. [CrossRef]
111. Crozier, I.; Richards, A.M.; Foy, S.G.; Ikram, H. Electrophysiological effects of atrial natriuretic peptide on the cardiac conduction system in man. *Pacing Clin. Electrophysiol.* **1993**, *16*, 738–742. [CrossRef]
112. Arakawa, M.; Miwa, H.; Noda, T.; Ito, Y.; Kambara, K.; Kagawa, K.; Nishigaki, K.; Kano, A.; Hirakawa, S. Alternations in atrial natriuretic peptide release after DC cardioversion of nonvalvular chronic atrial fibrillation. *Eur. Heart J.* **1995**, *16*, 977–985. [CrossRef] [PubMed]
113. Latini, R.; Masson, S.; Pirelli, S.; Barlera, S.; Pulitano, G.; Carbonieri, E.; Gulizia, M.; Vago, T.; Favero, C.; Zdunek, D.; et al. Circulating cardiovascular biomarkers in recurrent atrial fibrillation: Data from the GISSI-atrial fibrillation trial. *J. Intern. Med.* **2011**, *269*, 160–171. [CrossRef]
114. Zografos, T.A.; Katritsis, D.G. Natriuretic Peptides as Predictors of Atrial Fibrillation Recurrences Following Electrical Cardioversion. *Arrhythm. Electrophysiol. Rev.* **2013**, *2*, 109–114. [CrossRef] [PubMed]
115. Khan, M.S.; Siddiqi, T.J.; Usman, M.S.; Sreenivasan, J.; Fugar, S.; Riaz, H.; Murad, M.H.; Mookadam, F.; Figueredo, V.M. Does natriuretic peptide monitoring improve outcomes in heart failure patients? A systematic review and meta-analysis. *Int. J. Cardiol.* **2018**, *263*, 80–87. [CrossRef]
116. Sanders-van Wijk, S.; van Asselt, A.D.; Rickli, H.; Estlinbaum, W.; Erne, P.; Rickenbacher, P.; Vuillomenet, A.; Peter, M.; Pfisterer, M.E.; Brunner-La Rocca, H.P. Cost- effectiveness of N-terminal pro-B-type natriuretic-guided therapy in elderly heart failure patients: Results from TIME- CHF (Trial of Intensified versus Standard Medical Therapy in Elderly Patients with Congestive Heart Failure). *JACC Heart Fail.* **2013**, *1*, 64–71. [CrossRef] [PubMed]
117. Schou, M.; Gustafsson, F.; Videbaek, L.; Tuxen, C.; Keller, N.; Handberg, J.; Sejr Knudsen, A.; Espersen, G.; Markenvard, J.; Egstrup, K.; et al. Extended heart failure clinic follow-up in low-risk patients: A randomized clinical trial (NorthStar). *Eur. Heart J.* **2013**, *34*, 432–442. [CrossRef]
118. Maeder, M.T.; Rickenbacher, P.; Rickli, H.; Abbühl, H.; Gutmann, M.; Erne, P.; Vuilliomenet, A.; Peter, M.; Pfisterer, M.; Brunner-La Rocca, H.P. N-terminal pro brain natriuretic peptide-guided management in patients with heart failure and preserved ejection fraction: Findings from the Trial of Intensified versus standard medical therapy in elderly patients with congestive heart failure (TIME-CHF). *Eur. J. Heart Fail.* **2013**, *15*, 1148–1156. [CrossRef]
119. Troughton, R.W.; Frampton, C.M.; Brunner-La Rocca, H.P.; Pfisterer, M.; Eurlings, L.W.; Erntell, H.; Persson, H.; O'Connor, C.M.; Moertl, D.; Karlström, P.; et al. Effect of B-type natriuretic peptide-guided treatment of chronic heart failure on total mortality and hospitalization: An individual patient meta-analysis. *Eur. Heart J.* **2014**, *35*, 1559–1567. [CrossRef]
120. Mohiuddin, S.; Reeves, B.; Pufulete, M.; Maishman, R.; Dayer, M.; Macleod, J.; McDonagh, T.; Purdy, S.; Rogers, C.; Hollingworth, W. Model-based cost- effectiveness analysis of B-type natriuretic peptide-guided care in patients with heart failure. *BMJ Open* **2016**, *6*, e014010. [CrossRef] [PubMed]
121. Eurlings, L.W.; van Pol, P.E.; Kok, W.E.; van Wijk, S.; Lodewijks-van der Bolt, C.; Balk, A.H.; Lok, D.J.; Crijns, H.J.; van Kraaij, D.J.; de Jonge, N.; et al. Management of chronic heart failure guided by individual N-terminal pro-B- type natriuretic peptide targets: Results of the PRIMA (Can PRo-brain-natriuretic peptide guided therapy of chronic heart failure IMprove heart fAilure morbidity and mortality?) study. *J. Am. Coll. Cardiol.* **2010**, *56*, 2090–2100. [CrossRef] [PubMed]
122. Ibrahim, N.E.; Januzzi, J.L., Jr. The Future of Biomarker-Guided Therapy for Heart Failure After the Guiding Evidence-Based Therapy Using Biomarker Intensified Treatment in Heart Failure (GUIDE-IT) Study. *Curr. Heart Fail. Rep.* **2018**, *15*, 37–43. [CrossRef] [PubMed]
123. Shah, M.R.; Califf, R.M.; Nohria, A.; Bhapkar, M.; Bowers, M.; Mancini, D.M.; Fiuzat, M.; Stevenson, L.W.; O'Connor, C.M. The STARBRITE trial: A randomized.; pilot study of B-type natriuretic peptide- guided therapy in patients with advanced heart failure. *J. Card. Fail.* **2011**, *17*, 613–621. [CrossRef]
124. Sabatine, M.S.; Morrow, D.A.; de Lemos, J.A.; Omland, T.; Sloan, S.; Jarolim, P.; Solomon, S.D.; Pfeffer, M.A.; Braunwald, E. Evaluation of multiple biomarkers of cardiovascular stress for risk prediction and guiding medical therapy in patients with stable coronary disease. *Circulation* **2012**, *125*, 233–240. [CrossRef]
125. Arrigo, M.; Truong, Q.A.; Szymonifka, J.; Rivas-Lasarte, M.; Tolppanen, H.; Sadoune, M.; Gayat, E.; Cohen-Solal, A.; Ruschitzka, F.; Januzzi, J.L., Jr.; et al. Mid- regional pro-atrial natriuretic peptide to predict clinical course in heart failure patients undergoing cardiac resynchronization therapy. *Europace.* **2017**, *19*, 1848–1854. [CrossRef] [PubMed]

126. Wu, C.K.; Su, M.Y.; Lee, J.K.; Chiang, F.T.; Hwang, J.J.; Lin, J.L.; Chen, J.J.; Liu, F.T.; Tsai, C.T. Galectin-3 level and the severity of cardiac diastolic dysfunction using cellular and animal models and clinical indices. *Sci. Rep.* **2015**, *5*, 17007. [CrossRef]
127. Circulating NEP and NEP Inhibition Study in Heart Failure with Preserved Ejection Fraction (CNEPi). ClinicalTrials.gov Identifier:NCT03506412. Available online: https://clinicaltrials.gov/ct2/show/NCT03506412?cond=heart+failure+preserved+ejection+fraction&draw=2&rank=18 (accessed on 2 April 2019).
128. A Randomized, Double-blind Controlled Study Comparing LCZ696 to Medical Therapy for Comorbidities in HFpEF Patients (PARALLAX). ClinicalTrials.gov Identifier: NCT03066804. Available online: https://clinicaltrials.gov/ct2/show/NCT03066804?cond=heart+failure+preserved+ejection+fraction&draw=3&rank=27 (accessed on 2 April 2019).

© 2019 by the authors. Licensee MDPI, Basel, Switzerland. This article is an open access article distributed under the terms and conditions of the Creative Commons Attribution (CC BY) license (http://creativecommons.org/licenses/by/4.0/).

Review

Clinical Applications of Natriuretic Peptides in Heart Failure and Atrial Fibrillation

Masako Baba [1,2], Kentaro Yoshida [1,2,]* and Masaki Ieda [1]

[1] Department of Cardiology, Faculty of Medicine, University of Tsukuba, Tsukuba 305-8575, Japan; babamasako1010@yahoo.co.jp (M.B.); mieda@md.tsukuba.ac.jp (M.I.)
[2] Department of Cardiology, Ibaraki Prefectural Central Hospital, Kasama 309-1793, Japan
* Correspondence: kentaroyo@nifty.com; Tel.: +81-29-853-3142

Received: 12 April 2019; Accepted: 7 June 2019; Published: 10 June 2019

Abstract: Natriuretic peptides (NPs) have become important diagnostic and prognostic biomarkers in cardiovascular diseases, particularly in heart failure (HF). Diagnosis and management of coronary artery disease and atrial fibrillation (AF) can also be guided by NP levels. When interpreting NP levels, however, the caveat is that age, sex, body mass index, renal dysfunction, and race affect the clearance of NPs, resulting in different cut-off values in clinical practice. In AF, NP levels have been associated with incident AF in the general population, recurrences after catheter ablation, prediction of clinical prognosis, and the risk of stroke. In this article, we first review and summarize the current evidence and the roles of B-type NP and atrial NP in HF and coronary artery disease and then focus on the increasing utility of NPs in the diagnosis and management of and the research into AF.

Keywords: natriuretic peptides; heart failure; atrial fibrillation; remodeling

1. Introduction

Natriuretic peptide (NP) levels are now widely measured in clinical practice and have been extensively assessed in cardiovascular research throughout the world. B-type natriuretic peptide (BNP) and N-terminal proBNP (NT-proBNP) are the most commonly used to diagnose heart failure (HF) [1–3]. In addition, diagnosis and management of acute coronary syndrome (ACS) [4,5] and atrial fibrillation (AF) [6] can be guided by NP levels. Although the use of NP has been slow to permeate AF care compared with that for HF and ACS, several studies reported the utility of NPs for the diagnosis and treatment of AF over the last decade [7–9]. Recently, the 2016 European Society of Cardiology (ESC) guidelines on the management of AF recommended using NPs to further refine the risks of stroke and bleeding in AF patients as a class IIb recommendation with the level of evidence B [10]. In this article, we first review and summarize the current evidence and roles of BNP and atrial natriuretic peptide (ANP) in HF and ACS and then focus on the increasing utility of NPs in the diagnosis and management of and research into AF from the viewpoint of electrophysiologists routinely performing catheter ablation of AF.

2. Roles of NP: BNP and ANP

Both ANP and BNP are synthesized as pre-prohormones. ANP is primarily expressed and stored in the atrium. The primary stimulus for ANP release is atrial wall stretch resulting from increased intravascular volume [11]. ANP is translated into prepro-ANP that is cleaved into pro-ANP, which is stored in intracellular granules. The plasma level of ANP in healthy individuals is approximately 20 pg/mL and is evaluated to be 10–100-fold higher in patients with HF [12]. The half-life of ANP is approximately 2 min [13]. The clearance of ANP mainly occurs in the lung, liver, and kidney, with extraction ratios reported to be 24%, 30%, and 35%, respectively. In the kidney, a good correlation was shown between creatine clearance and ANP clearance ($r = 0.58$, $p < 0.05$) [14]. BNP is minimally stored in granules in the ventricles and secreted directly in large bursts following stimulation [3,15]. The plasma level of BNP in healthy individuals is approximately 3.5 pg/mL and is evaluated to be 100-fold higher in patients with HF [16]. The half-life of BNP is approximately 20 min [17]. Subsequently, the peptide is cleaved first into pro-BNP, then to biologically active BNP and the inactive NT-proBNP. BNP and NT-proBNP are secreted in equal concentrations, and the half-life of NT-proBNP is approximately 120 min. BNP clearance is dependent on neutral endopeptidase, and NT-proBNP clearance is dependent on direct renal filtration [18]. In normal subjects, although the BNP concentration is much lower than the ANP concentration, the BNP concentration is markedly increased in patients with HF in proportion to its severity. Thus, the BNP concentration is increased to a much greater degree than is the ANP concentration [17] (Table 1).

Table 1. Summary of some clinically relevant physiologic characteristics of B-type natriuretic peptide (BNP), N-terminal proBNP (NT-proBNP), atrial natriuretic peptide (ANP), and mid-regional proANP (MR-proANP) [11–16].

Characteristic	BNP	NT-proBNP	ANP	MR-proANP
Localization within heart	Atrial and ventricular	Same as BNP	Atrial	Same as ANP
Storage	Minimal	Same as BNP	In intracellular granules	Same as ANP
Basal cardiac secretion	(+)	Same as BNP	++	Same as ANP
Gene transcription response to stretch	Rapid	Same as BNP	Slow	Same as ANP
Half-life (min)	20	60-120	2	60–120
Biologically active	Yes	No	Yes	No
Clinical range	0–5000 pg/mL	0–35,000 pg/mL	0–2000 pg/mL	0–1000 pmol/L

BNP: B-type natriuretic peptide, NT-proBNP: N-terminal pro B-type natriuretic peptide, ANP: Atrial natriuretic peptide, MR-proANP: Mid-regional proANP.

Additionally, because ANP is labile and has a very short half-life, BNP is preferred for diagnostic and prognostic use in HF [19]. Tsutamoto et al. showed that in patients with chronic HF (left ventricular ejection fraction [EF] <45%), only the BNP level ($p < 0.0001$) was a significant independent predictor of mortality in patients with HF by Cox proportional hazard analysis, whereas the ANP level was not [20]. Some studies showed that in patients with HF, BNP is markedly increased in relation to HF severity and surpasses the levels of ANP [21,22]. Current guidelines have greatly affected the use of NP for the diagnosis and management of HF. The ESC guideline and the American Heart Association (ACCF/AHA) noted that measurements of BNP and NT-proBNP levels are useful (Class I) to support a clinical diagnosis of HF and to determine prognosis or disease severity in chronic HF and acutely decompensated HF [23,24].

3. Screening for Asymptomatic Patients

BNP and NT-proBNP can predict mortality and cardiovascular events in asymptomatic patients. McDonagh et al. studied a random sample of 1640 men and women aged 25–74 years and their four-year all-cause mortality rate. The median BNP in those who died was 16.9 (8.8–27) pg/mL compared with 7.8 (3.4–13) pg/mL in the survivors ($p < 0.0001$). One of the independent predictors of four-year all-cause mortality was BNP >17.9 pg/mL ($p = 0.006$) [25]. Similarly, in the large prospective study of the Framingham Offspring study, which included 3346 people without HF (mean follow-up of 5.2 years), those with a BNP above the 80th percentile (20.0 pg/mL for men and 23.3 pg/mL for women) were associated with multivariable-adjusted hazard ratios (HRs) of 1.62 for death, 1.76 for a first major cardiovascular event, 1.91 for AF, 1.99 for stroke or transient ischemic attack, and 3.07 for HF [26].

4. Diagnosis of Acute HF

Acute HF is often difficult to diagnose in the emergency department (ED). The symptoms are not specific and not sensitive. BNP and NT-proBNP are useful in establishing or excluding the diagnosis of acute HF. The large Breathing Not Properly Multinational Study, which first proved the efficacy of BNP, included 1586 patients with acute dyspnea in the ED. At a cut-off of 100 pg/mL, the diagnostic accuracy of BNP was 83.4%, whereas the negative predictive value of BNP at a cut-off of <50 pg/mL was 96% (area under the curve (AUC) 0.91) [3]. The PRIDE study applying NT-proBNP levels showed similar findings among 600 patients presenting to the ED with dyspnea, in which the cut-off level was set at 300 pg/mL, at 90% sensitivity and 85% specificity for the diagnosis of acute HF [1]. BNP is also useful for distinguishing between acute HF and acute respiratory deficiency syndrome (ARDS). In 80 ICU patients with acute hypoxemic respiratory failure, BNP offered good discriminatory performance for the diagnosis of ARDS or cardiogenic pulmonary edema (C-statistic, 0.80). At a cut-off point of ≤200 pg/mL, BNP provided specificity of 91% for ARDS, whereas at a cut-off point of ≥1200 pg/mL, BNP had a specificity of 92% for cardiogenic pulmonary edema [27].

5. Diagnosis of Chronic Ambulatory HF

BNP and NT-proBNP are useful in supporting or excluding the diagnosis of HF when the etiology of dyspnea is unclear. In a study including 250 patients with dyspnea, the BNP cut-off for the diagnosis of congestive HF was 80 pg/mL (95% accuracy), resulting in a satisfactory positive predictive value of 90% and negative predictive value of 98% [28]. In another study of 78 patients seen at a single HF clinic, BNP significantly increased according to different NYHA functional classes (class I: 21.6 ± 2.8 pg/mL, class II: 108.6 ± 16.3 pg/mL, class III: 197.1 ± 27.2 pg/mL, and class IV: 363.0 ± 67.8 pg/mL, $p < 0.0001$). A cut-off value of 107.5 pg/mL (75th percentile) was a significant predictor of clinical events, for which the relative HR was 1.492 (95% confidence interval (CI) 1.221–1.819) [29]. Using a cut-off of NT-proBNP of 125 pg/mL also had an excellent negative predictive power of 97% [2]. Because the mechanisms of chronic HF are more multifactorial with different underlying cardiac and non-cardiac diseases than those of acute HF, it may be difficult to determine a single cut-off value for the diagnosis of chronic HF. The ESC 2012 guideline mentioned that the sensitivity and specificity of BNP and NT-proBNP for the diagnosis of HF are lower in patients in the non-acute phase. The cut-off level of BNP for chronic HF was 35 pg/mL [30].

6. HF with Preserved versus Reduced EF (HFpEF vs. HFrEF)

Stretching of ventricular cardiomyocytes is the most important stimulus of BNP regulation [31], but LV diastolic wall stress also reflects an increased BNP [32]. Therefore, BNP can be used in the diagnosis of HFpEF. In 2042 community residents, the utility of BNP for the detection of diastolic dysfunction was limited, although that of BNP was valuable for the detection of systolic dysfunction [33]. In a comparison between HFrEF (EF ≤50%) and HFpEF (EF >50%) in 160 consecutive patients presenting with HF, the BNP level was significantly higher in those with HFrEF compared with those with HFpEF (267 (136–583) and 105 (64–146) pg/mL, respectively, $p < 0.001$) [32]. The Breathing Not Properly Multinational Study was a seven-center, prospective study including 1586 patients who presented with acute dyspnea. Congestive HF was diagnosed in 452 patients. In those with HFpEF (EF >45%), BNP level was significantly lower than in those with HFrEF (413 vs. 821 pg/mL, $p < 0.001$) [34]. Similarly, 1670 patients from the Korean Heart Failure registry with HFpEF (EF ≥50%) had significantly lower NT-proBNP levels than those with HFrEF (median 2723 vs. 5644 pg/mL, $p < 0.001$) [35]. Although the use of BNP alone results in relatively poor detection of diastolic dysfunction, its combination with the echocardiographic value of diastolic dysfunction such as from pulsed-wave Doppler examination of the mitral flow (E/A) might help reinforce the diagnosis of diastolic dysfunction [36].

7. Prognostication of HF

Measurements of BNP and NT-proBNP are also useful in the prognostication of HF. In 452 ambulatory patients with reduced EF (<35%) with three-year follow-up, patients with a BNP level of >130 pg/mL had a higher rate of sudden cardiac death [37]. The Rapid Emergency Department Heart Failure Outpatient Trial (REDHOT) study was a 10-center trial that included patients seen in the ED with shortness of breath. A BNP of >200 pg/mL was strongly predictive of the 90-day combined event rate (HF visits or admissions and mortality) [38]. In the large ADHERE (Acute Decompensated Heart Failure National Registry) study comprising 65,257 patients with acute decompensated HF, BNP at time of admission independently predicted in-hospital mortality [39]. In 599 patients with shortness of breath treated in the ED, the NT-proBNP cut-off point for predicting one-year mortality was 986 pg/mL, and this cut-off value was the single strongest predictor of death at one year (HR, 2.88, 95% CI, 1.64–5.06, $p < 0.001$) [40].

8. Prognostication of ACS

NPs are recognized as important predictors of cardiovascular events in patients with not only HF but also ACS. In 1996, Omland et al. published data of 131 patients with documented acute myocardial infarction. The median follow-up period was 1293 days, and BNP proved to be one of the powerful predictors of cardiovascular mortality (Cox regression of survival time, coefficient 0.69, SE 0.22, $p < 0.001$) [41]. In total, 438 patients presenting within 6 h of the onset of ST-elevation myocardial infarction were enrolled in the ENTIRE–TIMI-23 trial. Outcomes were assessed through 30 days. BNP was higher in patients who died compared with survivors (89 vs. 15 pg/mL, $p < 0.0001$). A BNP level of 80 pg/mL was associated with a seven-fold higher risk of mortality (odds ratio (OR), 7.2, 95% CI, 2.1–24.5, $p = 0.001$) [42]. Some studies showed that the prognostic value of BNP and NT-proBNP were superior to that of troponins. The adjusted ORs (95% CI) for death at 10 months in the second, third, and fourth quartiles of BNP were 3.8 (1.1–13.3), 4.0 (1.2–13.7), and 5.8 (1.7–19.7), respectively [4]. For two-year mortality, the OR applied to a doubling of the NT-proBNP level, 800 vs. 400 pg/mL, was 1.36 (1.04–1.76) [5]. Furthermore, the OR based on the NT-proBNP level at six months was higher than that at two days: 1.89 (1.14–3.14) vs. 1.29 (0.99–1.67). This suggested that the NT-proBNP level measured during a stable chronic phase is a better predictor of mortality than that measured during an acute phase [5].

9. Interpretations of NP Levels in Different Populations

Several factors increase the NP level: Renal dysfunction, age, and sex (female). Conversely obesity and flash pulmonary edema decrease NP level. Furthermore, NP levels differ substantially according to race/ethnicity.

In a reference sample of 911 healthy subjects (mean age 55 years, 62% women) from the Framingham Heart Study, the strongest predictors of higher NP levels were older age and female sex [43]. Similarly, in 2042 randomly selected community residents >44 years old, BNP increased significantly with age and was significantly higher in women than in men [44]. Framingham Study participants without HF were revealed to have mean BNP levels in lean (<25 kg/m^2), overweight (25 to 29.9 kg/m^2), and obese (≥30 kg/m^2) men of 21.4, 15.5, and 12.7 pg/mL, respectively (trend $p < 0.0001$) [45]. In 318 HF patients, levels of BNP were significantly lower in the obese than in the nonobese subjects (205 ± 22 vs. 335 ± 39 pg/mL, $p = 0.0007$), and multivariate regression analysis identified body mass index (BMI) as an independent negative correlate of BNP level [46]. In 1103 patients presenting to the ED with acute dyspnea, the NT-proBNP concentrations in the overweight and obese groups were significantly lower than that in the lean patients, regardless of the presence of acute HF ($p < 0.001$) [47]. There seemed to be a linear decrease in BNP levels with increasing BMI. In 316 systolic HF patients, the optimal BNP cut-off values for the prediction of death or urgent transplant in lean, overweight, and obese HF patients were 747, 380, and 332 pg/mL, respectively [48]. In obese patients (BMI >35 kg/m^2), a significantly lower

BNP cut-off level (<50 pg/mL) should be used to rule out HF [48]. The reason for the lower BNP in obese patients remains unclear. On the contrary, cut-off values of NPs for patients with an extremely low BMI have not been evaluated.

The BNP cut-off point for the diagnosis of HF may need to be raised when the estimated glomerular filtration rate (eGFR) is <60 mL/min/1.73 m^2. The BNP level in patients with an eGFR <60 mL/min/1.73 m^2 was approximately two- to four-fold greater than that in patients with an eGFR ≥60 mL/min/1.73 m^2 [49,50]. In the Breathing Not Properly Multinational Study including 1586 participants who presented with acute dyspnea with an eGFR <60 mL/min/1.73 m^2, BNP was influenced by renal function. The optimum cut-off points for BNP were 70.7, 104.3, 201.2, and 225.0 pg/mL for the eGFR categories of ≥90, 89 to 60, 59 to 30, and <30 mL/min/1.73 m^2, respectively [51] (Figure 1).

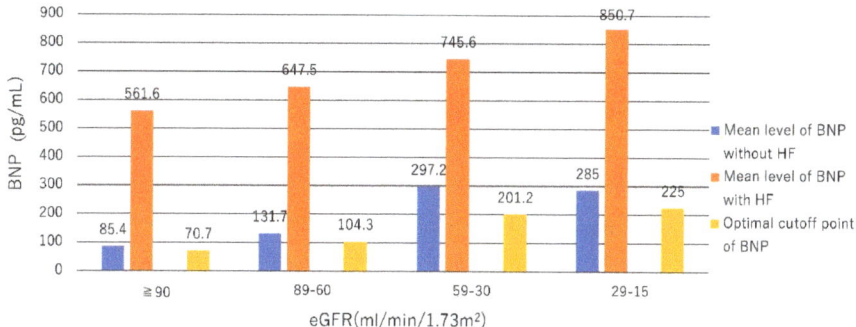

Figure 1. Correlations between B-type natriuretic peptide (BNP) and estimated glomerular filtration rate (eGFR) values. The level of BNP is influenced by renal function, especially when the eGFR is less than 60 mL/min/1.73 m^2. The BNP cut-off points for the diagnosis of heart failure (HF) may need to be raised in renal dysfunction [51].

As mentioned above, BNP clearance is dependent on neutral endopeptidase, and NT-proBNP clearance is dependent on direct renal filtration [18]. Therefore, the NT-proBNP level would seem to be more affected by renal dysfunction than would BNP [52]. A multi-ethnic cohort study showed that NT-proBNP levels differ substantially according to race/ethnicity. NT-proBNP levels were lowest in black (24 pg/mL) as compared with white (32 pg/mL) and Hispanic (30 pg/mL) patients ($p < 0.0001$) [53]. Another study revealed a similar trend for black (median 43 pg/mL), Chinese (43 pg/mL), Hispanic (53 pg/mL), and white (68 pg/mL) patients ($p = 0.0001$) [54].

10. NP-Guided Therapy

NP can help in the clinical management of HF. Several studies have shown the utility of NP-guided therapy. In the 2013 ACCF/AHA guideline, BNP (or NT-proBNP)-guided therapy is placed in the category of Class IIa, evidence level of B [24]. In a STARS-BNP trial including 220 patients with HF with New York Heart Association functional class II to III, patients were randomized to receive BNP-guided treatment with a goal of BNP levels of <100 pg/mL for the BNP group. At the end of the first three months, the mean dosages of angiotensin-converting enzyme inhibitors and beta-blockers were significantly higher in the BNP group ($p < 0.05$). By 15 months of follow-up, patients in the BNP-guided treatment group had a significantly lower number of events of HF-related death or readmission than the patients treated according to current guidelines (24% vs. 52%, $p < 0.001$) [55]. Similar effectiveness was also proved in the BATTLESCARRED trial using NT-proBNP-guided therapy in which 364 patients with HF admitted to a single hospital were randomly allocated 1:1:1 (stratified by age) to therapy guided by NT-proBNP levels or by intensive clinical management or according to usual care. Treatment strategies were applied for two years with a follow-up of three years. One-year mortality was less in both the NT-proBNP (9.1%) and clinically guided (9.1%) groups compared with the usual care group (18.9%, $p = 0.03$). Three-year mortality was selectively reduced in patients ≤75 years of age receiving NT-proBNP-guided treatment (15.5%) compared with their peers receiving either clinically managed treatment (30.9%, $p = 0.048$) or usual care (31.3%, $p = 0.021$) [56]. Conversely, BNP-guided therapy may be harmful in patients with HFpEF. In HFrEF patients, NT-pro or BNP-guided therapy compared with symptom-guided therapy resulted in lower mortality (HR, 0.78, 95% CI, 0.62–0.97, $p = 0.03$) and fewer HF admissions (HR, 0.80, 95% CI, 0.67–0.97, $p = 0.02$), whereas in HFpEF patients, renal failure provided the strongest interaction. Increased risk of (NT-pro) BNP-guided therapy was observed if renal failure was present ($p < 0.01$), and (NT-pro) BNP-guided therapy was beneficial only if none or one of the comorbidities, such as chronic obstructive pulmonary disease, diabetes, cardiovascular insult, or peripheral vascular disease, was present ($p < 0.01$). Additionally, (NT-pro) BNP-guided therapy may be inappropriate in HFpEF patients without hypertension ($p = 0.02$) [57]. Moreover, in elderly HF patients in the TIME-CHF trial, NT-proBNP-guided therapy resulted in a higher rate of survival and a lower rate of all-cause hospitalization in patients aged 60 to 70 years, but not in patients older than 75 years, by 18 months of follow-up after initial admission [58]. Taken together, in elderly and HFpEF patients, NP guided-therapy may not be beneficial compared with symptom-guided medication.

11. Mid-Regional proANP

ProANP is a polypeptide comprising 126 amino acids, with ANP consisting of amino acids 99-126. The N-terminal portion of proANP, termed proANP1-98 or NT-proANP, has a much longer half-life than ANP and has therefore been suggested to be a more reliable analyte for measurement than ANP. ProANP1-98 can be subjected to further fragmentation, and an immunoassay for mid-regional (MR) proANP (amino acids 53–90) was developed to measure the proANP level. In 325 healthy individuals, the range of MR-proANP was 9.6–313 pmol/L, and the median was 45 pmol/L [59].

The largest study to evaluate MR-proANP for the diagnosis of acute HF, the BACH (Biomarkers in Acute Heart Failure) trial, was a prospective, 15-center, international study including 1641 patients presenting to the ED with dyspnea. MR-proANP (≥120 pmol/L) provided a sensitivity of 97%, a negative predictive value of 97.4%, and AUC of 0.90 that proved noninferior to BNP (≥100 pg/mL) for the diagnosis of acute HF (accuracy difference 0.9%) [60]. Other studies have shown similar findings [61,62]. MR-proANP also has prognostic utility in acute HF and chronic HF. Although the utility of MR-proANP to diagnose acute HF was lower than that of BNP and NT-pro BNP (AUC 0.901 vs. 0.973 vs. 0.922, respectively), MR-proANP had better prognostic value for mortality than did BNP (AUC 0.668 vs. 0.604) at five years [63] (Figure 2).

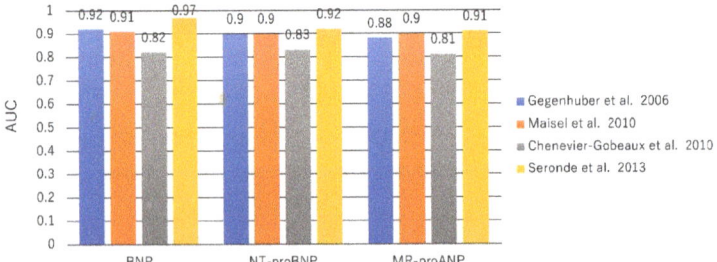

Figure 2. Area under the curve (AUC) for NP to diagnose acute HF. Similar values of BNP, NT-proBNP, and MR-proANP for diagnosis of acute HF [60–63].

In the GISSI-HF trial including 1237 patients with chronic and stable HF, MR-proANP and NT-pro BNP were measured at randomization and after three months. Changes in MR-proANP concentrations were related to mortality, whereas changes in NT-proBNP markers were not [64].

Moreover, MR-proANP may have utility as a screening tool in community populations. Although NT-proBNP and MR-proANP predicted incident HF during 14 months of follow-up, only MR-proANP predicted incident AF [65]. Similar to those of BNP and NT-proBNP, the level of MR-pro ANP is increased with age, decreased by a higher BMI, and influenced by race and sex [66].

12. AF and NPs

ANP is synthesized and secreted mainly by atrial cardiomyocytes in response to atrial dilatation, whereas BNP is produced chiefly in the ventricular myocardium in response to ventricular stretch and pressure overload [67]. In some patients in AF without HF, the level of ANP was normal, but that of BNP or NT-proBNP was elevated [6,68]. The reason for the elevated BNP and NT-proBNP levels was suggested to be due to the small amount of BNP that is also produced and secreted by atrial tissue [69]. Atrial dysrhythmia would also increase BNP secretion [70,71]. Asynchronous contraction of the atrial myocardium could produce a tethering effect of atrial myocardial fibers that may stimulate the secretion of BNP [71]. Furthermore, during AF, the elevated atrial pressure stretches the atrial wall (pressure overload), and loss of atrial contraction leads to an unfavorable alternation of the left ventricular filling pattern [72]. BNP decreased significantly 24 h after the restoration of sinus rhythm (SR) by cardioversion in patients with paroxysmal and persistent AF (from 95 to 28 pg/mL in paroxysmal AF and from 75 to 41 pg/mL in persistent AF) [73,74].

BNP is a more valuable marker for the diagnosis of LV diastolic function compared to ANP. Bakowski et al. investigated 42 patients with AF in whom SR was restored by cardioversion and maintained for at least 30 days. The average values of ANP during AF in patients with normal and impaired diastolic function were 167.3 ± 70.1 and 298.7 ± 83.6 pg/mL, respectively ($p < 0.001$), and those of BNP were 49.5 ± 14.7 and 145.6 ± 49.6 pg/mL, respectively ($p < 0.001$). An ANP value >220.7 pg/mL measured during AF identified patients with impaired LV diastolic function with 85% sensitivity and 90% specificity. A BNP value of >74.7 pg/mL proved to be 95% sensitive and 100% specific in the diagnosis of such patients [75]. BNP was a more specific and sensitive marker of impaired LV diastolic function than was ANP.

13. Incident AF in Community Studies

Some studies showed elevated BNP and NT-proBNP levels to be associated with increased AF incidence [9,26,76,77]. In three US community-based studies (ARIC, CHS, and FHS), including 18,556 participants overall, BNP and CRP were positively associated with AF incidence [78]. That finding was similar in the elderly population. In a community-based population of 5445 older patients in the Cardiovascular Health Study, NT-proBNP levels were strongly associated with prevalent AF. After a median follow-up of 10 years, the incidence of AF was 2.2 per 100 person-years [7]. Although BNP and NT-proBNP levels were highly predictive of incident AF, the cut-off levels were unclear. In the Framingham cohort, the correlation of NT-proANP with BNP was moderately high at 0.66, and after incorporation of both natriuretic peptides into the model, BNP emerged as the stronger biomarker [76].

14. Impact of Structural Heart Disease in AF patients

The cut-off level of NP to detect structural heart disease is different between SR and AF. In 793 patients with structural heart disease at a single center, NT-proBNP levels were 960 (IQR 359–2625) pg/mL for SR ($n = 591$) and 2491 (1443–4368) pg/mL for AF ($n = 202$) ($p < 0.001$). The areas under the ROC curve for NT-proBNP to detect structural heart disease were 0.79 for SR (95% CI, 0.77–0.82) and 0.78 for AF (95% CI, 0.72–0.84). NT-proBNP cut-off levels necessary to achieve a 1-in-100 false-negative rate were 27.5 (7.5–30.5) pg/mL for SR and 524 (253–662) pg/mL for AF [79].

15. HF and AF

Both AF and HF increase BNP and NT-proBNP levels, but these levels remain useful in the diagnosis of HF in patients with AF. In the PRIDE study, 600 patients presented to the ED with acute dyspnea. AF was associated with higher NT-proBNP in the dyspneic patients and particularly in those without acute HF [80]. The BASEL study randomly assigned 452 patients with AF and dyspnea to a diagnostic strategy with or without the use of BNP. BNP cut-off levels of 100 and 500 pg/mL for the diagnosis of HF were determined. If BNP was <100 pg/mL, HF was considered unlikely, whereas if BNP was >500 pg/mL, HF was considered likely. The use of BNP significantly reduced time to discharge (median eight days in the BNP group vs. 12 days in the control group, $p = 0.046$) and time to initiation of adequate therapy (median 51 min in the BNP group vs. 100 min in the control group, $p = 0.024$) [8]. In patients with both HF and AF, the higher cut-off levels of BNP and NT-proBNP should be used. The BACH study including 1445 patients with acute dyspnea showed that the diagnostic performance of BNP and NT-proBNP for acute HF was impaired by the presence of AF [81] (Figure 3).

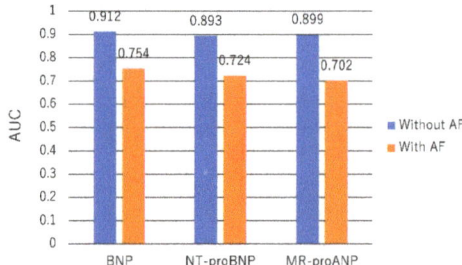

Figure 3. AUC for NP to diagnose acute HF in atrial fibrillation (AF). AUCs for BNP, NT-proBNP, and MR-proANP to diagnose acute HF are similarly reduced in the presence of AF [81].

Among 1431 patients without HF, permanent/paroxysmal AF was associated with significantly higher BNP levels ($p = 0.001$). Conversely, in patients with HF, BNP levels did not differ significantly between patients with and without AF ($p = 0.533$). A BNP cut-off value of 100 pg/mL had respective specificities of 40% and 79% for the diagnosis of acute HF in patients with and without AF. In patients with AF, a cut-off level of 200 pg/mL resulted in a marked improvement in specificity and positive likelihood ratio for diagnosing HF compared with the conventional cut-off level of 100 pg/mL, with little loss of sensitivity [82]. Another study showed that the BNP cut-off level for HF that maintained high sensitivity was 150 pg/mL for those with AF [83] (Figure 4).

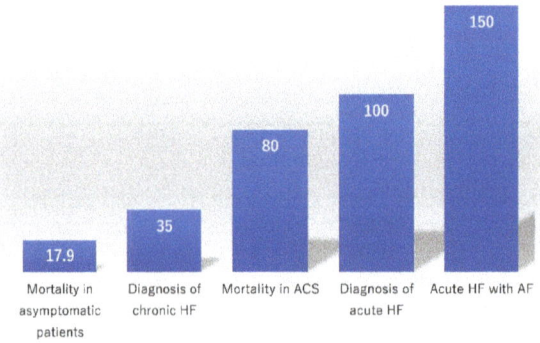

Figure 4. Cut-off points of BNP (pg/mL). The cut-off points of BNP vary among clinical settings [1–3,25,30,42,83]. ACS: Acute coronary syndrome.

Similarly, in 1941 elderly community-dwelling residents, NT-proBNP levels of patients with AF with and without HF were 744 pg/mL and 211 pg/mL, respectively. At the cut-off point of 125 pg/mL, sensitivity and specificity were 93% and 35%, respectively, and positive and negative predictive values were 51% and 86%, respectively [84].

NT-proBNP had a predictive value for adverse cardiovascular outcomes irrespective of AF status. In a large trial including 14,737 patients with HFrEF, NT-proBNP was associated with a risk of cardiovascular death or hospitalization for HF with and without AF. However, when the NT-proBNP level was >400 pg/mL, NT-proBNP had similar predictive value for adverse cardiovascular outcomes in patients with or without AF [85]. Even in patients with AF, BNP, and NT-proBNP are useful for the diagnosis and prediction of prognosis in patients with HF although their cut-off values should be offset.

16. AF Recurrence after Cardioversion or Pulmonary Vein Isolation

Baseline BNP and NT-proBNP values were found to be independent predictors for AF recurrence after cardioversion [86–88], but the cut-off levels differed between studies. Solheim et al. reported that at baseline, there were no differences in NT-proBNP levels (33.5 vs. 29.5 pmol/L, $p = 0.9$) between patients with AF recurrence and nonrecurrence after ablation. At long-term follow-up, there was a marked decrease in the NT-proBNP level at 22 ± 5 months after ablation in the successful ablation patients (7.0 vs. 17.5 pmol/L, $p < 0.05$). NT-proBNP correlated with LA volume both at baseline ($r = 0.71$, $p < 0.001$) and at follow-up ($r = 0.57$, $p < 0.001$). AF burden correlated with both NT-proBNP ($r = 0.47$, $p < 0.01$) and LA volume ($r = 0.52$, $p < 0.01$). A decrease in NT-pro-BNP of >25% from the baseline value could be useful as a marker of ablation success [89]. A meta-analysis of electronic databases including 10 studies suggested that both increased baseline BNP and NT-pro BNP levels, are associated with greater risk of AF recurrence after catheter ablation [90]. In another meta-analysis of 36 studies, compared with the nonrecurrence group, the recurrence group had increased pre-ablation levels of ANP, BNP, and NT-pro-BNP (standardized mean difference (95% CI): 0.37 (0.13–0.61), 0.77 (0.40–1.14), and 1.25 (0.64–1.87)) [91]. Deng et al. evaluated 1410 consecutive AF patients (68% male, 57.2 ± 11.6 years) undergoing AF ablation, during a mean follow-up of 20.7 ± 8.8 months. The cut-off value of BNP for AF recurrence was 237.45 pg/mL. Similar findings were evident in the subgroups of patients with paroxysmal or nonparoxysmal AF [92]. The NT-proBNP level at baseline was an independent predictor of AF recurrence ($p < 0.001$) after pulmonary vein isolation with a cut-off value of NT-proBNP of ≥ 423.2 pg/mL ($p = 0.002$) [68].

17. Stroke in AF Patients

BNP and NT-proBNP are also independent risk markers of stroke in AF patients. Anticoagulated AF patients with a high NT-proBNP level were associated with an increased risk of stroke [93,94]. In the RE-LY trial including 6189 patients, rates of stroke were independently related to levels of NT-proBNP (2.30%/year vs. 0.92%/year in the highest (>1402 pg/mL) versus lowest (<387 pg/mL) quartile groups, HR, 2.40 (95% CI, 1.41–4.07), $p = 0.0014$) [94]. The biomarker-based ABC stroke score (age, biomarkers, and clinical history of prior stroke) was recently shown to improve the prediction of stroke risk in patients with AF. In the ARISTOTLE trial including 18,201 patients with AF, adding NT-proBNP levels to the CHA2DS2-VASc score improved the C-statistic from 0.62 to 0.65 ($p = 0.0009$) for stroke or systemic embolism and from 0.59 to 0.69 for cardiac death ($p < 0.0001$) [95]. The biomarker-based ABC stroke score performed better than presently used scores such as the CHA2DS2-VASc and ATRIA scores [96,97].

18. MR-proANP in AF Patients

Due to the short half-life and lability of ANP, BNP is preferred for the diagnosis and management of AF. ANP is primarily a feature of atrial cardiomyocytes and may thus be a more appropriate biological marker of atrial changes. The more stable MR-proANP level may be more useful for the assessment of AF. In 632 consecutive patients presenting with acute dyspnea, the diagnostic accuracy of acute HF in AF patients was similar for MR-proANP (0.90, 95% CI 0.84–0.95) and NT-proBNP (0.89, 95% CI 0.81–0.96). MR-proANP strongly predicted one-year all-cause mortality (HR = 1.13 (1.09–1.17), per 100 pmol/L increase, $p < 0.001$) [98]. However, in the AMIO-CAT trial evaluating patients undergoing ablation for AF, patients with persistent AF had higher concentrations of both MR-proANP and NT-proBNP at baseline than those with paroxysmal AF. The NT-proBNP level was significantly associated with the incidence of documented AF/AT recurrence within the three-month blanking period after catheter ablation (HR, 1.84, 95% CI, 1.06–3.19, $p = 0.030$), but the MR-proANP level was not (HR, 2.87, 95% CI, 0.86–9.50, $p = 0.085$). The baseline MR-proANP and NT-proBNP levels were not associated with the recurrence of AF at six months after ablation (MR-proANP: OR, 4.40, 95% CI, 0.57–33.71, $p = 0.15$ and NT-proBNP: OR, 1.42, 95% CI, 0.59–3.41, $p = 0.15$) [99]. It is still unclear which is superior for AF management, MR-proANP, or NT-proBNP.

19. Depletion of ANP in AF Patients with Atrial Remodeling

Because the secretion of ANP is induced by stretching of the atrial wall, ANP is depleted in the atrium with advanced fibrosis, which leads to reduced ANP production capacity [100]. When AF converts to the longstanding form, the atria are characterized by a loss of myocytes and an increase in fibrous tissue [101,102]. Histological examination showed that in patients undergoing the maze procedure, preoperative ANP was significantly lower in the AF group than in the SR group. In the AF group, the messenger RNA expressions of ANP were lower, and collagen volumes were higher than those in the SR group [103]. Yoshida et al. reported that in patients with persistent AF and an enlarged LA undergoing ablation, the reduction of LA volume after ablation was greater in patients with a higher ANP level (73 vs. 50 pg/mL, $p = 0.02$). This finding indicated a relation between healthy atrial myocardium and preserved ANP secretion [104]. This hypothesis that ANP can serve as a marker of atrial integrity was further supported by another study performing longitudinal assessments of left atrial volume with cardiac computed tomography in patients with AF [105]. Yoshida et al. also proposed the original index ANP/BNP ratio, which may be more sensitive to a heart condition and better reflects atrial integrity than ANP or BNP alone. Patients with more severe HF (higher BNP) and more advanced atrial fibrosis (lower ANP) have a much lower ANP/BNP ratio than those without these conditions [106]. However, this interpretation of the ANP/BNP ratio needs validation in future studies, and assessment of the *MR-proANP*/BNP ratio is also of interest with respect to atrial remodeling in patients with HF and AF.

20. Conclusions

NP levels can greatly help in the clinical management of cardiovascular diseases. In patients with HF, NP has been established as a tool of diagnosis and prognostication, a guide to the management and monitoring of therapy, and a surrogate of the underlying disease and cut-off levels have been confirmed. Although NP is also useful for AF management, such as in screening for the new onset of incident AF and in predicting the success of cardioversions and pulmonary vein isolation, and the risk of stroke, we hope that further applications of NPs, particularly MR-proANP, to patients with AF will contribute to clarifying the complex mechanisms of AF.

Funding: This research received no external funding.

Conflicts of Interest: The authors declare no conflict of interest.

References

1. Januzzi, J.L., Jr.; Camargo, C.A.; Anwaruddin, S.; Baggish, A.L.; Chen, A.A.; Krauser, D.G.; Tung, R.; Cameron, R.; Nagurney, J.T.; Chae, C.U.; et al. The N-terminal Pro-BNP investigation of dyspnea in the emergency department (PRIDE) study. *Am. J. Cardiol.* **2005**, *95*, 948–954. [CrossRef] [PubMed]
2. Hildebrandt, P.; Collinson, P.O. Amino-terminal pro-B-type natriuretic peptide testing to assist the diagnostic evaluation of heart failure in symptomatic primary care patients. *Am. J. Cardiol.* **2008**, *101*, 25–28. [CrossRef]
3. Maisel, A.S.; Krishnaswamy, P.; Nowak, R.M.; McCord, J.; Hollander, J.E.; Duc, P.; Omland, T.; Storrow, A.B.; Abraham, W.T.; Wu, A.H.; et al. Rapid measurement of B-type natriuretic peptide in the emergency diagnosis of heart failure. *N. Engl. J. Med.* **2002**, *347*, 161–167. [CrossRef] [PubMed]
4. de Lemos, J.A.; Morrow, D.A.; Bentley, J.H.; Omland, T.; Sabatine, M.S.; McCabe, C.H.; Hall, C.; Cannon, C.P.; Braunwald, E. The prognostic value of B-type natriuretic peptide in patients with acute coronary syndromes. *N. Engl. J. Med.* **2001**, *345*, 1014–1021. [CrossRef] [PubMed]
5. Lindahl, B.; Lindback, J.; Jernberg, T.; Johnston, N.; Stridsberg, M.; Venge, P.; Wallentin, L. Serial analyses of N-terminal pro-B-type natriuretic peptide in patients with non-ST-segment elevation acute coronary syndromes: A Fragmin and fast Revascularisation during In Stability in Coronary artery disease (FRISC)-II substudy. *J. Am. Coll. Cardiol.* **2005**, *45*, 533–541. [CrossRef] [PubMed]
6. Ellinor, P.T.; Low, A.F.; Patton, K.K.; Shea, M.A.; Macrae, C.A. Discordant atrial natriuretic peptide and brain natriuretic peptide levels in lone atrial fibrillation. *J. Am. Coll. Cardiol.* **2005**, *45*, 82–86. [CrossRef] [PubMed]
7. Patton, K.K.; Ellinor, P.T.; Heckbert, S.R.; Christenson, R.H.; DeFilippi, C.; Gottdiener, J.S.; Kronmal, R.A. N-terminal pro-B-type natriuretic peptide is a major predictor of the development of atrial fibrillation: The Cardiovascular Health Study. *Circulation* **2009**, *120*, 1768–1774. [CrossRef]
8. Breidthardt, T.; Noveanu, M.; Cayir, S.; Viglino, M.; Laule, K.; Hochholzer, W.; Reichlin, T.; Potocki, M.; Christ, M.; Mueller, C. The use of B-type natriuretic peptide in the management of patients with atrial fibrillation and dyspnea. *Int. J. Cardiol.* **2009**, *136*, 193–199. [CrossRef]
9. Patton, K.K.; Heckbert, S.R.; Alonso, A.; Bahrami, H.; Lima, J.A.; Burke, G.; Kronmal, R.A. N-terminal pro-B-type natriuretic peptide as a predictor of incident atrial fibrillation in the Multi-Ethnic Study of Atherosclerosis: The effects of age, sex and ethnicity. *Heart* **2013**, *99*, 1832–1836. [CrossRef]
10. Kirchhof, P.; Benussi, S.; Kotecha, D.; Ahlsson, A.; Atar, D.; Casadei, B.; Castella, M.; Diener, H.C.; Heidbuchel, H.; Hendriks, J.; et al. 2016 ESC Guidelines for the management of atrial fibrillation developed in collaboration with EACTS. *Europace* **2016**, *18*, 1609–1678. [CrossRef]
11. Edwards, B.S.; Zimmerman, R.S.; Schwab, T.R.; Heublein, D.M.; Burnett, J.C., Jr. Atrial stretch, not pressure, is the principal determinant controlling the acute release of atrial natriuretic factor. *Circ. Res.* **1988**, *62*, 191–195. [CrossRef] [PubMed]
12. Cody, R.J.; Atlas, S.A.; Laragh, J.H.; Kubo, S.H.; Covit, A.B.; Ryman, K.S.; Shaknovich, A.; Pondolfino, K.; Clark, M.; Camargo, M.J.; et al. Atrial natriuretic factor in normal subjects and heart failure patients. Plasma levels and renal, hormonal, and hemodynamic responses to peptide infusion. *J. Clin. Investig.* **1986**, *78*, 1362–1374. [CrossRef] [PubMed]
13. Nakao, K.; Sugawara, A.; Morii, N.; Sakamoto, M.; Yamada, T.; Itoh, H.; Shiono, S.; Saito, Y.; Nishimura, K.; Ban, T.; et al. The pharmacokinetics of alpha-human atrial natriuretic polypeptide in healthy subjects. *Eur. J. Clin. Pharmacol.* **1986**, *31*, 101–103. [CrossRef] [PubMed]
14. Hollister, A.S.; Rodeheffer, R.J.; White, F.J.; Potts, J.R.; Imada, T.; Inagami, T. Clearance of atrial natriuretic factor by lung, liver, and kidney in human subjects and the dog. *J. Clin. Investig.* **1989**, *83*, 623–628. [CrossRef] [PubMed]
15. Hall, C. Essential biochemistry and physiology of (NT-pro)BNP. *Eur. J. Heart Fail.* **2004**, *6*, 257–260. [CrossRef] [PubMed]
16. Abassi, Z.; Karram, T.; Ellaham, S.; Winaver, J.; Hoffman, A. Implications of the natriuretic peptide system in the pathogenesis of heart failure: Diagnostic and therapeutic importance. *Pharmacol. Ther.* **2004**, *102*, 223–241. [CrossRef] [PubMed]
17. Mukoyama, M.; Nakao, K.; Hosoda, K.; Suga, S.; Saito, Y.; Ogawa, Y.; Shirakami, G.; Jougasaki, M.; Obata, K.; Yasue, H.; et al. Brain natriuretic peptide as a novel cardiac hormone in humans. Evidence for an exquisite dual natriuretic peptide system, atrial natriuretic peptide and brain natriuretic peptide. *J. Clin. Investig.* **1991**, *87*, 1402–1412. [CrossRef] [PubMed]

18. Daniels, L.B.; Maisel, A.S. Natriuretic peptides. *J. Am. Coll. Cardiol.* **2007**, *50*, 2357–2368. [CrossRef] [PubMed]
19. Lerman, A.; Gibbons, R.J.; Rodeheffer, R.J.; Bailey, K.R.; McKinley, L.J.; Heublein, D.M.; Burnett, J.C., Jr. Circulating N-terminal atrial natriuretic peptide as a marker for symptomless left-ventricular dysfunction. *Lancet* **1993**, *341*, 1105–1109. [CrossRef]
20. Tsutamoto, T.; Wada, A.; Maeda, K.; Hisanaga, T.; Maeda, Y.; Fukai, D.; Ohnishi, M.; Sugimoto, Y.; Kinoshita, M. Attenuation of compensation of endogenous cardiac natriuretic peptide system in chronic heart failure: Prognostic role of plasma brain natriuretic peptide concentration in patients with chronic symptomatic left ventricular dysfunction. *Circulation* **1997**, *96*, 509–516. [CrossRef]
21. Mukoyama, M.; Nakao, K.; Saito, Y.; Ogawa, Y.; Hosoda, K.; Suga, S.; Shirakami, G.; Jougasaki, M.; Imura, H. Human brain natriuretic peptide, a novel cardiac hormone. *Lancet* **1990**, *335*, 801–802. [CrossRef]
22. Mukoyama, M.; Nakao, K.; Saito, Y.; Ogawa, Y.; Hosoda, K.; Suga, S.; Shirakami, G.; Jougasaki, M.; Imura, H. Increased human brain natriuretic peptide in congestive heart failure. *N. Engl. J. Med.* **1990**, *323*, 757–758. [PubMed]
23. Ponikowski, P.; Voors, A.A.; Anker, S.D.; Bueno, H.; Cleland, J.G.F.; Coats, A.J.S.; Falk, V.; Gonzalez-Juanatey, J.R.; Harjola, V.P.; Jankowska, E.A.; et al. 2016 ESC Guidelines for the diagnosis and treatment of acute and chronic heart failure: The Task Force for the diagnosis and treatment of acute and chronic heart failure of the European Society of Cardiology (ESC)Developed with the special contribution of the Heart Failure Association (HFA) of the ESC. *Eur. Heart J.* **2016**, *37*, 2129–2200. [PubMed]
24. Yancy, C.W.; Jessup, M.; Bozkurt, B.; Butler, J.; Casey, D.E., Jr.; Drazner, M.H.; Fonarow, G.C.; Geraci, S.A.; Horwich, T.; Januzzi, J.L.; et al. 2013 ACCF/AHA guideline for the management of heart failure: A report of the American College of Cardiology Foundation/American Heart Association Task Force on Practice Guidelines. *J. Am. Coll. Cardiol.* **2013**, *62*, e147–e239. [CrossRef]
25. McDonagh, T.A.; Cunningham, A.D.; Morrison, C.E.; McMurray, J.J.; Ford, I.; Morton, J.J.; Dargie, H.J. Left ventricular dysfunction, natriuretic peptides, and mortality in an urban population. *Heart* **2001**, *86*, 21–26. [CrossRef] [PubMed]
26. Wang, T.J.; Larson, M.G.; Levy, D.; Benjamin, E.J.; Leip, E.P.; Omland, T.; Wolf, P.A.; Vasan, R.S. Plasma natriuretic peptide levels and the risk of cardiovascular events and death. *N. Engl. J. Med.* **2004**, *350*, 655–663. [CrossRef] [PubMed]
27. Karmpaliotis, D.; Kirtane, A.J.; Ruisi, C.P.; Polonsky, T.; Malhotra, A.; Talmor, D.; Kosmidou, I.; Jarolim, P.; de Lemos, J.A.; Sabatine, M.S.; et al. Diagnostic and prognostic utility of brain natriuretic Peptide in subjects admitted to the ICU with hypoxic respiratory failure due to noncardiogenic and cardiogenic pulmonary edema. *Chest* **2007**, *131*, 964–971. [CrossRef] [PubMed]
28. Dao, Q.; Krishnaswamy, P.; Kazanegra, R.; Harrison, A.; Amirnovin, R.; Lenert, L.; Clopton, P.; Alberto, J.; Hlavin, P.; Maisel, A.S. Utility of B-type natriuretic peptide in the diagnosis of congestive heart failure in an urgent-care setting. *J. Am. Coll. Cardiol.* **2001**, *37*, 379–385. [CrossRef]
29. Koglin, J.; Pehlivanli, S.; Schwaiblmair, M.; Vogeser, M.; Cremer, P.; vonScheidt, W. Role of brain natriuretic peptide in risk stratification of patients with congestive heart failure. *J. Am. Coll. Cardiol.* **2001**, *38*, 1934–1941. [CrossRef]
30. McMurray, J.J.; Adamopoulos, S.; Anker, S.D.; Auricchio, A.; Bohm, M.; Dickstein, K.; Falk, V.; Filippatos, G.; Fonseca, C.; Gomez-Sanchez, M.A.; et al. ESC Guidelines for the diagnosis and treatment of acute and chronic heart failure 2012: The Task Force for the Diagnosis and Treatment of Acute and Chronic Heart Failure 2012 of the European Society of Cardiology. Developed in collaboration with the Heart Failure Association (HFA) of the ESC. *Eur. Heart J.* **2012**, *33*, 1787–1847.
31. Tokola, H.; Hautala, N.; Marttila, M.; Magga, J.; Pikkarainen, S.; Kerkela, R.; Vuolteenaho, O.; Ruskoaho, H. Mechanical load-induced alterations in B-type natriuretic peptide gene expression. *Can. J. Physiol. Pharmacol.* **2001**, *79*, 646–653. [CrossRef] [PubMed]
32. Iwanaga, Y.; Nishi, I.; Furuichi, S.; Noguchi, T.; Sase, K.; Kihara, Y.; Goto, Y.; Nonogi, H. B-type natriuretic peptide strongly reflects diastolic wall stress in patients with chronic heart failure: Comparison between systolic and diastolic heart failure. *J. Am. Coll. Cardiol.* **2006**, *47*, 742–748. [CrossRef] [PubMed]
33. Redfield, M.M.; Rodeheffer, R.J.; Jacobsen, S.J.; Mahoney, D.W.; Bailey, K.R.; Burnett, J.C., Jr. Plasma brain natriuretic peptide to detect preclinical ventricular systolic or diastolic dysfunction: A community-based study. *Circulation* **2004**, *109*, 3176–3181. [CrossRef] [PubMed]

34. Maisel, A.S.; McCord, J.; Nowak, R.M.; Hollander, J.E.; Wu, A.H.B.; Duc, P.; Omland, T.; Storrow, A.B.; Krishnaswamy, P.; Abraham, W.T.; et al. Bedside B-Type natriuretic peptide in the emergency diagnosis of heart failure with reduced or preserved ejection fraction. *J. Am. Coll. Cardiol.* **2003**, *41*, 2010–2017. [CrossRef]
35. Kang, S.H.; Park, J.J.; Choi, D.J.; Yoon, C.H.; Oh, I.Y.; Kang, S.M.; Yoo, B.S.; Jeon, E.S.; Kim, J.J.; Cho, M.C.; et al. Prognostic value of NT-proBNP in heart failure with preserved versus reduced EF. *Heart* **2015**, *101*, 1881–1888. [CrossRef]
36. Lubien, E.; DeMaria, A.; Krishnaswamy, P.; Clopton, P.; Koon, J.; Kazanegra, R.; Gardetto, N.; Wanner, E.; Maisel, A.S. Utility of B-natriuretic peptide in detecting diastolic dysfunction: Comparison with Doppler velocity recordings. *Circulation* **2002**, *105*, 595–601. [CrossRef]
37. Berger, R.; Huelsman, M.; Strecker, K.; Bojic, A.; Moser, P.; Stanek, B.; Pacher, R. B-type natriuretic peptide predicts sudden death in patients with chronic heart failure. *Circulation* **2002**, *105*, 2392–2397. [CrossRef] [PubMed]
38. Maisel, A.; Hollander, J.E.; Guss, D.; McCullough, P.; Nowak, R.; Green, G.; Saltzberg, M.; Ellison, S.R.; Bhalla, M.A.; Bhalla, V.; et al. Primary results of the Rapid Emergency Department Heart Failure Outpatient Trial (REDHOT). A multicenter study of B-type natriuretic peptide levels, emergency department decision making, and outcomes in patients presenting with shortness of breath. *J. Am. Coll. Cardiol.* **2004**, *44*, 1328–1333. [CrossRef]
39. Fonarow, G.C.; Peacock, W.F.; Phillips, C.O.; Givertz, M.M.; Lopatin, M.; ADHERE Scientific Advisory Committee and Investigators. Admission B-type natriuretic peptide levels and in-hospital mortality in acute decompensated heart failure. *J. Am. Coll. Cardiol.* **2007**, *49*, 1943–1950. [CrossRef]
40. Januzzi, J.L., Jr.; Sakhuja, R.; O'Donoghue, M.; Baggish, A.L.; Anwaruddin, S.; Chae, C.U.; Cameron, R.; Krauser, D.G.; Tung, R.; Camargo, C.A., Jr.; et al. Utility of amino-terminal pro-brain natriuretic peptide testing for prediction of 1-year mortality in patients with dyspnea treated in the emergency department. *Arch. Intern. Med.* **2006**, *166*, 315–320. [CrossRef]
41. Omland, T.; Aakvaag, A.; Bonarjee, V.V.; Caidahl, K.; Lie, R.T.; Nilsen, D.W.; Sundsfjord, J.A.; Dickstein, K. Plasma brain natriuretic peptide as an indicator of left ventricular systolic function and long-term survival after acute myocardial infarction. Comparison with plasma atrial natriuretic peptide and N-terminal proatrial natriuretic peptide. *Circulation* **1996**, *93*, 1963–1969. [CrossRef] [PubMed]
42. Mega, J.L.; Morrow, D.A.; De Lemos, J.A.; Sabatine, M.S.; Murphy, S.A.; Rifai, N.; Gibson, C.M.; Antman, E.M.; Braunwald, E. B-type natriuretic peptide at presentation and prognosis in patients with ST-segment elevation myocardial infarction: An ENTIRE-TIMI-23 substudy. *J. Am. Coll. Cardiol.* **2004**, *44*, 335–339. [CrossRef] [PubMed]
43. Wang, T.J.; Larson, M.G.; Levy, D.; Leip, E.P.; Benjamin, E.J.; Wilson, P.W.; Sutherland, P.; Omland, T.; Vasan, R.S. Impact of age and sex on plasma natriuretic peptide levels in healthy adults. *Am. J. Cardiol.* **2002**, *90*, 254–258. [CrossRef]
44. Redfield, M.M.; Rodeheffer, R.J.; Jacobsen, S.J.; Mahoney, D.W.; Bailey, K.R.; Burnett, J.C. Plasma brain natriuretic peptide concentration: Impact of age and gender. *J. Am. Coll. Cardiol.* **2002**, *40*, 976–982. [CrossRef]
45. Wang, T.J.; Larson, M.G.; Levy, D.; Benjamin, E.J.; Leip, E.P.; Wilson, P.W.; Vasan, R.S. Impact of obesity on plasma natriuretic peptide levels. *Circulation* **2004**, *109*, 594–600. [CrossRef] [PubMed]
46. Mehra, M.R.; Uber, P.A.; Park, M.H.; Scott, R.L.; Ventura, H.O.; Harris, B.C.; Frohlich, E.D. Obesity and suppressed B-type natriuretic peptide levels in heart failure. *J. Am. Coll. Cardiol.* **2004**, *43*, 1590–1595. [CrossRef] [PubMed]
47. Bayes-Genis, A.; Lloyd-Jones, D.M.; van Kimmenade, R.R.; Lainchbury, J.G.; Richards, A.M.; Ordonez-Llanos, J.; Santalo, M.; Pinto, Y.M.; Januzzi, J.L., Jr. Effect of body mass index on diagnostic and prognostic usefulness of amino-terminal pro-brain natriuretic peptide in patients with acute dyspnea. *Arch. Intern. Med.* **2007**, *167*, 400–407. [CrossRef] [PubMed]
48. Horwich, T.B.; Hamilton, M.A.; Fonarow, G.C. B-type natriuretic peptide levels in obese patients with advanced heart failure. *J. Am. Coll. Cardiol.* **2006**, *47*, 85–90. [CrossRef]
49. Tsutamoto, T.; Wada, A.; Sakai, H.; Ishikawa, C.; Tanaka, T.; Hayashi, M.; Fujii, M.; Yamamoto, T.; Dohke, T.; Ohnishi, M.; et al. Relationship between renal function and plasma brain natriuretic peptide in patients with heart failure. *J. Am. Coll. Cardiol.* **2006**, *47*, 582–586. [CrossRef]

50. Forfia, P.R.; Watkins, S.P.; Rame, J.E.; Stewart, K.J.; Shapiro, E.P. Relationship between B-type natriuretic peptides and pulmonary capillary wedge pressure in the intensive care unit. *J. Am. Coll. Cardiol.* **2005**, *45*, 1667–1671. [CrossRef]
51. McCullough, P.A.; Duc, P.; Omland, T.; McCord, J.; Nowak, R.M.; Hollander, J.E.; Herrmann, H.C.; Steg, P.G.; Westheim, A.; Knudsen, C.W.; et al. B-type natriuretic peptide and renal function in the diagnosis of heart failure: An analysis from the Breathing Not Properly Multinational Study. *Am. J. Kidney Dis.* **2003**, *41*, 571–579. [CrossRef]
52. Anwaruddin, S.; Lloyd-Jones, D.M.; Baggish, A.; Chen, A.; Krauser, D.; Tung, R.; Chae, C.; Januzzi, J.L., Jr. Renal function, congestive heart failure, and amino-terminal pro-brain natriuretic peptide measurement: Results from the ProBNP Investigation of Dyspnea in the Emergency Department (PRIDE) Study. *J. Am. Coll. Cardiol.* **2006**, *47*, 91–97. [CrossRef] [PubMed]
53. Gupta, D.K.; de Lemos, J.A.; Ayers, C.R.; Berry, J.D.; Wang, T.J. Racial Differences in Natriuretic Peptide Levels: The Dallas Heart Study. *JACC Heart Fail.* **2015**, *3*, 513–519. [CrossRef] [PubMed]
54. Gupta, D.K.; Daniels, L.B.; Cheng, S.; deFilippi, C.R.; Criqui, M.H.; Maisel, A.S.; Lima, J.A.; Bahrami, H.; Greenland, P.; Cushman, M.; et al. Differences in Natriuretic Peptide Levels by Race/Ethnicity (From the Multi-Ethnic Study of Atherosclerosis). *Am. J. Cardiol.* **2017**, *120*, 1008–1015. [CrossRef] [PubMed]
55. Jourdain, P.; Jondeau, G.; Funck, F.; Gueffet, P.; Le Helloco, A.; Donal, E.; Aupetit, J.F.; Aumont, M.C.; Galinier, M.; Eicher, J.C.; et al. Plasma brain natriuretic peptide-guided therapy to improve outcome in heart failure: The STARS-BNP Multicenter Study. *J. Am. Coll. Cardiol.* **2007**, *49*, 1733–1739. [CrossRef]
56. Lainchbury, J.G.; Troughton, R.W.; Strangman, K.M.; Frampton, C.M.; Pilbrow, A.; Yandle, T.G.; Hamid, A.K.; Nicholls, M.G.; Richards, A.M. N-terminal pro-B-type natriuretic peptide-guided treatment for chronic heart failure: Results from the BATTLESCARRED (NT-proBNP-Assisted Treatment To Lessen Serial Cardiac Readmissions and Death) trial. *J. Am. Coll. Cardiol.* **2009**, *55*, 53–60. [CrossRef]
57. Brunner-La Rocca, H.P.; Eurlings, L.; Richards, A.M.; Januzzi, J.L.; Pfisterer, M.E.; Dahlstrom, U.; Pinto, Y.M.; Karlstrom, P.; Erntell, H.; Berger, R.; et al. Which heart failure patients profit from natriuretic peptide guided therapy? A meta-analysis from individual patient data of randomized trials. *Eur. J. Heart Fail.* **2015**, *17*, 1252–1261. [CrossRef]
58. Pfisterer, M.; Buser, P.; Rickli, H.; Gutmann, M.; Erne, P.; Rickenbacher, P.; Vuillomenet, A.; Jeker, U.; Dubach, P.; Beer, H.; et al. BNP-guided vs symptom-guided heart failure therapy: The Trial of Intensified vs Standard Medical Therapy in Elderly Patients With Congestive Heart Failure (TIME-CHF) randomized trial. *JAMA* **2009**, *301*, 383–392. [CrossRef]
59. Morgenthaler, N.G.; Struck, J.; Thomas, B.; Bergmann, A. Immunoluminometric assay for the midregion of pro-atrial natriuretic peptide in human plasma. *Clin. Chem.* **2004**, *50*, 234–236. [CrossRef]
60. Maisel, A.; Mueller, C.; Nowak, R.; Peacock, W.F.; Landsberg, J.W.; Ponikowski, P.; Mockel, M.; Hogan, C.; Wu, A.H.; Richards, M.; et al. Mid-region pro-hormone markers for diagnosis and prognosis in acute dyspnea: Results from the BACH (Biomarkers in Acute Heart Failure) trial. *J. Am. Coll. Cardiol.* **2010**, *55*, 2062–2076. [CrossRef]
61. Gegenhuber, A.; Struck, J.; Poelz, W.; Pacher, R.; Morgenthaler, N.G.; Bergmann, A.; Haltmayer, M.; Mueller, T. Midregional pro-A-type natriuretic peptide measurements for diagnosis of acute destabilized heart failure in short-of-breath patients: Comparison with B-type natriuretic peptide (BNP) and amino-terminal proBNP. *Clin. Chem.* **2006**, *52*, 827–831. [CrossRef] [PubMed]
62. Chenevier-Gobeaux, C.; Guerin, S.; Andre, S.; Ray, P.; Cynober, L.; Gestin, S.; Pourriat, J.L.; Claessens, Y.E. Midregional pro-atrial natriuretic peptide for the diagnosis of cardiac-related dyspnea according to renal function in the emergency department: A comparison with B-type natriuretic peptide (BNP) and N-terminal proBNP. *Clin. Chem.* **2010**, *56*, 1708–1717. [CrossRef] [PubMed]
63. Seronde, M.F.; Gayat, E.; Logeart, D.; Lassus, J.; Laribi, S.; Boukef, R.; Sibellas, F.; Launay, J.M.; Manivet, P.; Sadoune, M.; et al. Comparison of the diagnostic and prognostic values of B-type and atrial-type natriuretic peptides in acute heart failure. *Int. J. Cardiol.* **2013**, *168*, 3404–3411. [CrossRef] [PubMed]
64. Masson, S.; Latini, R.; Carbonieri, E.; Moretti, L.; Rossi, M.G.; Ciricugno, S.; Milani, V.; Marchioli, R.; Struck, J.; Bergmann, A.; et al. The predictive value of stable precursor fragments of vasoactive peptides in patients with chronic heart failure: Data from the GISSI-heart failure (GISSI-HF) trial. *Eur. J. Heart Fail.* **2010**, *12*, 338–347. [CrossRef] [PubMed]

65. Smith, J.G.; Newton-Cheh, C.; Almgren, P.; Struck, J.; Morgenthaler, N.G.; Bergmann, A.; Platonov, P.G.; Hedblad, B.; Engstrom, G.; Wang, T.J.; et al. Assessment of conventional cardiovascular risk factors and multiple biomarkers for the prediction of incident heart failure and atrial fibrillation. *J. Am. Coll. Cardiol.* **2010**, *56*, 1712–1719. [CrossRef]
66. Daniels, L.B.; Clopton, P.; Potocki, M.; Mueller, C.; McCord, J.; Richards, M.; Hartmann, O.; Anand, I.S.; Wu, A.H.; Nowak, R.; et al. Influence of age, race, sex, and body mass index on interpretation of midregional pro atrial natriuretic peptide for the diagnosis of acute heart failure: Results from the BACH multinational study. *Eur. J. Heart Fail.* **2012**, *14*, 22–31. [CrossRef] [PubMed]
67. Burke, M.A.; Cotts, W.G. Interpretation of B-type natriuretic peptide in cardiac disease and other comorbid conditions. *Heart Fail. Rev.* **2007**, *12*, 23–36. [CrossRef]
68. Fan, J.; Cao, H.; Su, L.; Ling, Z.; Liu, Z.; Lan, X.; Xu, Y.; Chen, W.; Yin, Y. NT-proBNP, but not ANP and C-reactive protein, is predictive of paroxysmal atrial fibrillation in patients undergoing pulmonary vein isolation. *J. Interv. Card. Electrophysiol.* **2012**, *33*, 93–100. [CrossRef]
69. Martinez-Rumayor, A.; Richards, A.M.; Burnett, J.C.; Januzzi, J.L., Jr. Biology of the natriuretic peptides. *Am. J. Cardiol.* **2008**, *101*, 3–8. [CrossRef]
70. Arima, M.; Kanoh, T.; Kawano, Y.; Oigawa, T.; Yamagami, S.; Matsuda, S. Plasma levels of brain natriuretic peptide increase in patients with idiopathic bilateral atrial dilatation. *Cardiology* **2002**, *97*, 12–17. [CrossRef]
71. Inoue, S.; Murakami, Y.; Sano, K.; Katoh, H.; Shimada, T. Atrium as a source of brain natriuretic polypeptide in patients with atrial fibrillation. *J. Card. Fail.* **2000**, *6*, 92–96. [CrossRef]
72. Bai, M.; Yang, J.; Li, Y. Serum N-terminal-pro-brain natriuretic peptide level and its clinical implications in patients with atrial fibrillation. *Clin. Cardiol.* **2009**, *32*, E1–E5. [CrossRef] [PubMed]
73. Jourdain, P.; Bellorini, M.; Funck, F.; Fulla, Y.; Guillard, N.; Loiret, J.; Thebault, B.; Sadeg, N.; Desnos, M. Short-term effects of sinus rhythm restoration in patients with lone atrial fibrillation: A hormonal study. *Eur. J. Heart Fail.* **2002**, *4*, 263–267. [CrossRef]
74. Wozakowska-Kaplon, B. Effect of sinus rhythm restoration on plasma brain natriuretic peptide in patients with atrial fibrillation. *Am. J. Cardiol.* **2004**, *93*, 1555–1558. [CrossRef] [PubMed]
75. Bakowski, D.; Wozakowska-Kaplon, B.; Opolski, G. The influence of left ventricle diastolic function on natriuretic peptides levels in patients with atrial fibrillation. *Pacing Clin. Electrophysiol.* **2009**, *32*, 745–752. [CrossRef] [PubMed]
76. Schnabel, R.B.; Larson, M.G.; Yamamoto, J.F.; Sullivan, L.M.; Pencina, M.J.; Meigs, J.B.; Tofler, G.H.; Selhub, J.; Jacques, P.F.; Wolf, P.A.; et al. Relations of biomarkers of distinct pathophysiological pathways and atrial fibrillation incidence in the community. *Circulation* **2010**, *121*, 200–207. [CrossRef]
77. Svennberg, E.; Lindahl, B.; Berglund, L.; Eggers, K.M.; Venge, P.; Zethelius, B.; Rosenqvist, M.; Lind, L.; Hijazi, Z. NT-proBNP is a powerful predictor for incident atrial fibrillation—Validation of a multimarker approach. *Int. J. Cardiol.* **2016**, *223*, 74–81. [CrossRef]
78. Sinner, M.F.; Stepas, K.A.; Moser, C.B.; Krijthe, B.P.; Aspelund, T.; Sotoodehnia, N.; Fontes, J.D.; Janssens, A.C.; Kronmal, R.A.; Magnani, J.W.; et al. B-type natriuretic peptide and C-reactive protein in the prediction of atrial fibrillation risk: The CHARGE-AF Consortium of community-based cohort studies. *Europace* **2014**, *16*, 1426–1433. [CrossRef]
79. Shelton, R.J.; Clark, A.L.; Goode, K.; Rigby, A.S.; Cleland, J.G. The diagnostic utility of N-terminal pro-B-type natriuretic peptide for the detection of major structural heart disease in patients with atrial fibrillation. *Eur. Heart J.* **2006**, *27*, 2353–2361. [CrossRef]
80. Morello, A.; Lloyd-Jones, D.M.; Chae, C.U.; van Kimmenade, R.R.; Chen, A.C.; Baggish, A.L.; O'Donoghue, M.; Lee-Lewandrowski, E.; Januzzi, J.L., Jr. Association of atrial fibrillation and amino-terminal pro-brain natriuretic peptide concentrations in dyspneic subjects with and without acute heart failure: Results from the ProBNP Investigation of Dyspnea in the Emergency Department (PRIDE) study. *Am. Heart J.* **2007**, *153*, 90–97. [CrossRef]
81. Richards, M.; Di Somma, S.; Mueller, C.; Nowak, R.; Peacock, W.F.; Ponikowski, P.; Mockel, M.; Hogan, C.; Wu, A.H.; Clopton, P.; et al. Atrial fibrillation impairs the diagnostic performance of cardiac natriuretic peptides in dyspneic patients: Results from the BACH Study (Biomarkers in ACute Heart Failure). *JACC Heart Fail.* **2013**, *1*, 192–199. [CrossRef] [PubMed]

82. Knudsen, C.W.; Omland, T.; Clopton, P.; Westheim, A.; Wu, A.H.; Duc, P.; McCord, J.; Nowak, R.M.; Hollander, J.E.; Storrow, A.B.; et al. Impact of atrial fibrillation on the diagnostic performance of B-type natriuretic peptide concentration in dyspneic patients: An analysis from the breathing not properly multinational study. *J. Am. Coll. Cardiol.* **2005**, *46*, 838–844. [CrossRef] [PubMed]
83. Rogers, R.K.; Stoddard, G.J.; Greene, T.; Michaels, A.D.; Fernandez, G.; Freeman, A.; Nord, J.; Stehlik, J. Usefulness of adjusting for clinical covariates to improve the ability of B-type natriuretic peptide to distinguish cardiac from noncardiac dyspnea. *Am. J. Cardiol.* **2009**, *104*, 689–694. [CrossRef] [PubMed]
84. van Doorn, S.; Geersing, G.J.; Kievit, R.F.; van Mourik, Y.; Bertens, L.C.; van Riet, E.E.S.; Boonman-de Winter, L.J.; Moons, K.G.M.; Hoes, A.W.; Rutten, F.H. Opportunistic screening for heart failure with natriuretic peptides in patients with atrial fibrillation: A meta-analysis of individual participant data of four screening studies. *Heart* **2018**, *104*, 1236–1237. [CrossRef] [PubMed]
85. Kristensen, S.L.; Jhund, P.S.; Mogensen, U.M.; Rorth, R.; Abraham, W.T.; Desai, A.; Dickstein, K.; Rouleau, J.L.; Zile, M.R.; Swedberg, K.; et al. Prognostic Value of N-Terminal Pro-B-Type Natriuretic Peptide Levels in Heart Failure Patients With and Without Atrial Fibrillation. *Circ. Heart Fail.* **2017**, *10*, e004409. [CrossRef]
86. Kallergis, E.M.; Manios, E.G.; Kanoupakis, E.M.; Mavrakis, H.E.; Goudis, C.A.; Maliaraki, N.E.; Saloustros, I.G.; Milathianaki, M.E.; Chlouverakis, G.I.; Vardas, P.E. Effect of sinus rhythm restoration after electrical cardioversion on apelin and brain natriuretic Peptide prohormone levels in patients with persistent atrial fibrillation. *Am. J. Cardiol.* **2010**, *105*, 90–94. [CrossRef] [PubMed]
87. Beck-da-Silva, L.; de Bold, A.; Fraser, M.; Williams, K.; Haddad, H. Brain natriuretic peptide predicts successful cardioversion in patients with atrial fibrillation and maintenance of sinus rhythm. *Can. J. Cardiol.* **2004**, *20*, 1245–1248.
88. Lellouche, N.; Berthier, R.; Mekontso-Dessap, A.; Braconnier, F.; Monin, J.L.; Duval, A.M.; Dubois-Rande, J.L.; Gueret, P.; Garot, J. Usefulness of plasma B-type natriuretic peptide in predicting recurrence of atrial fibrillation one year after external cardioversion. *Am. J. Cardiol.* **2005**, *95*, 1380–1382. [CrossRef]
89. Solheim, E.; Off, M.K.; Hoff, P.I.; De Bortoli, A.; Schuster, P.; Ohm, O.J.; Chen, J. N-terminal pro-B-type natriuretic peptide level at long-term follow-up after atrial fibrillation ablation: A marker of reverse atrial remodelling and successful ablation. *J. Interv. Card. Electrophysiol.* **2012**, *34*, 129–136. [CrossRef]
90. Zhang, Y.; Chen, A.; Song, L.; Li, M.; Chen, Y.; He, B. Association Between Baseline Natriuretic Peptides and Atrial Fibrillation Recurrence After Catheter Ablation. *Int. Heart J.* **2016**, *57*, 183–189. [CrossRef]
91. Jiang, H.; Wang, W.; Wang, C.; Xie, X.; Hou, Y. Association of pre-ablation level of potential blood markers with atrial fibrillation recurrence after catheter ablation: A meta-analysis. *Europace* **2017**, *19*, 392–400. [CrossRef] [PubMed]
92. Deng, H.; Shantsila, A.; Guo, P.; Zhan, X.; Fang, X.; Liao, H.; Liu, Y.; Wei, W.; Fu, L.; Wu, S.; et al. Multiple biomarkers and arrhythmia outcome following catheter ablation of atrial fibrillation: The Guangzhou Atrial Fibrillation Project. *J. Arrhythm.* **2018**, *34*, 617–625. [CrossRef] [PubMed]
93. Roldan, V.; Vilchez, J.A.; Manzano-Fernandez, S.; Jover, E.; Galvez, J.; Puche, C.M.; Valdes, M.; Vicente, V.; Lip, G.Y.; Marin, F. Usefulness of N-terminal pro-B-type natriuretic Peptide levels for stroke risk prediction in anticoagulated patients with atrial fibrillation. *Stroke* **2014**, *45*, 696–701. [CrossRef] [PubMed]
94. Hijazi, Z.; Oldgren, J.; Andersson, U.; Connolly, S.J.; Ezekowitz, M.D.; Hohnloser, S.H.; Reilly, P.A.; Vinereanu, D.; Siegbahn, A.; Yusuf, S.; et al. Cardiac biomarkers are associated with an increased risk of stroke and death in patients with atrial fibrillation: A Randomized Evaluation of Long-term Anticoagulation Therapy (RE-LY) substudy. *Circulation* **2012**, *125*, 1605–1616. [CrossRef] [PubMed]
95. Hijazi, Z.; Wallentin, L.; Siegbahn, A.; Andersson, U.; Christersson, C.; Ezekowitz, J.; Gersh, B.J.; Hanna, M.; Hohnloser, S.; Horowitz, J.; et al. N-terminal pro-B-type natriuretic peptide for risk assessment in patients with atrial fibrillation: Insights from the ARISTOTLE Trial (Apixaban for the Prevention of Stroke in Subjects With Atrial Fibrillation). *J. Am. Coll. Cardiol.* **2013**, *61*, 2274–2284. [CrossRef] [PubMed]
96. Oldgren, J.; Hijazi, Z.; Lindback, J.; Alexander, J.H.; Connolly, S.J.; Eikelboom, J.W.; Ezekowitz, M.D.; Granger, C.B.; Hylek, E.M.; Lopes, R.D.; et al. Performance and Validation of a Novel Biomarker-Based Stroke Risk Score for Atrial Fibrillation. *Circulation* **2016**, *134*, 1697–1707. [CrossRef] [PubMed]
97. Hijazi, Z.; Lindback, J.; Alexander, J.H.; Hanna, M.; Held, C.; Hylek, E.M.; Lopes, R.D.; Oldgren, J.; Siegbahn, A.; Stewart, R.A.; et al. The ABC (age, biomarkers, clinical history) stroke risk score: A biomarker-based risk score for predicting stroke in atrial fibrillation. *Eur. Heart J.* **2016**, *37*, 1582–1590. [CrossRef] [PubMed]

98. Eckstein, J.; Potocki, M.; Murray, K.; Breidthardt, T.; Ziller, R.; Mosimann, T.; Klima, T.; Hoeller, R.; Moehring, B.; Sou, S.M.; et al. Direct comparison of mid-regional pro-atrial natriuretic peptide with N-terminal pro B-type natriuretic peptide in the diagnosis of patients with atrial fibrillation and dyspnoea. *Heart* **2012**, *98*, 1518–1522. [CrossRef]
99. Darkner, S.; Goetze, J.P.; Chen, X.; Henningsen, K.; Pehrson, S.; Svendsen, J.H. Natriuretic Propeptides as Markers of Atrial Fibrillation Burden and Recurrence (from the AMIO-CAT Trial). *Am. J. Cardiol.* **2017**, *120*, 1309–1315. [CrossRef]
100. van den Berg, M.P.; van Gelder, I.C.; van Veldhuisen, D.J. Depletion of atrial natriuretic peptide during longstanding atrial fibrillation. *Europace* **2004**, *6*, 433–437. [CrossRef]
101. Davies, M.J.; Pomerance, A. Pathology of atrial fibrillation in man. *Br. Heart J.* **1972**, *34*, 520–525. [CrossRef] [PubMed]
102. Seino, Y.; Shimai, S.; Ibuki, C.; Itoh, K.; Takano, T.; Hayakawa, H. Disturbed secretion of atrial natriuretic peptide in patients with persistent atrial standstill: Endocrinologic silence. *J. Am. Coll. Cardiol.* **1991**, *18*, 459–463. [CrossRef]
103. Yoshihara, F.; Nishikimi, T.; Sasako, Y.; Hino, J.; Kobayashi, J.; Minatoya, K.; Bando, K.; Kosakai, Y.; Horio, T.; Suga, S.-I.; et al. Plasma atrial natriuretic peptide concentration inversely correlates with left atrial collagen volume fraction in patients with atrial fibrillation. *J. Am. Coll. Cardiol.* **2002**, *39*, 288–294. [CrossRef]
104. Yoshida, K.; Tada, H.; Ogata, K.; Sekiguchi, Y.; Inaba, T.; Ito, Y.; Sato, Y.; Sato, A.; Seo, Y.; Kandori, A.; et al. Electrogram organization predicts left atrial reverse remodeling after the restoration of sinus rhythm by catheter ablation in patients with persistent atrial fibrillation. *Heart Rhythm.* **2012**, *9*, 1769–1778. [CrossRef] [PubMed]
105. Nakanishi, K.; Fukuda, S.; Yamashita, H.; Kosaka, M.; Shirai, N.; Tanaka, A.; Yoshikawa, J.; Shimada, K. Pre-procedural serum atrial natriuretic peptide levels predict left atrial reverse remodeling after catheter ablation in patients with atrial fibrillation. *JACC Clin. Electrophysiol.* **2016**, *2*, 151–158. [CrossRef]
106. Ogawa, K.; Yoshida, K.; Uehara, Y.; Ebine, M.; Kimata, A.; Nishina, H.; Takeyasu, N.; Noguchi, Y.; Ieda, M.; Aonuma, K.; et al. Mechanistic implication of decreased plasma atrial natriuretic peptide level for transient rise in the atrial capture threshold early after ICD or CRT-D implantation. *J. Interv. Card. Electrophysiol.* **2018**, *53*, 131–140. [CrossRef] [PubMed]

© 2019 by the authors. Licensee MDPI, Basel, Switzerland. This article is an open access article distributed under the terms and conditions of the Creative Commons Attribution (CC BY) license (http://creativecommons.org/licenses/by/4.0/).

Review

BNP as a Major Player in the Heart-Kidney Connection

Ryuji Okamoto [1,*], Yusuf Ali [1], Ryotaro Hashizume [2], Noboru Suzuki [3] and Masaaki Ito [1]

1. Department of Cardiology and Nephrology, Mie University Graduate School of Medicine, 2-174 Edobashi, Tsu, Mie 514-8507, Japan
2. Department of Pathology and Matrix Biology, Mie University Graduate School of Medicine, 2-174 Edobashi, Tsu, Mie 514-8507, Japan
3. Department of Animal Genomics, Functional Genomics Institute, Mie University Life Science Research Center, 2-174 Edobashi, Tsu, Mie 514-8507, Japan
* Correspondence: ryuji@clin.medic.mie-u.ac.jp; Tel.: +81-59-231-5015; Fax: +81-59-231-5201

Received: 22 June 2019; Accepted: 17 July 2019; Published: 22 July 2019

Abstract: Brain natriuretic peptide (BNP) is an important biomarker for patients with heart failure, hypertension and cardiac hypertrophy. Although it is known that BNP levels are relatively higher in patients with chronic kidney disease and no heart disease, the mechanism remains unknown. Here, we review the functions and the roles of BNP in the heart-kidney interaction. In addition, we discuss the relevant molecular mechanisms that suggest BNP is protective against chronic kidney diseases and heart failure, especially in terms of the counterparts of the renin-angiotensin-aldosterone system (RAAS). The renal medulla has been reported to express depressor substances. The extract of the papillary tips from kidneys may induce the expression and secretion of BNP from cardiomyocytes. A better understanding of these processes will help accelerate pharmacological treatments for heart-kidney disease.

Keywords: natriuretic peptide; cardiorenal syndrome; vasopressor; vasodilator; kidney; medulla; renin-angiotensin-aldosterone system

1. Introduction

The Brain natriuretic peptide (BNP), which is a component of the natriuretic peptide (NP) system and also known as B-type NP, is mainly secreted from the cardiomyocytes in response to cardiac stretch and ischemia, and plays an important role in cardiorenal protection [1–4]. The renoprotective effects of BNP include the inhibition of sodium reabsorption in the proximal tubule and the distal nephron, and the improvement of the glomerular filtration rate (GFR) and renal plasma flow (RPF) with respect to vasodilatation by inhibiting multiple plasma vasoconstrictors [5–7]. Furthermore, BNP infusion inhibits the cardiac and renal sympathetic tones [8] and the renin-angiotensin-aldosterone system (RAAS) [9], in addition to decreasing the endothelin release [10]. NPs mediate their functions through interactions with specific surface receptors on target cells. At present, there are three distinct natriuretic peptide receptors (NPR) that have been reported and which include NPR-A, NPR-B, and NPR-C. NPR-A and NPR-B stimulate guanylyl cyclase, mediating their effect via the activation of the second messenger, cyclic guanosine monophosphate (cGMP). In the kidney, NPs cause a relaxation of mesangial cells, which increases GFR and reduces fractional sodium reabsorption in the renal tubules [7,11–13].

Congestive heart failure (CHF) is a complex syndrome characterized by sodium and water retention through the activation of different neurohormonal systems, such as the RAAS and the sympathetic nervous system (SNS), and also importantly, the NP system. Several experimental and clinical studies have implicated BNP in the pathophysiology of the unbalanced cardiorenal axis in

CHF. In CHF patients, BNP plasma levels are in excess of 100 pg/mL [14]. Due to BNP's longer half-life, it has been shown to have greater stability and offer a better understanding of the disease progression diagnostically as compared to atrial NP (ANP) in terms of improving cardiovascular function [15–18]. BNP plasma levels are also elevated in chronic kidney disease (CKD) patients. Underlying causes include renal dysfunction, diminished neprilysin (NEP) activity in the kidney and associated cardiovascular pathophysiology. Previous studies reported finding markedly high plasma BNP levels in patients with renal impairments [19–21].

Growing evidence suggests that in CHF, coronary artery disease and/or left ventricular hypertrophy etc., are increasingly associated with CKD patients. More recently, the cardiorenal syndrome (CRS), which is a complex pathophysiological condition that involves an association between acute heart failure (AHF) or CHF and renal impairment, has received much more attention. However, this condition is more than just a simultaneous cardiac and renal disease [22]. A recent French prospective study on 507 AHF patients demonstrated that BNP and the BNP prohormone were higher in AHF patients with renal dysfunction (CRS patients) as compared to those with normal renal function [23].

BNP acts as a compensating agent in the early stages of disease progression by inducing natriuresis and dieresis, and reducing RAAS and SNS. Similar to that seen with severe disease states like HF or CRS and despite high levels, endogenous BNP becomes resistant and is no longer able to compensate for volume overload in such cases. Thus, the supportive role of BNP in counteracting the unwanted effects of activated RAAS and SNS in these diseases provides the rationale for using this peptide as a potential therapeutic agent [13,24]. As it has been reported that the renal medulla expresses depressor substances, an extract of the papillary tips from kidneys might be able to induce the expression and secretion of endogenous BNP from cardiomyocytes [25]. A better understanding of these processes could potentially accelerate pharmacological treatments for CRS. Here, we review the functions and the roles of BNP in the heart-kidney interaction. In addition, we discuss the relevant molecular mechanisms for the protective effect of BNP against CKD and HF, especially in terms of the counterparts of the RAA system.

2. Biochemical Characteristics of BNP

Human BNP mRNA is translated to preproBNP of 134 amino acids, from which proBNP of 108 amino acids is processed and cleaved by the serine proteases corin and/or furin, yielding a biological inactive amino-terminal fragment, 76-amino acid proBNP (NT-proBNP) and an active carboxy-terminal fragment, 32-amino acid BNP [6,26]. It has been considered that, after maturation, both NT-proBNP and BNP are secreted at a 1:1 molar ratio from the heart. However, recent studies have shown that a certain amount of proBNP is also secreted without its cleavage from the heart [27]. Therefore, we should take into consideration the possibility that the increase in NT-proBNP is partly due to the elevation of proBNP when we see patients with an elevation of NT-proBNP. In contrast, BNP is increased without additional cardiac stress in patients treated with angiotensin-receptor/Neprilysin inhibitor (ARNi) due to its inhibitory effects of BNP degradation. In this review, we will not discuss them in detail here. Both NT-proBNP and BNP are equally useful in the differential diagnosis of heart failure [28,29], although the NT-proBNP and BNP assay allows for complementary information during ARNi treatments [30]. The gene structure and the post-translational processing of BNP and the diversity of circulating BNP-related peptides have been reviewed in detail [2,6,31].

3. The Function of BNP in Kidneys

It has been well established that BNP plays important roles in the kidneys, and provides multiple beneficial effects involving renal function [3]. Abnormalities or alterations of this system may contribute to renal impairment including tubular damage as a consequence of other CV disorders [32]. On the other hand, BNP plasma levels are affected by renal function, although the mechanism remains unclear. Thus, during renal failure, these are not considered to be an ideal choice as hemodynamic biomarkers [33,34]. The elevated levels of BNP may be the result of an increased cardiac release in

CKD patients. In CKD patients, the increase in circulatory blood volume, the elevation in BP due to volume overload and arterial stiffness, and the cardiac hypertrophy and HF etc., can contribute to the elevation in BNP. The elevation of BNP in CKD patients is partly due to the impaired clearance of BNP from the kidneys.

The functions of NPs are mediated by their interactions with specific surface receptors on the target cells. In the kidney, BNP increases GFR by relaxing the mesangial cells and inhibits the tubular fractional reabsorption of sodium [7,35] (Figure 1). BNP also decreases vascular resistance by relaxing vascular smooth muscle cells, while it has no effect on vascular permeability, unlike ANP.

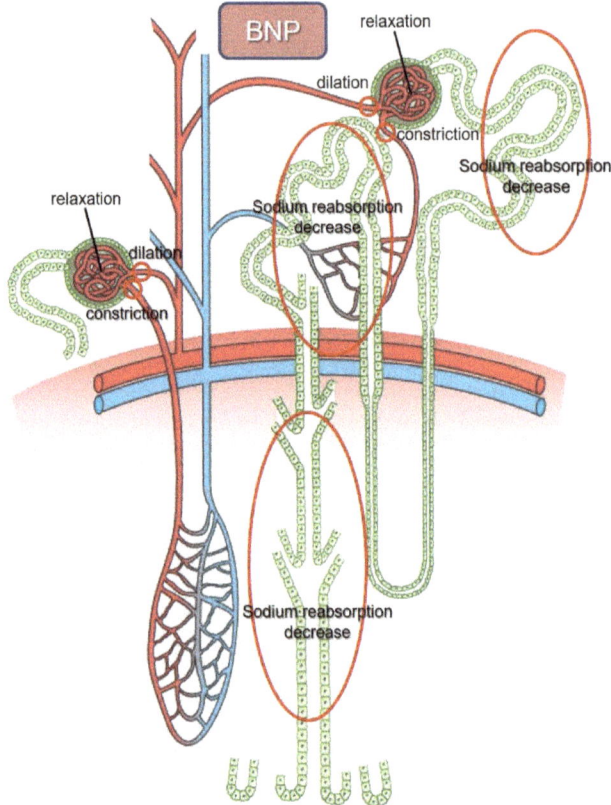

Figure 1. The interaction site of BNP in the nephron. BNP: brain natriuretic peptide.

Table 1 shows a sequence of pioneer studies that examined the injection of BNP in humans [5,9,35–52]. Among them, Jensen et al. performed a well-organized study of BNP infusion in healthy men in 1998 [35]. The maximum concentration in the plasma that was reached was 199 pmol/L (688.6 pg/mL) after a dose of 4 pmol/kg/min for 60 min. The authors observed that there was an increase in the urinary flow rate and GFR (about +60% and +5% respectively) and that there was inhibition of the renin secretion (−24%). In addition, the fractional excretion of sodium was increased (+140%). The fractional reabsorption of sodium was decreased under the infusion (30 to 60 min after the start of infusion). The measurement of the clearance using lithium, which is assumed to only be reabsorbed in the proximal tubule and to the same degree as sodium and water, demonstrated that the tubular site of action occurred both in the proximal tubules (−7%) and distal nephron (−5%). There were no changes in the blood pressure, heart rate and aldosterone concentration. There was a decrease in the

ANF concentration (−16%). These results suggest that the infusion of BNP within the physiological range, which can be observed in patients with HF, induces an increase in the GFR and the inhibition of sodium excretion, which leads to an increase in both the urine volume and the sodium excretion without affecting the blood pressure and heart rate in healthy subjects. These results are similar to other studies in terms of urination and sodium excretion irrespective of the minor differences in the protocol and the results (Table 1). Sodium reabsorption appears to decrease in the distal nephron earlier than that observed in the proximal tubule [41]. Interestingly, a trial of BNP infusion (2 pmol/kg/min for 60 min) in patients with heart failure and reduced ejection fraction (HFrEF) due to previous myocardial infarction or dilated cardiomyopathy showed that the impaired natriuretic response due to the reduced responsiveness in the distal nephron in patients with HFrEF was comparable to that found in healthy control subjects [5] and the results reported in another previous study [41]. This finding may be of interest, as this is in line with another previous report that found that a reduced reabsorption of sodium occurs in the distal nephron earlier than in the proximal tubules in patients with mild CKD [53]. In the distal nephron, the inner medullary collecting duct is the most prominent target of ANP [13]. It has been reported that BNP is colocalized with ANP in the distal tubules, while CNP is observed in the proximal tubules in human kidneys [54]. Thus, the adaptive recovery from impaired reabsorption of sodium in the distal nephron and feasibility for lowering the blood pressure suggest these can be therapeutic targets if we can determine the range of nesiritide in patients with CHF or/and CKD [55,56]. Moreover, as the NPRs are predominantly expressed in the distal part of the renal tubules [57], this can be another possible reason for their impaired natriuretic effect in advanced HF in which the proximal reabsorption of sodium is greatly enhanced [58]. In an experimental CHF model, BNP enhances the renal diuretic and natriuretic actions of loop diuretics while at the same time it also reduces the diuretic-induced aldosterone production [59].

BNP infusion may play an important role in preventing the development of CKD. McKie et al. recently examined BNP's role in the pathophysiology of cardiorenal dysfunction and found that in asymptomatic systolic HF patients, chronic subcutaneous BNP therapy for 12 weeks improved renal function along with favourable hemodynamic effects in response to volume expansion [60]. A recent study also investigated the prophylactic effects of early BNP administration on contrast-induced nephropathy (CIN) in CKD patients undergoing elective percutaneous coronary intervention or coronary angiography [61]. BNP effectively decreased the incidence of CIN in patients with CKD, as was shown by the improved estimated GFR, cystatin C, and serum creatinine compared to the control group. In addition, the BNP group showed a faster recovery. Thus, exogenous BNP served as a prophylactic agent for attenuating the CIN incidence in CKD patients [61]. A retrospective study analyzed the effect of nesiritide on renal function and its clinical safety in 328 patients with decompensated HFpEF (dHFpEF) and concluded that nesiritide can be safely administered without negatively impacting the long-term renal function in these patients [56]. The GFR and creatinine remained stable at 1-month post-nesiritide infusion, whereas there was a significant deterioration of kidney function (GFR and creatinine) observed in the control subjects. In addition, their multivariate analysis showed that nesiritide was an important predictor of renal function at 1 month [56].

The vascular effects of NPs are site-specific [62]. Importantly, in the kidney's vasculature, while NPs relax the afferent arterioles by acting as vasodilators, they act as vasoconstrictive agents on the efferent arterioles, thereby causing the GFR to be increased [63]. Therefore, since it is presumed that the renal blood flow may increase or decrease or even remain unchanged in response to exogenous NPs, it can be accompanied by a similar discrepant result regarding the renal plasma flow [13].

In CKD patients, impaired renal function restricts the use of NPs, as plasma BNP levels are elevated to ~200 pg/mL in CKD patients without HF. Whether these elevated BNP levels in CKD promote the activation of the NP system and have an effect on the target organ still remains unclear. Downregulation of NPR-A and upregulation of NPR-C expression in the renal medulla and renal cortex, respectively, may be responsible for the resistance of NPs including BNP in CKD, thereby resulting in its limited use in treating CKD patients with HF [64,65].

Table 1. The effects of brain natriuretic peptide (BNP) infusion on renal function, the renin angiotensin II aldosterone system and hemodynamics in normal subjects and patients with heart failure and hypertension.

Study/Reference	Dosage of BNP pmol/kg/min	GFR	RPF	Urine Volume	Urine Na	Urine cGMP	Urine Aldo	Plasma cGMP	PRA	AngII	Plasma Aldo	MAP	HR	CO	SVR	PCWP
Normal Subjects																
McGregor. J. Clin. Endocrinol. Metab. 1990 [36]	2				↑↔*				↑		→	↑↔	↑↔			
Yoshimura. Circulation. 1991 [37]	30			↔	↑				↑↔*		→	→	↔	←	↑	
Holmes. J. Clin. Endocrinol. Metab. 1993 [9]	2			↔	↑	↑			↓↔*		→	↑↔	↔			
Cheung. Clin. Sci. 1994 [38]	0.4	↔	↑		↑			↑		↔		↔	↔			
Florkowski. Am. J. Physiol. 1994 [39]	2(+ANP(2))**	↔	↔	↔	↑	↑		↑	↑↔	↔	↔	↔	↔	↔	↔	
La Villa. J. Clin. Endocrinol. Metab. 1994 [40]	4	←		↔	↑	↑	→		↑↔			↔	↔			
La Villa. Hypertension. 1995 [41]	0.25 and 0.5	↔			↑	↑		↑	↑↔*	→	→	↔	↔			
Lazzeri. Cardiology. 1995 [42]	4, 8, 10, 12				↑			↑				↔	←	↔	↔	
Hunt. J Clin. Endocrinol. Metab. 1996 [43]	2	↔		↑	↑			↑	↓↔*		→	↑↔	←			↓
Yasue. J Card. Fail. 1996 [44]	30	↑	↓***	↑	↑			↑	→	↔	↔	↑↔	↔			
Jensen. Am. J. Phy.1998 [35]	1,2 and 4	↑	↔	↔	↔	↔		↑	→	↔	↔	↔	↔			
Jensen. Clin. Sci. 1999 [5]	2	↑		↑	↑			↑				←	←	↓↔*	↔	
van der Zander. Am. J. Physiol. 2003 [45]	4	↑		↔	↑			↑				←	←			
Summarized		←	↔	↑	↑	↑		↑	↓↔	↔	→	↔	↔	↔	↔	↓
Patients with Heart Failure																
Yoshimura. Circulation. 1991 [37]	30	↔		↑	↑			↑	↔		→	→	←	←	→	→
Marcus. Circulation. 1996 [46]	1 to 30	↔		←	↔			↑				→	←	←	→	→
Yasue. J. Card. Fail. 1996 [44]	30	↔		↔	↑			↑	↔→		↔→	↑↔	↔			→
Lainchbury. HTN 1997 [47]	3.3	↔		↔	↑			↑				←	←	↑↔	→	→
Abraham. J. Cardiac. fail. 1998 [48]	7.5, 15	↔	↔	↔	↑↔		↑	↑	↔→		↔	↔	↔	←	→	
Jensen. Clin. Sci. 1999 [5]	2	↔	↔	↔	←	↑		↑	→			←	↔			
Wang. Am. J. Trans. Res. 2016 [49]	A bolus followed by 2 to 6 for 72 h	→										→				↓
Summarized		↔	↔	↔	↔	↑		↑	↔	↔	↔	←	↔	←	→	→
Patients with Hypertension																
Richards. J. Hypertension. 1993 [50]	2	↔		←	←	↑↔		↑	↑↔#		→	↔	↔	↔	↔	
Lazzeri. Am. J. Hypertension. 1995 [51]	4	←		←	←	←			↑↔		→	↑↔	←			
Pidgeon. Hypertension. 1996 [52]	2			←	↔			↑	↔		→	↔				
Summarized		←		←	←	↑		↑	↔		→	↔	↔	↔	↔	

ANP, atrial natriuretic peptide; cGMP, cyclic guanosine monophosphate; CO, cardiac output; GFR, glomerular filtration rate; HR, heart rate; MAP, mean arterial pressure; PCWP, pulmonary capillary wedge pressure; PRA, plasma renin activity; RPF, renal plasma flow; SVR, systemic vascular resistance. * marginal significance. ** BNP(2)+ANP(2) vs ANP(2) alone. *** after the infusion. # Plasma renin concentration.

4. BNP Clearance from Kidneys

BNP clearance is associated with two major pathways. The first involves binding to the NPR-C receptor, while the second one is involved with the degradation by neprilysin (NEP), a zinc metallopeptidase. NPR-C, which is devoid of any guanylyl cyclase activity and coupled to the adenylyl cyclase inhibition [66], plays an important role as a clearance receptor for BNP in addition to ANP and CNP. In humans, it has been demonstrated that NPR-C mRNA is expressed in a variety of tissues such as the atria, kidney, lung, mesentery, placenta, adrenal, heart, cerebral cortex, cerebellum and aortic smooth muscle and endothelial cells [67,68]. ANP has been shown to be degraded mainly in the lung, liver and kidney. The binding ability of BNP with NPR-C in these organs is much weaker than ANP and CNP [69], which suggests a diminished removal of BNP by NPR-C-mediated internalization, and a long half-life due to the smaller amount of degradation. BNP is also degraded by a neutral endopeptidase known as NEP, which is mainly expressed at high levels at the luminal side of the renal proximal tubules in the kidneys [70] in addition to other tissues such as the heart, lungs, liver and vascular smooth muscle and endothelial cells. NEP extensively degrades BNP in the rat renal membrane in collaboration with a renal protease. In the human renal membrane, however, not all BNPs are degraded by NEP [71]. Moreover, the NEP mediated degradation of BNP is slower [72]. These results suggest that it is difficult to clear BNP from the kidneys in humans. Despite the fact that BNP degradation is mediated by both the NPR-C and NEP pathways, the precise role of each process with regard to the BNP concentrations remains undetermined [72]. In addition to NEP, BNP has also been reported to be degraded by dipeptidyl peptidase-4 and insulin-degrading enzyme [73,74]. In CHF patients, a marked decrease in urinary excretion of NPs was observed in conjunction with elevated plasma levels [75]. However, another report showed that there was an increase of NT-proBNP in fresh urine from HF patients [76].

5. BNP System vs. RAA System

Sympathetic stimulation activates the RAAS, including the augmentation of angiotensin II (Ang II) and aldosterone production, as a result of the elevation of vasopressin and norepinephrine plasma levels, as well as the secretion of renin. SNS and RAAS augment the cardiac output by increasing heart rate, contractibility, preload and afterload at the expense of increased oxygen consumption [77]. Inhibiting the SNS through beta-blockers and antagonizing the RAAS through the angiotensin-converting enzyme (ACE) inhibitors or angiotensin receptor blockers (ARBs), and mineralocorticoid receptor antagonists, may be insufficient for certain neurohormonal abnormalities [78]. As a response to volume overload or cardiac stretching, secretion of BNP from the cardiac chambers manifests the heart as one of the endocrine organs in order to maintain the salt balance by interacting with the RAAS, SNS and the kidneys (Figure 2). Continual or excessive activation of the RAAS and SNS leads to the development of CHF. Furthermore, BNP is also activated in order to resist the actions attributed to the diuretic, natriuretic and vasorelaxant effects of BNP. As a result, this counterbalances the neurohormonal activation in CHF or hypertension leading to beneficial effects, such as natriuresis, vasodilation, and anti-cardiac remodelling (Figure 2) [79–81].

BNP has been shown to directly inhibit renin production from the kidney before affecting RBF or GFR [82], similar to ANP [83]. This result indicates that BNP directly inhibits the tubuloglomerular feedback response that is activated by salt over intake.

Experimental studies using rat cardiomyocytes reported that both endogenous and exogenous BNP reduced aldosterone synthase (CYP11B2) mRNA expression, which may lead to the inhibition of the RAAS and attenuation of cardiac hypertrophy and fibrosis [84]. Furthermore, in cultured primary human adrenocortical cells, it has been reported that BNP opposed the Ang II-stimulated biosynthesis of aldosterone due to decreased expression of both CYP11B2 and CYP11B1, which are the most important synthetic enzymes of aldosterone [85]. In contrast, BNP failed to directly inhibit the production of catecholamine and the synthesis of tyrosine hydroxylase, a dopamine synthetic enzyme, in rat adrenal pheochromocytoma cells [86]. Infusion of BNP induced a sympatho-inhibitory effect in normal subjects and inhibited renal sympathetic nervous activity in patients with CHF but not in healthy subjects [8]. However, the findings of another study that found that there were reductions in systemic and right-sided cardiac pressures in HF patients infused with BNP without any changes in the renin, aldosterone and norepinephrine plasma levels suggests that there is a RAAS or SNS independent direct vasorelaxant effect of BNP [47]. Both systems appear to influence each other, with BNP counteracting RAAS by inhibiting the renin secretion and CYP11B2 expression through cGMP, while the RAAS blockade, in turn, activates BNP. This suggests these may exert a synergistic effect in HF [87]. Thus, NP clearly interacts with the RAAS, and is inversely correlated with the plasma Ang II levels in certain physiologic conditions [88]. Furthermore, since the plasma levels of both hormonal systems are augmented in HF, this suggests that they counterbalance each other [89].

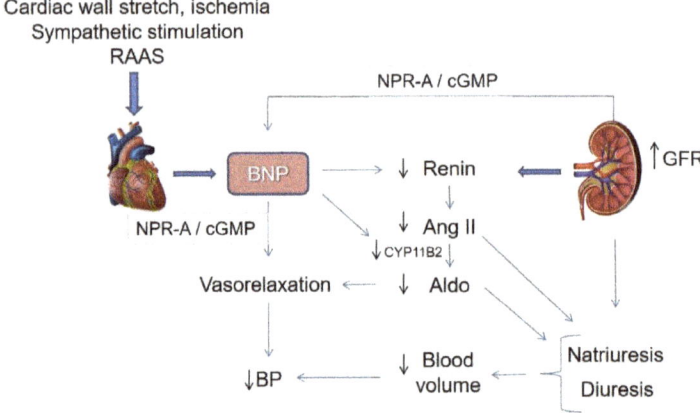

Figure 2. BNP as a counterregulatory system of the renin angiotensin II aldosterone system (RAAS). Aldo, aldosterone; Ang II, angiotensin II; BP, blood pressure; BNP, brain natriuretic peptide; CYP11B2, cytochrome P450 family 11 subfamily B member 2 (aldosterone synthase); cGMP, cyclic guanosine monophosphate; GFR, glomerular filtration rate; NPR-A, natriuretic peptide receptor A.

6. NPs Augmentation Combined with RAAS Blockade: Dual-Acting Angiotensin-Receptor/Neprilysin Inhibitors (ARNi)

Although BNP is progressively activated in HF, its response may often be insufficient to counteract the sodium retention and vascular constriction due to activation of RAAS and SNS. Therefore, more thorough approaches were undertaken both experimentally and clinically in an attempt to minimize the dysregulation of these neurohormonal systems. Ongoing strategies in promoting NP include synthesis or using agonists to increase its bioactivity and inhibition of NEP to reduce its catabolism [90,91]. Nesiritide, a recombinant BNP approved by the US Food and Drug Administration (FDA) in 2001, has been shown to promote clinical improvements in the management of CHF [92]. However, it has been reported to worsen renal function and increase the mortality rate in a meta-analysis [93]. Moreover, the severe hypotension and short half-life made these agents, including nesiritide, carperitide and ularitide, clinically imperfect. NEP inhibitors (NEPi) alone leads to activation of RAAS and attenuated Ang II degradation. Again the NEPi and ACE inhibitor combination predisposes a high risk of angioedema [94]. Therefore, the use of an ARB would be the optimal method of RAAS inhibition for use with NEPi. The first angiotensin receptor–neprilysin inhibitor (ARNI, LCZ696), developed by combining an ARB (Valsartan) with a NEPi (Sacubitril), was a major advance in the therapies for HF [95,96]. The combination of sacubitril and valsartan augments the beneficial effects of NPs and inhibits the harmful effects of Ang II. It preserves the ACE mechanism for bradykinin degradation and protects from angioedema formation [97]. Animal studies have shown that the NEPi and RAAS inhibitor combination reduced proteinuria and prevented kidney damage [98], also improved cardiac remodelling, fibrosis, and hypertrophy [99]. The recent UK HARP-III (United Kingdom Heart and Renal Protection-III), study on 414 CKD patients with an eGFR 20 to 60 mL/min/1.73 m^2 has demonstrated that over 12 months, the combination of sacubitril and valsartan was well tolerated and had similar effects on kidney function and albuminuria compared to ARB irbesartan, with a BP and cardiac biomarker lowering effect, suggested that this combination could have the potential for reducing cardiovascular risk in CKD [100].

LCZ696 lowered BP more effectively than valsartan in hypertensive patients [101]. The PARAMOUNT (Prospective comparison of ARNI with ARB on Management Of heart failUre with preserved ejectioN fracTion, HFpEF) trial demonstrated that compared to valsartan, LCZ696 improved the overall clinical status and decreased atrial pressure and elevated GFR as well [102]. In HF patients with reduced ejection fraction (HFrEF), LCZ696 was reported to be more effective than enalapril in reducing hospitalization and cardiovascular and sudden death, preventing HF progression and improving renal function, as well as quality of life, in the Prospective comparison of ARNI with the ACE inhibitor to Determine Impact on Global Mortality and Morbidity in Heart Failure (PARADIGM-HF) trial [103–106]. Based on the myriads of favourable results of this trial, LCZ696 has been approved by the FDA in 2015 for treating HFrEF. ARNI has been shown in several studies including the PARADIGM-HF trial, to improve kidney function compared with RAAS inhibitor in HF [102,106,107]. Recently, an experimental study using a mouse myocardial infarction (MI) model and an in vitro mouse peritoneal macrophage, demonstrated that LCZ696 was associated with a better balance between the RAA and NP systems, and attenuated cardiac rupture following MI, suggesting that LCZ696 by its dual regulating mechanisms inhibited the inflammation and degradation response of macrophages and that early treatment with LCZ696 might have a cardioprotective effect after MI [108].

In spite of having numerous benefits of ARNi so far, however, some concerns raised that limits its clinical use, such as the elevation of bradykinin levels with ARB, the development of angioedema [109], and the chance of inducing Alzheimer disease by blocking the breakdown of amyloid-β [110]. Importantly, since sacubitril is predominantly excreted in the kidney in patients with renal impairment, it may accumulate and induce the development of hypotension particularly in patients with borderline BP [74].

7. Vasopressor and Vasodepressor Derived from Kidneys

Kidneys play an important role in regulating blood pressure. In addition to volume control via urination with balanced water and electrolytes excretion, the kidneys possess several substances that are used to control blood pressure [111] (Figure 3). The most well-known substance is derived from the renal cortex. Tigerstedt and Bergman demonstrated that the injection of cortical extracts from rabbit kidneys caused a blood pressure elevation in the recipient rabbits, which led to the subsequent discovery of renin, a powerful vasopressor [112].

Endothelin-1 (ET-1), a 21 amino acid vasoconstrictor, is also produced by the endothelial cells, tubular cells and inner medullary collecting duct cells [113]. In physiological states, increased salt intake induces the tubular production of ET-1, which inhibits epithelial sodium channels and causes a reduced sodium reabsorption rate that leads to natriuresis. However, in pathophysiological states, ET-1 has been considered to be a vasopressor due to the development of glomerular and interstitial kidney and vascular diseases [114].

Prostaglandin (PG) is a lipid mediator involved in a variety of physiological and pathological processes in the kidneys [115]. PGE2 is the most abundant renal arachidonic acid metabolite. PGE2 is produced by the microsomal PGE synthase in the macula densa, distal convoluted tubule, collecting duct and renal medullary interstitial cells (RMIC) [116]. Although there are still conflicting conclusions, PGE2 is generally considered to be a vasodepressor due to its diuretic effect [117].

Tissue kallikrein is mainly expressed in the submandibular gland, pancreas and kidneys. Kallikreins are primarily produced in the distal convoluted tubule [118,119], the cortical collecting tubule [119,120], and possibly in the collecting tubule in the inner medulla [120]. Tissue kallikrein releases kinins from low and high molecular-weight kininogen. Kinins are inactivated kininases and the best-known kininase is the angiotensin-converting enzyme (ACE). In humans, tissue kallikrein releases Lys-bradykinin (kallidin). The kallikrein-kinin system increases the renal blood flow leading to more urination, water and sodium excretion [121] in collaboration with its direct effect on the distal nephron [122]. It has been reported that a reduced urinary kallikrein is associated with the development of hypertension [123–125]. Thus, kallikrein is considered to be a vasodepressor.

It has been reported by several independent groups that the renal medulla possesses a depressor substance that acts as a counterpart to RAAS [126,127]. This depressor agent is produced from RMICs. RMICs contain intracellular endocrine granules, which are reduced in size in proportion to blood pressure and renal blood flow [128]. Although the inclusions have not been correctly identified, electron microscopy studies have revealed that the contents consist of free fatty acids and PGs. However, the depressor substance does not seem to be a PG, nitric oxide or a platelet-activating factor [126,129]. The renal papillary RMICs have abundant granules when a kidney is clipped, and they have been shown to degranulate after unclipping in a one-kidney and one-clip hypertensive rat model [130]. It remains unknown whether the renal papillary tip contains the same substance the researchers have been trying to identify in the renal medulla, as most researchers did not distinguish between the papillary tip and the entire inner medulla.

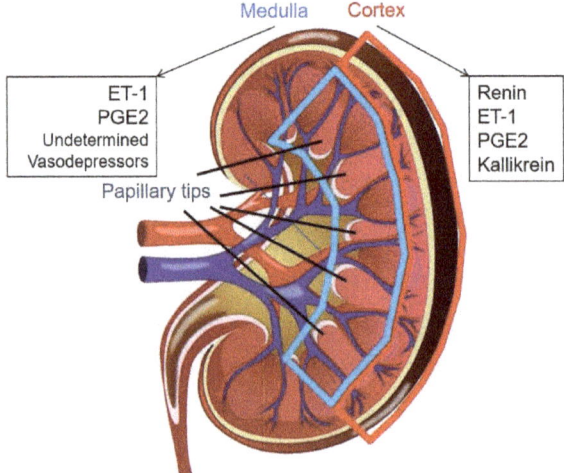

Figure 3. The vasopressors and vasodepressors derived from the kidneys. ET-1, endothelin 1; PGE2, prostaglandin E2.

8. Renal Papillary Tip May Contribute to the Expression of BNP in Cardiomyocytes

We recently developed a BNP reporter mouse carrying tdTomato under the mouse promoter of the 1136-bp fragment of the mouse *NPPB* gene from −1000 to +136 and demonstrated that this promoter was specifically activated in the papillary tips of the kidneys and was not accompanied by the BNP mRNA expression [25]. No evidence has been found that shows the existence of BNP isoforms or other nucleotide expressions apart from BNP and tdTomato. After the treatment with the extract from the renal papillary tip, both the expression and the secretion of BNP unexpectedly increased in the primary cultured neonatal cardiomyocytes. Although it is possible that artefacts due to contamination could occur, we found that there was no change in the expression of Ang II, ET-1, and type A, B and C NPs between the papillary tip and other portions of the kidneys. Even though its mechanism remains unknown, we initially evaluated elderly female mice as ageing and the female sex contribute to the expression of BNP in both normal subjects and patients with CHF [131–134]. However, we observed a similar activation of the BNP promoter in the papillary tips from young adults and/or male adult mice, although this was not recognized in neonatal mice [25].

The pBNP-tdTomato-positive cells were interstitial cells and were not proliferative. The papillary medulla has been reported to possess the ability to decrease blood pressure due to its vasodilatory activity [126,127]. To evaluate this activity of the papillary medulla from the kidneys, we injected an extract of the papillary tip intraperitoneally into stroke-prone spontaneously hypertensive rats (SHR-SPs). Intraperitoneal injection of the papillary extract reduced blood pressure from 210 mmHg to 165 mmHg and this was accompanied by an increase in serum BNP and urinary cGMP production in SHR-SP rats. Furthermore, the treatment with the papillary extract from rats with heart failure due to myocardial infarction significantly induced BNP expression in cardiomyocytes [25].

9. Conclusions

BNP plays an important role as a major player in the heart-kidney connection via its inhibitory effect on the RAAS, especially in the heart and the kidneys. Kidneys possess several substances involved in regulating the blood pressure in addition to volume control via urination. Furthermore, the papillary tips may play important roles in regulating the BNP expression from cardiomyocytes. Additional investigations will need to be undertaken in order to determine the relationship between the renal depressor system and BNP regulation, especially in terms of cardiovascular diseases, such as heart failure, hypertension and CKD.

Funding: This work is supported in part by Grants-in-Aid for Scientific Research from the Ministry of Education, Culture, Sports, Science and Technology, Japan (No. 19K08578 to R.O.) and the Okasan-Kato Foundation (to R.O.).

Acknowledgments: We thank Rie Ito and Miho Hisamura for their excellent technical assistance. The Department of Cardiology and Nephrology, Mie University Graduate School of Medicine, received research grants from Bristol-Myers Squibb, MSD K.K., Pfizer Japan Inc., Takeda Pharmaceutical Co., Ltd., Astellas Pharma Inc., Daiichi Sankyo Pharmaceutical Co., Ltd., Genzyme Japan, Shionogi & Co., Ltd., Sumitomo Dainippon Pharma Co., Ltd., Mitsubishi Tanabe Corporation, Otsuka Pharmaceutical Co., Ltd., Bayer Yakuhin, Ltd., AstraZeneca K.K., and Boehringer Ingelheim Co., Ltd. Masaaki Ito received lecture fees from Daiichi Sankyo Co. Pharmaceutical Co., Ltd., Mitsubishi Tanabe Corporation, Bayer Yakuhin, Ltd. and Takeda Pharmaceutical Co., Ltd.

Conflicts of Interest: The authors declare no conflict of interest.

Abbreviations

ANP	atrial natriuretic peptide
BNP	brain natriuretic peptide
cGMP	cyclic guanosine monophosphate
CHF	congestive heart failure
CKD	chronic kidney disease
CRS	cardiorenal syndrome
GFR	glomerular filtration rate
NP	natriuretic peptide
NPR	natriuretic peptide receptor
RAAS	renin-angiotensin-aldosterone system
RMIC	renal medullary interstitial cell
RPF	renal plasma flow
SNS	sympathetic nervous system

References

1. Sudoh, T.; Kangawa, K.; Minamino, N.; Matsuo, H. A new natriuretic peptide in porcine brain. *Nature* **1988**, *332*, 78–81. [CrossRef] [PubMed]
2. LaPointe, M.C. Molecular regulation of the brain natriuretic peptide gene. *Peptides* **2005**, *26*, 944–956. [CrossRef] [PubMed]

3. Volpe, M. Natriuretic peptides and cardio-renal disease. *Int. J. Cardiol.* **2014**, *176*, 630–639. [CrossRef] [PubMed]
4. Sabatine, M.S.; Morrow, D.A.; de Lemos, J.A.; Omland, T.; Desai, M.Y.; Tanasijevic, M.; Hall, C.; McCabe, C.H.; Braunwald, E. Acute changes in circulating natriuretic peptide levels in relation to myocardial ischemia. *J. Am. Coll. Cardiol.* **2004**, *44*, 1988–1995. [CrossRef] [PubMed]
5. Jensen, K.T.; Eiskjaer, H.; Carstens, J.; Pedersen, E.B. Renal effects of brain natriuretic peptide in patients with congestive heart failure. *Clin. Sci.* **1999**, *96*, 5–15. [CrossRef] [PubMed]
6. Nishikimi, T.; Kuwahara, K.; Nakao, K. Current biochemistry, molecular biology, and clinical relevance of natriuretic peptides. *J. Cardiol.* **2011**, *57*, 131–140. [CrossRef]
7. Levin, E.R.; Gardner, D.G.; Samson, W.K. Natriuretic peptides. *N. Engl. J. Med.* **1998**, *339*, 321–328.
8. Brunner-La Rocca, H.P.; Kaye, D.M.; Woods, R.L.; Hastings, J.; Esler, M.D. Effects of intravenous brain natriuretic peptide on regional sympathetic activity in patients with chronic heart failure as compared with healthy control subjects. *J. Am. Coll. Cardiol.* **2001**, *37*, 1221–1227. [CrossRef]
9. Holmes, S.J.; Espiner, E.A.; Richards, A.M.; Yandle, T.G.; Frampton, C. Renal, endocrine, and hemodynamic effects of human brain natriuretic peptide in normal man. *J. Clin. Endocrinol. Metab.* **1993**, *76*, 91–96.
10. Kohno, M.; Yokokawa, K.; Horio, T.; Yasunari, K.; Murakawa, K.; Takeda, T. Atrial and brain natriuretic peptides inhibit the endothelin-1 secretory response to angiotensin II in porcine aorta. *Circ. Res.* **1992**, *70*, 241–247. [CrossRef]
11. De Arriba, G.; Barrio, V.; Olivera, A.; Rodriguez-Puyol, D.; Lopez-Novoa, J.M. Atrial natriuretic peptide inhibits angiotensin II-induced contraction of isolated glomeruli and cultured glomerular mesangial cells of rats: The role of calcium. *J. Lab. Clin. Med.* **1988**, *111*, 466–474. [PubMed]
12. Ballermann, B.J.; Hoover, R.L.; Karnovsky, M.J.; Brenner, B.M. Physiologic regulation of atrial natriuretic peptide receptors in rat renal glomeruli. *J. Clin. Investig.* **1985**, *76*, 2049–2056. [CrossRef] [PubMed]
13. Houben, A.J.; van der Zander, K.; de Leeuw, P.W. Vascular and renal actions of brain natriuretic peptide in man: Physiology and pharmacology. *Fundam. Clin. Pharmacol.* **2005**, *19*, 411–419. [CrossRef] [PubMed]
14. Doust, J.; Lehman, R.; Glasziou, P. The role of BNP testing in heart failure. *Am. Fam. Physician* **2006**, *74*, 1893–1898. [PubMed]
15. Brenner, B.M.; Ballermann, B.J.; Gunning, M.E.; Zeidel, M.L. Diverse biological actions of atrial natriuretic peptide. *Physiol. Rev.* **1990**, *70*, 665–699. [CrossRef] [PubMed]
16. Rademaker, M.T.; Richards, A.M. Cardiac natriuretic peptides for cardiac health. *Clin. Sci.* **2005**, *108*, 23–36. [CrossRef] [PubMed]
17. Semenov, A.G.; Katrukha, A.G. Analytical Issues with Natriuretic Peptides—Has this been Overly Simplified? *EJIFCC* **2016**, *27*, 189–207.
18. Kitakaze, M.; Asakura, M.; Kim, J.; Shintani, Y.; Asanuma, H.; Hamasaki, T.; Seguchi, O.; Myoishi, M.; Minamino, T.; Ohara, T.; et al. Human atrial natriuretic peptide and nicorandil as adjuncts to reperfusion treatment for acute myocardial infarction (J-WIND): Two randomised trials. *Lancet* **2007**, *370*, 1483–1493. [CrossRef]
19. Spanaus, K.S.; Kronenberg, F.; Ritz, E.; Schlapbach, R.; Fliser, D.; Hersberger, M.; Kollerits, B.; Konig, P.; von Eckardstein, A. B-type natriuretic peptide concentrations predict the progression of nondiabetic chronic kidney disease: The Mild-to-Moderate Kidney Disease Study. *Clin. Chem.* **2007**, *53*, 1264–1272. [CrossRef]
20. Van Kimmenade, R.R.; Januzzi, J.L., Jr.; Bakker, J.A.; Houben, A.J.; Rennenberg, R.; Kroon, A.A.; Crijns, H.J.; van Dieijen-Visser, M.P.; de Leeuw, P.W.; Pinto, Y.M. Renal clearance of B-type natriuretic peptide and amino terminal pro-B-type natriuretic peptide a mechanistic study in hypertensive subjects. *J. Am. Coll. Cardiol.* **2009**, *53*, 884–890. [CrossRef]
21. Takase, H.; Dohi, Y. Kidney function crucially affects B-type natriuretic peptide (BNP), N-terminal proBNP and their relationship. *Eur. J. Clin. Investig.* **2014**, *44*, 303–308. [CrossRef] [PubMed]
22. Niizuma, S.; Iwanaga, Y.; Yahata, T.; Miyazaki, S. Renocardiovascular Biomarkers: From the Perspective of Managing Chronic Kidney Disease and Cardiovascular Disease. *Front. Cardiovasc. Med.* **2017**, *4*, 10. [CrossRef] [PubMed]
23. Dos Reis, D.; Fraticelli, L.; Bassand, A.; Manzo-Silberman, S.; Peschanski, N.; Charpentier, S.; Elbaz, M.; Savary, D.; Bonnefoy-Cudraz, E.; Laribi, S.; et al. Impact of renal dysfunction on the management and outcome of acute heart failure: Results from the French prospective, multicentre, DeFSSICA survey. *BMJ Open* **2019**, *9*, e022776. [CrossRef] [PubMed]

24. Volpe, M.; Carnovali, M.; Mastromarino, V. The natriuretic peptides system in the pathophysiology of heart failure: From molecular basis to treatment. *Clin. Sci.* **2016**, *130*, 57–77. [CrossRef] [PubMed]
25. Goto, I.; Okamoto, R.; Hashizume, R.; Suzuki, N.; Ito, R.; Yamanaka, K.; Saito, H.; Kiyonari, H.; Tawara, I.; Kageyama, Y.; et al. Renal papillary tip extract stimulates BNP production and excretion from cardiomyocytes. *PLoS ONE* **2018**, *13*, e0197078. [CrossRef] [PubMed]
26. Clerico, A.; Passino, C.; Franzini, M.; Emdin, M. Cardiac biomarker testing in the clinical laboratory: Where do we stand? General overview of the methodology with special emphasis on natriuretic peptides. *Clin. Chim. Acta* **2015**, *443*, 17–24. [CrossRef] [PubMed]
27. Waldo, S.W.; Beede, J.; Isakson, S.; Villard-Saussine, S.; Fareh, J.; Clopton, P.; Fitzgerald, R.L.; Maisel, A.S. Pro-B-type natriuretic peptide levels in acute decompensated heart failure. *J. Am. Coll. Cardiol.* **2008**, *51*, 1874–1882. [CrossRef]
28. Santaguida, P.L.; Don-Wauchope, A.C.; Oremus, M.; McKelvie, R.; Ali, U.; Hill, S.A.; Balion, C.; Booth, R.A.; Brown, J.A.; Bustamam, A.; et al. BNP and NT-proBNP as prognostic markers in persons with acute decompensated heart failure: A systematic review. *Heart Fail. Rev.* **2014**, *19*, 453–470. [CrossRef]
29. Mueller, T.; Gegenhuber, A.; Poelz, W.; Haltmayer, M. Head-to-head comparison of the diagnostic utility of BNP and NT-proBNP in symptomatic and asymptomatic structural heart disease. *Clin. Chim. Acta* **2004**, *341*, 41–48. [CrossRef]
30. Clerico, A.; Zaninotto, M.; Passino, C.; Plebani, M. New issues on measurement of B-type natriuretic peptides. *Clin. Chem. Lab. Med.* **2017**, *56*, 32–39. [CrossRef]
31. Kuwahara, K.; Nakagawa, Y.; Nishikimi, T. Cutting Edge of Brain Natriuretic Peptide (BNP) Research- The Diversity of BNP Immunoreactivity and Its Clinical Relevance. *Circ. J.* **2018**, *82*, 2455–2461. [CrossRef] [PubMed]
32. Rubattu, S.; Forte, M.; Marchitti, S.; Volpe, M. Molecular Implications of Natriuretic Peptides in the Protection from Hypertension and Target Organ Damage Development. *Int. J. Mol. Sci.* **2019**, *20*, 798. [CrossRef] [PubMed]
33. Santos-Araujo, C.; Leite-Moreira, A.; Pestana, M. Clinical value of natriuretic peptides in chronic kidney disease. *Nefrologia* **2015**, *35*, 227–233. [CrossRef]
34. Vanderheyden, M.; Bartunek, J.; Filippatos, G.; Goethals, M.; Vlem, B.V.; Maisel, A. Cardiovascular disease in patients with chronic renal impairment: Role of natriuretic peptides. *Congest. Heart Fail.* **2008**, *14*, 38–42. [CrossRef]
35. Jensen, K.T.; Carstens, J.; Pedersen, E.B. Effect of BNP on renal hemodynamics, tubular function and vasoactive hormones in humans. *Am. J. Physiol.* **1998**, *274*, F63–F72. [CrossRef] [PubMed]
36. McGregor, A.; Richards, M.; Espiner, E.; Yandle, T.; Ikram, H. Brain natriuretic peptide administered to man: Actions and metabolism. *J. Clin. Endocrinol. Metab.* **1990**, *70*, 1103–1107. [CrossRef] [PubMed]
37. Yoshimura, M.; Yasue, H.; Morita, E.; Sakaino, N.; Jougasaki, M.; Kurose, M.; Mukoyama, M.; Saito, Y.; Nakao, K.; Imura, H. Hemodynamic, renal, and hormonal responses to brain natriuretic peptide infusion in patients with congestive heart failure. *Circulation* **1991**, *84*, 1581–1588. [CrossRef]
38. Cheung, B.M.; Dickerson, J.E.; Ashby, M.J.; Brown, M.J.; Brown, J. Effects of physiological increments in human alpha-atrial natriuretic peptide and human brain natriuretic peptide in normal male subjects. *Clin. Sci.* **1994**, *86*, 723–730. [CrossRef]
39. Florkowski, C.M.; Richards, A.M.; Espiner, E.A.; Yandle, T.G.; Frampton, C. Renal, endocrine, and hemodynamic interactions of atrial and brain natriuretic peptides in normal men. *Am. J. Physiol.* **1994**, *266*, R1244–R1250. [CrossRef]
40. La Villa, G.; Fronzaroli, C.; Lazzeri, C.; Porciani, C.; Bandinelli, R.; Vena, S.; Messeri, G.; Franchi, F. Cardiovascular and renal effects of low dose brain natriuretic peptide infusion in man. *J. Clin. Endocrinol. Metab.* **1994**, *78*, 1166–1171.
41. La Villa, G.; Stefani, L.; Lazzeri, C.; Zurli, C.; Guerra, C.T.; Barletta, G.; Bandinelli, R.; Strazzulla, G.; Franchi, F. Acute effects of physiological increments of brain natriuretic peptide in humans. *Hypertension* **1995**, *26*, 628–633. [CrossRef] [PubMed]
42. Lazzeri, C.; La Villa, G.; Bisi, G.; Boddi, V.; Messeri, G.; Strazzulla, G.; Franchi, F. Cardiovascular function during brain natriuretic peptide infusion in man. *Cardiology* **1995**, *86*, 396–401. [CrossRef] [PubMed]

43. Hunt, P.J.; Espiner, E.A.; Nicholls, M.G.; Richards, A.M.; Yandle, T.G. Differing biological effects of equimolar atrial and brain natriuretic peptide infusions in normal man. *J. Clin. Endocrinol. Metab.* **1996**, *81*, 3871–3876. [PubMed]
44. Yasue, H.; Yoshimura, M. Natriuretic peptides in the treatment of heart failure. *J. Card. Fail.* **1996**, *2*, S277–S285. [CrossRef]
45. Van der Zander, K.; Houben, A.J.; Hofstra, L.; Kroon, A.A.; de Leeuw, P.W. Hemodynamic and renal effects of low-dose brain natriuretic peptide infusion in humans: A randomized, placebo-controlled crossover study. *Am. J. Physiol. Heart Circ. Physiol.* **2003**, *285*, H1206–H1212. [CrossRef] [PubMed]
46. Marcus, L.S.; Hart, D.; Packer, M.; Yushak, M.; Medina, N.; Danziger, R.S.; Heitjan, D.F.; Katz, S.D. Hemodynamic and renal excretory effects of human brain natriuretic peptide infusion in patients with congestive heart failure. A double-blind, placebo-controlled, randomized crossover trial. *Circulation* **1996**, *94*, 3184–3189. [CrossRef] [PubMed]
47. Lainchbury, J.G.; Richards, A.M.; Nicholls, M.G.; Hunt, P.J.; Ikram, H.; Espiner, E.A.; Yandle, T.G.; Begg, E. The effects of pathophysiological increments in brain natriuretic peptide in left ventricular systolic dysfunction. *Hypertension* **1997**, *30*, 398–404. [CrossRef]
48. Abraham, W.T.; Lowes, B.D.; Ferguson, D.A.; Odom, J.; Kim, J.K.; Robertson, A.D.; Bristow, M.R.; Schrier, R.W. Systemic hemodynamic, neurohormonal, and renal effects of a steady-state infusion of human brain natriuretic peptide in patients with hemodynamically decompensated heart failure. *J. Card. Fail.* **1998**, *4*, 37–44. [CrossRef]
49. Wang, Y.; Gu, X.; Fan, W.; Fan, Y.; Li, W.; Fu, X. Effects of recombinant human brain natriuretic peptide on renal function in patients with acute heart failure following myocardial infarction. *Am. J. Transl. Res.* **2016**, *8*, 239–245.
50. Richards, A.M.; Crozier, I.G.; Holmes, S.J.; Espiner, E.A.; Yandle, T.G.; Frampton, C. Brain natriuretic peptide: Natriuretic and endocrine effects in essential hypertension. *J. Hypertens.* **1993**, *11*, 163–170. [CrossRef]
51. Lazzeri, C.; Franchi, F.; Porciani, C.; Fronzaroli, C.; Casini Raggi, V.; De Feo, M.L.; Mannelli, M.; Cersosimo, R.M.; La Villa, G. Systemic hemodynamics and renal function during brain natriuretic peptide infusion in patients with essential hypertension. *Am. J. Hypertens.* **1995**, *8*, 799–807. [CrossRef]
52. Pidgeon, G.B.; Richards, A.M.; Nicholls, M.G.; Espiner, E.A.; Yandle, T.G.; Frampton, C. Differing metabolism and bioactivity of atrial and brain natriuretic peptides in essential hypertension. *Hypertension* **1996**, *27*, 906–913. [CrossRef] [PubMed]
53. Kamper, A.L.; Holstein-Rathlou, N.H.; Leyssac, P.P.; Strandgaard, S. Lithium clearance in chronic nephropathy. *Clin. Sci.* **1989**, *77*, 311–318. [CrossRef] [PubMed]
54. Totsune, K.; Takahashi, K.; Murakami, O.; Satoh, F.; Sone, M.; Saito, T.; Sasano, H.; Mouri, T.; Abe, K. Natriuretic peptides in the human kidney. *Hypertension* **1994**, *24*, 758–762. [CrossRef] [PubMed]
55. Gottlieb, S.S.; Stebbins, A.; Voors, A.A.; Hasselblad, V.; Ezekowitz, J.A.; Califf, R.M.; O'Connor, C.M.; Starling, R.C.; Hernandez, A.F. Effects of nesiritide and predictors of urine output in acute decompensated heart failure: Results from ASCEND-HF (acute study of clinical effectiveness of nesiritide and decompensated heart failure). *J. Am. Coll. Cardiol.* **2013**, *62*, 1177–1183. [CrossRef] [PubMed]
56. Kelesidis, I.; Mazurek, J.; Khullar, P.; Saeed, W.; Vittorio, T.; Zolty, R. The effect of nesiritide on renal function and other clinical parameters in patients with decompensated heart failure and preserved ejection fraction. *Congest. Heart Fail.* **2012**, *18*, 158–164. [CrossRef] [PubMed]
57. Fontoura, B.M.; Nussenzveig, D.R.; Pelton, K.M.; Maack, T. Atrial natriuretic factor receptors in cultured renomedullary interstitial cells. *Am. J. Physiol.* **1990**, *258*, C692–C699. [CrossRef] [PubMed]
58. Verbrugge, F.H.; Dupont, M.; Steels, P.; Grieten, L.; Swennen, Q.; Tang, W.H.; Mullens, W. The kidney in congestive heart failure: 'Are natriuresis, sodium, and diuretics really the good, the bad and the ugly?'. *Eur. J. Heart Fail.* **2014**, *16*, 133–142. [CrossRef]
59. Cataliotti, A.; Boerrigter, G.; Costello-Boerrigter, L.C.; Schirger, J.A.; Tsuruda, T.; Heublein, D.M.; Chen, H.H.; Malatino, L.S.; Burnett, J.C., Jr. Brain natriuretic peptide enhances renal actions of furosemide and suppresses furosemide-induced aldosterone activation in experimental heart failure. *Circulation* **2004**, *109*, 1680–1685. [CrossRef]
60. McKie, P.M.; Schirger, J.A.; Benike, S.L.; Harstad, L.K.; Slusser, J.P.; Hodge, D.O.; Redfield, M.M.; Burnett, J.C., Jr.; Chen, H.H. Chronic subcutaneous brain natriuretic peptide therapy in asymptomatic systolic heart failure. *Eur. J. Heart Fail.* **2016**, *18*, 433–441. [CrossRef]

61. Liu, J.; Xie, Y.; He, F.; Gao, Z.; Hao, Y.; Zu, X.; Chang, L.; Li, Y. Recombinant Brain Natriuretic Peptide for the Prevention of Contrast-Induced Nephropathy in Patients with Chronic Kidney Disease Undergoing Nonemergent Percutaneous Coronary Intervention or Coronary Angiography: A Randomized Controlled Trial. *Biomed Res. Int.* **2016**, *2016*, 5985327. [CrossRef] [PubMed]
62. Woods, R.L.; Jones, M.J. Atrial, B-type, and C-type natriuretic peptides cause mesenteric vasoconstriction in conscious dogs. *Am. J. Physiol.* **1999**, *276*, R1443–R1452. [CrossRef] [PubMed]
63. Marin-Grez, M.; Fleming, J.T.; Steinhausen, M. Atrial natriuretic peptide causes pre-glomerular vasodilatation and post-glomerular vasoconstriction in rat kidney. *Nature* **1986**, *324*, 473–476. [CrossRef] [PubMed]
64. Santos-Araujo, C.; Roncon-Albuquerque, R., Jr.; Moreira-Rodrigues, M.; Henriques-Coelho, T.; Quelhas-Santos, J.; Faria, B.; Sampaio-Maia, B.; Leite-Moreira, A.F.; Pestana, M. Local modulation of the natriuretic peptide system in the rat remnant kidney. *Nephrol. Dial. Transplant.* **2009**, *24*, 1774–1782. [CrossRef] [PubMed]
65. Sackner-Bernstein, J.D.; Skopicki, H.A.; Aaronson, K.D. Risk of worsening renal function with nesiritide in patients with acutely decompensated heart failure. *Circulation* **2005**, *111*, 1487–1491. [CrossRef] [PubMed]
66. Anand-Srivastava, M.B. Natriuretic peptide receptor-C signaling and regulation. *Peptides* **2005**, *26*, 1044–1059. [CrossRef] [PubMed]
67. Porter, J.G.; Arfsten, A.; Fuller, F.; Miller, J.A.; Gregory, L.C.; Lewicki, J.A. Isolation and functional expression of the human atrial natriuretic peptide clearance receptor cDNA. *Biochem. Biophys. Res. Commun.* **1990**, *171*, 796–803. [CrossRef]
68. Wilcox, J.N.; Augustine, A.; Goeddel, D.V.; Lowe, D.G. Differential regional expression of three natriuretic peptide receptor genes within primate tissues. *Mol. Cell. Biol.* **1991**, *11*, 3454–3462. [CrossRef]
69. He, X.; Chow, D.; Martick, M.M.; Garcia, K.C. Allosteric activation of a spring-loaded natriuretic peptide receptor dimer by hormone. *Science* **2001**, *293*, 1657–1662. [CrossRef]
70. Jalal, F.; Dehbi, M.; Berteloot, A.; Crine, P. Biosynthesis and polarized distribution of neutral endopeptidase in primary cultures of kidney proximal tubule cells. *Biochem. J.* **1994**, *302*, 669–674. [CrossRef]
71. Dickey, D.M.; Potter, L.R. Human B-type natriuretic peptide is not degraded by meprin A. *Biochem. Pharmacol.* **2010**, *80*, 1007–1011. [CrossRef] [PubMed]
72. Smith, M.W.; Espiner, E.A.; Yandle, T.G.; Charles, C.J.; Richards, A.M. Delayed metabolism of human brain natriuretic peptide reflects resistance to neutral endopeptidase. *J. Endocrinol.* **2000**, *167*, 239–246. [CrossRef] [PubMed]
73. Ralat, L.A.; Guo, Q.; Ren, M.; Funke, T.; Dickey, D.M.; Potter, L.R.; Tang, W.J. Insulin-degrading enzyme modulates the natriuretic peptide-mediated signaling response. *J. Biol. Chem.* **2011**, *286*, 4670–4679. [CrossRef] [PubMed]
74. Fu, S.; Ping, P.; Wang, F.; Luo, L. Synthesis, secretion, function, metabolism and application of natriuretic peptides in heart failure. *J. Biol. Eng.* **2018**, *12*, 2. [CrossRef] [PubMed]
75. Linssen, G.C.; Damman, K.; Hillege, H.L.; Navis, G.; van Veldhuisen, D.J.; Voors, A.A. Urinary N-terminal prohormone brain natriuretic peptide excretion in patients with chronic heart failure. *Circulation* **2009**, *120*, 35–41. [CrossRef] [PubMed]
76. Jungbauer, C.G.; Buchner, S.; Birner, C.; Resch, M.; Heinicke, N.; Debl, K.; Buesing, M.; Biermeier, D.; Schmitz, G.; Riegger, G.; et al. N-terminal pro-brain natriuretic peptide from fresh urine for the biochemical detection of heart failure and left ventricular dysfunction. *Eur. J. Heart Fail.* **2010**, *12*, 331–337. [CrossRef] [PubMed]
77. Wong, P.C.; Guo, J.; Zhang, A. The renal and cardiovascular effects of natriuretic peptides. *Adv. Physiol. Educ.* **2017**, *41*, 179–185. [CrossRef]
78. Volpe, M.; Battistoni, A.; Mastromarino, V. Natriuretic peptides and volume handling in heart failure: The paradigm of a new treatment. *Eur. J. Heart Fail.* **2016**, *18*, 442–444. [CrossRef]
79. Von Lueder, T.G.; Sangaralingham, S.J.; Wang, B.H.; Kompa, A.R.; Atar, D.; Burnett, J.C., Jr.; Krum, H. Renin-angiotensin blockade combined with natriuretic peptide system augmentation: Novel therapeutic concepts to combat heart failure. *Circ. Heart Fail.* **2013**, *6*, 594–605. [CrossRef]
80. Corti, R.; Burnett, J.C., Jr.; Rouleau, J.L.; Ruschitzka, F.; Luscher, T.F. Vasopeptidase inhibitors: A new therapeutic concept in cardiovascular disease? *Circulation* **2001**, *104*, 1856–1862. [CrossRef]

81. Tokudome, T.; Kishimoto, I.; Horio, T.; Arai, Y.; Schwenke, D.O.; Hino, J.; Okano, I.; Kawano, Y.; Kohno, M.; Miyazato, M.; et al. Regulator of G-protein signaling subtype 4 mediates antihypertrophic effect of locally secreted natriuretic peptides in the heart. *Circulation* **2008**, *117*, 2329–2339. [CrossRef] [PubMed]
82. Akabane, S.; Matsushima, Y.; Matsuo, H.; Kawamura, M.; Imanishi, M.; Omae, T. Effects of brain natriuretic peptide on renin secretion in normal and hypertonic saline-infused kidney. *Eur. J. Pharmacol.* **1991**, *198*, 143–148. [CrossRef]
83. Kurtz, A.; Della Bruna, R.; Pfeilschifter, J.; Taugner, R.; Bauer, C. Atrial natriuretic peptide inhibits renin release from juxtaglomerular cells by a cGMP-mediated process. *Proc. Natl. Acad. Sci. USA* **1986**, *83*, 4769–4773. [CrossRef] [PubMed]
84. Ito, T.; Yoshimura, M.; Nakamura, S.; Nakayama, M.; Shimasaki, Y.; Harada, E.; Mizuno, Y.; Yamamuro, M.; Harada, M.; Saito, Y.; et al. Inhibitory effect of natriuretic peptides on aldosterone synthase gene expression in cultured neonatal rat cardiocytes. *Circulation* **2003**, *107*, 807–810. [CrossRef] [PubMed]
85. Liang, F.; Kapoun, A.M.; Lam, A.; Damm, D.L.; Quan, D.; O'Connell, M.; Protter, A.A. B-Type natriuretic peptide inhibited angiotensin II-stimulated cholesterol biosynthesis, cholesterol transfer, and steroidogenesis in primary human adrenocortical cells. *Endocrinology* **2007**, *148*, 3722–3729. [CrossRef] [PubMed]
86. Takekoshi, K.; Ishii, K.; Isobe, K.; Nomura, F.; Nammoku, T.; Nakai, T. Effects of natriuretic peptides (ANP, BNP, CNP) on catecholamine synthesis and TH mRNA levels in PC12 cells. *Life Sci.* **2000**, *66*, PL303–PL311. [CrossRef]
87. Han, B.; Hasin, Y. Cardiovascular effects of natriuretic peptides and their interrelation with endothelin-1. *Cardiovasc. Drugs Ther.* **2003**, *17*, 41–52. [CrossRef] [PubMed]
88. Johnston, C.I.; Hodsman, P.G.; Kohzuki, M.; Casley, D.J.; Fabris, B.; Phillips, P.A. Interaction between atrial natriuretic peptide and the renin angiotensin aldosterone system. Endogenous antagonists. *Am. J. Med.* **1989**, *87*, 24S–28S. [PubMed]
89. Rossi, F.; Mascolo, A.; Mollace, V. The pathophysiological role of natriuretic peptide-RAAS cross talk in heart failure. *Int. J. Cardiol.* **2017**, *226*, 121–125. [CrossRef] [PubMed]
90. Filippatos, G.; Farmakis, D.; Parissis, J.; Lekakis, J. Drug therapy for patients with systolic heart failure after the PARADIGM-HF trial: In need of a new paradigm of LCZ696 implementation in clinical practice. *BMC Med.* **2015**, *13*, 35. [CrossRef] [PubMed]
91. Bayes-Genis, A.; Barallat, J.; Galan, A.; de Antonio, M.; Domingo, M.; Zamora, E.; Urrutia, A.; Lupon, J. Soluble neprilysin is predictive of cardiovascular death and heart failure hospitalization in heart failure patients. *J. Am. Coll. Cardiol.* **2015**, *65*, 657–665. [CrossRef] [PubMed]
92. Colucci, W.S.; Elkayam, U.; Horton, D.P.; Abraham, W.T.; Bourge, R.C.; Johnson, A.D.; Wagoner, L.E.; Givertz, M.M.; Liang, C.S.; Neibaur, M.; et al. Intravenous nesiritide, a natriuretic peptide, in the treatment of decompensated congestive heart failure. Nesiritide Study Group. *N. Engl. J. Med.* **2000**, *343*, 246–253. [CrossRef] [PubMed]
93. Partovian, C.; Li, S.X.; Xu, X.; Lin, H.; Strait, K.M.; Hwa, J.; Krumholz, H.M. Patterns of change in nesiritide use in patients with heart failure: How hospitals react to new information. *JACC Heart Fail.* **2013**, *1*, 318–324. [CrossRef]
94. Kostis, J.B.; Packer, M.; Black, H.R.; Schmieder, R.; Henry, D.; Levy, E. Omapatrilat and enalapril in patients with hypertension: The Omapatrilat Cardiovascular Treatment vs. Enalapril (OCTAVE) trial. *Am. J. Hypertens.* **2004**, *17*, 103–111. [CrossRef] [PubMed]
95. McCormack, P.L. Sacubitril/Valsartan: A Review in Chronic Heart Failure with Reduced Ejection Fraction. *Drugs* **2016**, *76*, 387–396. [CrossRef] [PubMed]
96. Volpe, M.; Rubattu, S.; Battistoni, A. ARNi: A Novel Approach to Counteract Cardiovascular Diseases. *Int. J. Mol. Sci.* **2019**, *20*, 2092. [CrossRef] [PubMed]
97. Braunwald, E. The path to an angiotensin receptor antagonist-neprilysin inhibitor in the treatment of heart failure. *J. Am. Coll. Cardiol.* **2015**, *65*, 1029–1041. [CrossRef]
98. Roksnoer, L.C.; van Veghel, R.; van Groningen, M.C.; de Vries, R.; Garrelds, I.M.; Bhaggoe, U.M.; van Gool, J.M.; Friesema, E.C.; Leijten, F.P.; Hoorn, E.J.; et al. Blood pressure-independent renoprotection in diabetic rats treated with AT1 receptor-neprilysin inhibition compared with AT1 receptor blockade alone. *Clin. Sci.* **2016**, *130*, 1209–1220. [CrossRef]

99. Von Lueder, T.G.; Wang, B.H.; Kompa, A.R.; Huang, L.; Webb, R.; Jordaan, P.; Atar, D.; Krum, H. Angiotensin receptor neprilysin inhibitor LCZ696 attenuates cardiac remodeling and dysfunction after myocardial infarction by reducing cardiac fibrosis and hypertrophy. *Circ. Heart Fail.* **2015**, *8*, 71–78. [CrossRef]
100. Haynes, R.; Judge, P.K.; Staplin, N.; Herrington, W.G.; Storey, B.C.; Bethel, A.; Bowman, L.; Brunskill, N.; Cockwell, P.; Hill, M.; et al. Effects of Sacubitril/Valsartan Versus Irbesartan in Patients With Chronic Kidney Disease. *Circulation* **2018**, *138*, 1505–1514. [CrossRef]
101. Ruilope, L.M.; Dukat, A.; Bohm, M.; Lacourciere, Y.; Gong, J.; Lefkowitz, M.P. Blood-pressure reduction with LCZ696, a novel dual-acting inhibitor of the angiotensin II receptor and neprilysin: A randomised, double-blind, placebo-controlled, active comparator study. *Lancet* **2010**, *375*, 1255–1266. [CrossRef]
102. Solomon, S.D.; Zile, M.; Pieske, B.; Voors, A.; Shah, A.; Kraigher-Krainer, E.; Shi, V.; Bransford, T.; Takeuchi, M.; Gong, J.; et al. The angiotensin receptor neprilysin inhibitor LCZ696 in heart failure with preserved ejection fraction: A phase 2 double-blind randomised controlled trial. *Lancet* **2012**, *380*, 1387–1395. [CrossRef]
103. McMurray, J.J.; Packer, M.; Desai, A.S.; Gong, J.; Lefkowitz, M.P.; Rizkala, A.R.; Rouleau, J.L.; Shi, V.C.; Solomon, S.D.; Swedberg, K.; et al. Angiotensin-neprilysin inhibition versus enalapril in heart failure. *N. Engl. J. Med.* **2014**, *371*, 993–1004. [CrossRef] [PubMed]
104. Vardeny, O.; Miller, R.; Solomon, S.D. Combined neprilysin and renin-angiotensin system inhibition for the treatment of heart failure. *JACC Heart Fail.* **2014**, *2*, 663–670. [CrossRef] [PubMed]
105. Desai, A.S.; McMurray, J.J.; Packer, M.; Swedberg, K.; Rouleau, J.L.; Chen, F.; Gong, J.; Rizkala, A.R.; Brahimi, A.; Claggett, B.; et al. Effect of the angiotensin-receptor-neprilysin inhibitor LCZ696 compared with enalapril on mode of death in heart failure patients. *Eur. Heart J.* **2015**, *36*, 1990–1997. [CrossRef] [PubMed]
106. Packer, M.; McMurray, J.J.; Desai, A.S.; Gong, J.; Lefkowitz, M.P.; Rizkala, A.R.; Rouleau, J.L.; Shi, V.C.; Solomon, S.D.; Swedberg, K.; et al. Angiotensin receptor neprilysin inhibition compared with enalapril on the risk of clinical progression in surviving patients with heart failure. *Circulation* **2015**, *131*, 54–61. [CrossRef]
107. Voors, A.A.; Gori, M.; Liu, L.C.; Claggett, B.; Zile, M.R.; Pieske, B.; McMurray, J.J.; Packer, M.; Shi, V.; Lefkowitz, M.P.; et al. Renal effects of the angiotensin receptor neprilysin inhibitor LCZ696 in patients with heart failure and preserved ejection fraction. *Eur. J. Heart Fail.* **2015**, *17*, 510–517. [CrossRef]
108. Ishii, M.; Kaikita, K.; Sato, K.; Sueta, D.; Fujisue, K.; Arima, Y.; Oimatsu, Y.; Mitsuse, T.; Onoue, Y.; Araki, S.; et al. Cardioprotective Effects of LCZ696 (Sacubitril/Valsartan) After Experimental Acute Myocardial Infarction. *JACC Basic Transl. Sci.* **2017**, *2*, 655–668. [CrossRef]
109. Campbell, D.J.; Krum, H.; Esler, M.D. Losartan increases bradykinin levels in hypertensive humans. *Circulation* **2005**, *111*, 315–320. [CrossRef]
110. Grimm, M.O.; Mett, J.; Stahlmann, C.P.; Haupenthal, V.J.; Zimmer, V.C.; Hartmann, T. Neprilysin and Abeta Clearance: Impact of the APP Intracellular Domain in NEP Regulation and Implications in Alzheimer's Disease. *Front. Aging Neurosci.* **2013**, *5*, 98. [CrossRef]
111. Kuwahara, M.; Marumo, F. Biosynthesis of hormones in renal tubular and interstitial cells. *Nihon Rinsho* **1995**, *53*, 1873–1878. [PubMed]
112. Marks, L.S.; Maxwell, M.H. Tigerstedt and the discovery of renin. An historical note. *Hypertension* **1979**, *1*, 384–388. [CrossRef] [PubMed]
113. De Miguel, C.; Speed, J.S.; Kasztan, M.; Gohar, E.Y.; Pollock, D.M. Endothelin-1 and the kidney: New perspectives and recent findings. *Curr. Opin. Nephrol. Hypertens.* **2016**, *25*, 35–41. [CrossRef] [PubMed]
114. Boesen, E.I. Endothelin receptors, renal effects and blood pressure. *Curr. Opin. Pharmacol.* **2015**, *21*, 25–34. [CrossRef]
115. Nasrallah, R.; Hassouneh, R.; Hebert, R.L. PGE2, Kidney Disease, and Cardiovascular Risk: Beyond Hypertension and Diabetes. *J. Am. Soc. Nephrol.* **2016**, *27*, 666–676. [CrossRef]
116. Li, Y.; Xia, W.; Zhao, F.; Wen, Z.; Zhang, A.; Huang, S.; Jia, Z.; Zhang, Y. Prostaglandins in the pathogenesis of kidney diseases. *Oncotarget* **2018**, *9*, 26586–26602. [CrossRef]
117. Li, Y.; Wei, Y.; Zheng, F.; Guan, Y.; Zhang, X. Prostaglandin E2 in the Regulation of Water Transport in Renal Collecting Ducts. *Int. J. Mol. Sci.* **2017**, *18*, 2539. [CrossRef]
118. Tomita, K.; Endou, H.; Sakai, F. Localization of kallikrein-like activity along a single nephron in rabbits. *Pflug. Arch.* **1981**, *389*, 91–95. [CrossRef]
119. Omata, K.; Carretero, O.A.; Scicli, A.G.; Jackson, B.A. Localization of active and inactive kallikrein (kininogenase activity) in the microdissected rabbit nephron. *Kidney Int.* **1982**, *22*, 602–607. [CrossRef]

120. Proud, D.; Perkins, M.; Pierce, J.V.; Yates, K.N.; Highet, P.F.; Herring, P.L.; Mangkornkanok/Mark, M.; Bahu, R.; Carone, F.; Pisano, J.J. Characterization and localization of human renal kininogen. *J. Biol. Chem.* **1981**, *256*, 10634–10639.
121. Rhaleb, N.E.; Yang, X.P.; Carretero, O.A. The kallikrein-kinin system as a regulator of cardiovascular and renal function. *Compr. Physiol.* **2011**, *1*, 971–993. [PubMed]
122. Kauker, M.L. Bradykinin action on the efflux of luminal 22Na in the rat nephron. *J. Pharmacol. Exp. Ther.* **1980**, *214*, 119–123. [PubMed]
123. Sinaiko, A.R.; Glasser, R.J.; Gillum, R.F.; Prineas, R.J. Urinary kallikrein excretion in grade school children with high and low blood pressure. *J. Pediatr.* **1982**, *100*, 938–940. [CrossRef]
124. Sharma, J.N.; Narayanan, P. The kallikrein-kinin pathways in hypertension and diabetes. *Prog. Drug Res.* **2014**, *69*, 15–36. [PubMed]
125. Wollheim, E.; Peterknecht, S.; Dees, C.; Wiener, A.; Wollheim, C.B. Defect in the excretion of a vasoactive polypeptide fraction A possible genetic marker of primary hypertension. *Hypertension* **1981**, *3*, 574–579. [CrossRef] [PubMed]
126. Thomas, C.J.; Woods, R.L.; Evans, R.G.; Alcorn, D.; Christy, I.J.; Anderson, W.P. Evidence for a renomedullary vasodepressor hormone. *Clin. Exp. Pharmacol. Physiol.* **1996**, *23*, 777–785. [CrossRef] [PubMed]
127. Muirhead, E.E. Renal vasodepressor mechanisms: The medullipin system. *J. Hypertens. Suppl.* **1993**, *11*, S53–S58. [CrossRef] [PubMed]
128. Maric, C.; Harris, P.J.; Alcorn, D. Changes in mean arterial pressure predict degranulation of renomedullary interstitial cells. *Clin. Exp. Pharmacol. Physiol.* **2002**, *29*, 1055–1059. [CrossRef]
129. Glodny, B.; Pauli, G.F. The vasodepressor function of the kidney: Prostaglandin E2 is not the principal vasodepressor lipid of the renal medulla. *Acta Physiol.* **2006**, *187*, 419–430. [CrossRef]
130. Pitcock, J.A.; Brown, P.S.; Byers, W.; Brooks, B.; Muirhead, E.E. Degranulation of renomedullary interstitial cells during reversal of hypertension. *Hypertension* **1981**, *3*, II-75. [CrossRef]
131. Wang, T.J.; Larson, M.G.; Levy, D.; Leip, E.P.; Benjamin, E.J.; Wilson, P.W.; Sutherland, P.; Omland, T.; Vasan, R.S. Impact of age and sex on plasma natriuretic peptide levels in healthy adults. *Am. J. Cardiol.* **2002**, *90*, 254–258. [CrossRef]
132. Redfield, M.M.; Rodeheffer, R.J.; Jacobsen, S.J.; Mahoney, D.W.; Bailey, K.R.; Burnett, J.C., Jr. Plasma brain natriuretic peptide concentration: Impact of age and gender. *J. Am. Coll. Cardiol.* **2002**, *40*, 976–982. [CrossRef]
133. Sayama, H.; Nakamura, Y.; Saito, N.; Kinoshita, M. Why is the concentration of plasma brain natriuretic peptide in elderly inpatients greater than normal? *Coron. Artery Dis.* **1999**, *10*, 537–540. [CrossRef] [PubMed]
134. Sobhani, K.; Nieves Castro, D.K.; Fu, Q.; Gottlieb, R.A.; Van Eyk, J.E.; Merz, C.N.B. Sex differences in ischemic heart disease and heart failure biomarkers. *Biol. Sex Differ.* **2018**, *9*, 43. [CrossRef] [PubMed]

© 2019 by the authors. Licensee MDPI, Basel, Switzerland. This article is an open access article distributed under the terms and conditions of the Creative Commons Attribution (CC BY) license (http://creativecommons.org/licenses/by/4.0/).

Review

BNP and NT-proBNP as Diagnostic Biomarkers for Cardiac Dysfunction in Both Clinical and Forensic Medicine

Zhipeng Cao, Yuqing Jia and Baoli Zhu *

Department of Forensic Pathology, School of Forensic Medicine, China Medical University, Shenyang 110122, China; zpcao@cmu.edu.cn (Z.C.); jia214@hotmail.com (Y.J.)
* Correspondence: zhu1127@hotmail.com; Tel.: +86-24-3193-9433

Received: 14 March 2019; Accepted: 11 April 2019; Published: 12 April 2019

Abstract: Currently, brain natriuretic peptide (BNP) and *N*-terminal proBNP (NT-proBNP) are widely used as diagnostic biomarkers for heart failure (HF) and cardiac dysfunction in clinical medicine. They are also used as postmortem biomarkers reflecting cardiac function of the deceased before death in forensic medicine. Several previous studies have reviewed BNP and NT-proBNP in clinical medicine, however, few articles have reviewed their application in forensic medicine. The present article reviews the biological features, the research and application status, and the future research prospects of BNP and NT-proBNP in both clinical medicine and forensic medicine, thereby providing valuable assistance for clinicians and forensic pathologists.

Keywords: BNP; NT-proBNP; heart failure; cardiac dysfunction; forensic medicine; postmortem biochemistry

1. Introduction

More than 26 million people all over the world are suffering from heart failure (HF) and cardiac dysfunction, which are currently serious global public health problems. The global burden of HF and cardiac dysfunction is increasing rapidly and substantially with the aging of the population [1–6]. Due to high morbidity and mortality, the diagnosis of HF and cardiac dysfunction is extremely important in both clinical and forensic medicine [7–10]. For inpatients, the diagnosis of HF and cardiac dysfunction can be combined with clinically assisted examinations, such as electrocardiography or echocardiography. However, for the deceased examined by forensic pathologists, the diagnosis of HF or the evaluation of cardiac function after death is very difficult due to the lack of clinical medical records of the deceased and unavailability of assisted examinations. Postmortem assessment and diagnosis, especially for HF or cardiac dysfunction of the deceased without typical visible morphological changes, are extremely challenging [10].

Brain natriuretic peptide (BNP) and *N*-terminal proBNP (NT-proBNP) are widely used as significant indicators for the clinical diagnosis of HF and cardiac dysfunction [11–17]. In recent years, many forensic studies have demonstrated that BNP and NT-proBNP could be used to reflect the cardiac function of the deceased before their death through extensive animal experiments and postmortem specimens, and they could also be used as postmortem biomarkers for the diagnosis of HF or cardiac dysfunction in forensic medicine [9,10,18–20]. However, few articles have reviewed application of BNP and NT-proBNP in forensic medicine. For this purpose, this article reviews the biological features, the clinical and forensic research, and the application status of BNP and NT-proBNP, as well as their future research prospects in order to provide valuable assistance for clinicians and forensic pathologists.

2. Biological Features of BNP and NT-proBNP

The natriuretic peptide family mainly includes atrial natriuretic peptide (ANP), which is mostly synthesized and secreted by atrial myocytes, BNP, and C-type natriuretic peptide (CNP) [21]. BNP was originally isolated from pig brain tissue in 1988 and was named brain natriuretic peptide, but subsequent studies have shown that its synthesis and secretion are mainly in ventricular myocytes [22].

2.1. Structure, Synthesis, and Secretion of BNP and NT-proBNP

BNP is mainly synthesized and secreted by myocytes in the left ventricle (LV) as a response to myocytes stretched by pressure overload or volume expansion of the ventricle [12,23–26]. The structure of BNP is highly conserved among different species, and the difference between different species is in the length and amino acid composition of the N-terminal and C-terminal tail chains [27]. Human BNP is a 32 amino acid polypeptide containing a 17 amino acid ring structure with a disulfide bond connecting two cysteine residues [28,29]. The human gene encoding BNP is located on chromosome 1, and the mRNA encoding BNP contains an unstable repeat TATTTAT sequence [28,30,31]. Instead of storage in normal physiological myocardial tissue, the transcription of BNP mRNA and the synthesis and secretion of BNP protein occur in an explosive way and are rapidly released into surrounding tissues after myocardial synthesis [30,32]. Under pathological conditions, the unstable mRNA can rapidly synthesize a 134 amino acid BNP precursor (pre-proBNP) and remove the N-terminal 26 amino acid signal peptide to form a 108 amino acid BNP (proBNP), and then, proBNP is split by the proNP convertases, corin or furin, into an inactive 76-amino acid NT-proBNP and an active 32-amino acid BNP [24,33]. Both the biologically active BNP and NT-proBNP could be found in plasma [34,35].

2.2. Receptors of Natriuretic Peptides

There are three membrane-bound natriuretic peptide receptors (NPR) for natriuretic peptides, namely NPR-A, NPR-B, and NPR-C. NPR-A is abundant in the vascular endothelium system and some other organs such as kidney and brain [12,34]. NPR-A receptor is the main effector of both ANP and BNP actions, whereas the NPR-B receptor mediates CNP effects. The cyclic guanylate monophosphate (cGMP) levels increase after activation of NPR-A and NPR-B [33,36]. After binding with NPR-A, BNP mediates its biological activities working against the renin–angiotensin–aldosterone system (RAAS) and sympathetic nervous system, improving the glomerular filtration rate and filtration fraction and having diuretic, natriuretic, and vasodilatory effects [37,38].

2.3. Degradation of BNP and NT-proBNP

NPR-C is considered by the majority of physiological data to be the receptor mediating internalization and degradation process of clearing natriuretic peptides from the extracellular environment [39]. In addition to NPR-C receptors involved in the degradation of BNP, neutral endopeptidase (NEP), dipeptidyl peptidase-IV (DPPIV), and insulin degrading enzyme (IDE) are also associated with the clearance of BNP under physiological conditions, which leads to an approximate half-life of 20 min for BNP and 90–120 min for NT-proBNP [12,34,39,40]. In 2015, the first of a new class of drugs was approved by the Food and Drug Administration (FDA) of America; it was a sodium supramolecular complex with an equal ratio of the angiotensin receptor blocker valsartan and the neprilysin inhibitor prodrug sacubitril, and it has been proven to be able to successfully cut down mortality in patients suffering from heart failure with reduced ejection fraction (HFrEF) [33,41,42].

3. Regulation of BNP Gene Expression

The synthesis and secretion of BNP can be induced by mechanical stress, systemic ischemia and hypoxia, neurohumoral factors, and more. However, the exact mechanism of complete regulation remains unclear. It is now generally accepted that mechanical stretch is the main cause of BNP rise in the myocardium. After mechanical stress acts on cardiomyocytes, BNP may be induced by an endothelin (ET)-independent or an ET-dependent pathway [36,43] (Figure 1).

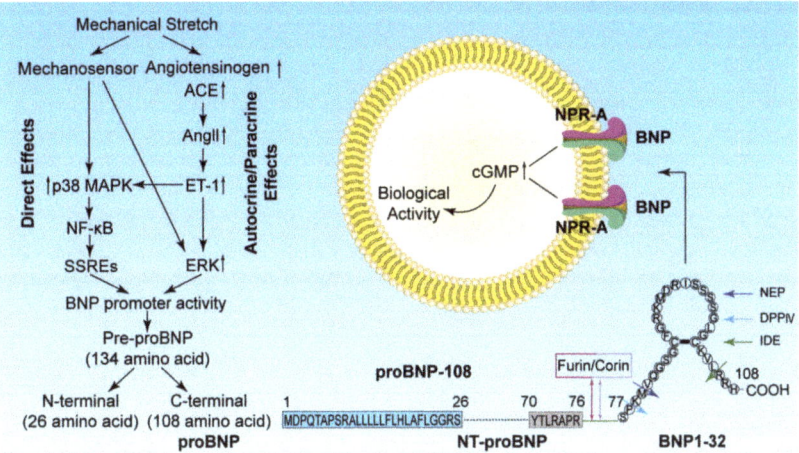

Figure 1. Diagrammatic sketch of mechanical stretch inducing brain natriuretic peptide (BNP) signal transduction events, structural processing, receptor binding, cleavage processing, and degradation enzymes (modified from [36,43] with permission).

3.1. ET-Independent Pathway (Direct Effects)

Mechanical stress signals act on mechanosensors. Then, signaling from the extracellular matrix through integrin activates the mitogen-activated protein kinase (MAPK) signaling pathway, thereby activating the BNP promoter [43]. BNP production induced by mechanical stress is mainly dependent on p38 MAPK, which is a subtype of MAPK. The activated p38 MAPK continues to activate its downstream nuclear factor kappa B (NF-κB), which binds NF-κB to shear stress-responsive elements (SSREs) in the BNP gene promoter, thereby enabling BNP gene promoter activation [43]. p38 MAPK has four subtypes: α, β, γ, and δ. Among them, p38α induces BNP gene transcription through activator protein-1 (AP-1), while p38β regulates BNP gene expression through ET-1-induced transcription factor GATA-4 [33,44]. GATA-4 and many other transcriptional regulators, such as nuclear factor of activated T-cells, myocardin, serum response factor, and more, have been shown to be transcriptional effectors that regulate the transcription of BNP [33].

3.2. ET-Dependent Pathway (Autocrine/Paracrine Effects)

While stress receptors activate intracellular kinases, mechanical stress stimulates the formation of angiotensin II (Ang II) and ET-1 complexes, which activate BNP gene activation via p38 MAPK and extracellular signal regulated kinase (ERK) signaling pathways [45,46]. Ang II is an octapeptide substance produced by the hydrolysis of angiotensin I (Ang I) under the actions of angiotensin enzyme and is the main response factor of the renin-angiotensin system [47]. Animal studies have demonstrated that the BNP mRNA level in the left ventricle of rats increased to 4.5 times that of the control group after Ang II was injected into rats for 6 h and increased to 1.8 times after two weeks. While, the Ang II type 1 receptor (AT1R) antagonist was administered, the BNP mRNA level in the left ventricle of rats was significantly reduced, which may be related to the decrease of aldosterone. This indicated that Ang II induced BNP production by binding to AT1R [47]. Ang II has also been proven to be able to promote the synthesis of BNP during myocardial fibrosis by inducing ET-1 gene expression [48]. ET is currently the most potent long-acting vasoconstrictor. It is produced by both endothelial cells and cardiomyocytes, and has three isomeric peptides, of which ET-1 conducts the very potent vasoconstriction and smooth muscle contraction by binding to ET-A receptor [49]. ET-1 is also a major cause of cardiovascular disease and has been reported to activate the NF-κB transcription factor, which is mediated by the phosphorylation of p38 MAPK, and also to activate the GATA-4 transcription factor, which regulates the expression of BNP [46].

3.3. Other Factors

Some other factors have also been reported to regulate BNP expression but may not be the dominant ones. Natriuretic peptides are frequently increased in primary aldosteronism patients. Aldosterone has been widely proven to be able to active NF-κB, and Ang II is reported to stimulate the synthesis of aldosterone, which can also be suppressed by BNP [50–52]. Ang II and aldosterone often collaborate in pathological conditions to induce cardiac fibrosis, hypertrophy of cardiomyocytes, and cardiac remodeling [53]. Thyroid hormone and its receptor levels are decreased in patients with HF and myocardial infarction animal models, suggesting that BNP mediates the pathophysiological mechanism of thyroxine involved in HF and myocardial infarction. Thyroid hormone may trigger hypertrophy in cardiac myocytes, and BNP gene, as a target of thyroid hormone action, increases under the action of thyroid hormone including BNP promoter activity, BNP mRNA, and BNP protein expression levels [54,55]. In cardiac allograft rejection, activated T lymphocytes produce inflammatory factors such as tumor necrosis factor, IL-1, and IL-6, which also selectively upregulate BNP secretion [56].

Various stimuli that cause cardiac hypertrophy, ischemia, and hypoxic damage, such as growth factors, adrenergic receptor agonists (catecholamines), thyroid hormone, Ca^{2+}, and more, may act on BNP promoter elements through a variety of signaling pathways and affect the activity of its promoter. The activation and transmission of these signaling pathways are different but could cooperate with each other [56].

4. BNP and NT-proBNP as Clinical Biomarkers for the Diagnosis of HF

HF is a multifactorial systemic disease affecting approximately 1 to 2% of the adult population. Cases of HF can currently be divided into HFrEF and "heart failure with normal or preserved ejection fraction" (HFnEF or HFpEF), depending on the ejection fraction (EF) [57,58]. According to the guidelines of the American College of Cardiology Foundation/American Heart Association (ACCF/AHA) and the European Society of Cardiology (ESC), BNP and NT-proBNP are considered to be the most valuable and reliable biomarkers for diagnosing HF and cardiac dysfunction. They are also responsible for the determination of the severity, guiding the relevant treatment strategies, and assessing the prognosis of heart disease [59–64].

4.1. Clinical Cutoffs of BNP and NT-proBNP

The ESC guidelines for the diagnosis and treatment of acute and chronic HF in 2016 recommends that all patients with suspected acute HF should have their plasma natriuretic peptide levels (BNP and NT-proBNP) tested to help identify acute HF. The upper limit of normal in the non-acute setting for BNP is 35 pg/mL and for NT-proBNP is 125 pg/mL, while in acute setting, the cutoff value for BNP is 100 pg/mL and for NT-proBNP is 300 pg/mL [59]. BNP levels can help clinicians distinguish the cause of dyspnea due to HF or other causes. If BNP < 100 pg/mL, HF is considered unlikely and alternative causes of dyspnea are pursued. If BNP is between 100 and 500 pg/mL, clinical judgment should be used to diagnose HF. If BNP is >500 pg/mL, HF or cardiac dysfunction is considered possible and rapid therapy for HF is suggested [65]. Based on International Collaborative of NT-proBNP (ICON) study, age-dependent cutoffs of NT-proBNP may be more useful for the diagnosis of HF. Acute HF could be excluded with a general age-independent cutoff of 300 pg/mL. However, HF should be diagnosed for patients who are less than 50 years old with NT-proBNP levels > 450 pg/mL, patients who are between 50 and 75 years old with NT-proBNP levels > 900 pg/mL, and patients who are more than 75 years old with NT-proBNP levels > 1800 pg/mL [66].

4.2. Diagnostic Role in a Failing Heart

HF and cardiac dysfunction—caused by various causes, such as ischemic heart disease, different types of arrhythmia, and cardiomyopathy—can lead to an increase in BNP and NT-proBNP [29,65,67–73].

Acute ischemic heart disease is associated with an elevation of BNP levels, which might reflect the severity of LV dysfunction, and studies have suggested using natriuretic peptide levels as a guide to institute more aggressive treatments for ischemic heart diseases aimed at reducing ventricular wall stress [29]. In patients with stable coronary heart disease, both BNP and NT-proBNP are strong predictors of adverse cardiovascular events [67]. BNP and NT-proBNP were evaluated—along with myocardial injury markers cardiac troponin T (cTnT), myoglobin, and creatine kinase MB (CK-MB)—in acute myocardial infarction patients. NT-proBNP, which remained elevated on average for 12 weeks, might be a better diagnostic biomarker than BNP [32,70]. BNP and NT-proBNP are highly sensitive and specific indicators of the size of a myocardial infarction, and they are also valuable markers for predicting the prognosis and severity of ischemic heart disease in patients with acute coronary syndrome [68,69].

Apart from ischemic heart diseases, BNP and NT-proBNP were also reported to be related to arrhythmias and cardiomyopathies. Both BNP and NT-proBNP were found to be increased in atrial fibrillation patients [65]. BNP mRNA and its protein are demonstrated to increase as early as 10 min after transient lethal ventricular arrhythmias in animal experiment [74]. BNP and NT-proBNP correlated directly with left ventricular end-diastolic dimension (LVEDD) and left ventricular volumes and were inversely correlated with left ventricular ejection fraction (LVEF) in patients with dilated cardiomyopathy and hypertrophic cardiomyopathy [71–73]. The levels of BNP are significantly high in Takotsubo cardiomyopathy, and early BNP/cTnT and BNP/CK-MB ratios help differentiate Takotsubo cardiomyopathy from acute myocardial infarction (AMI) with greater accuracy than BNP alone [75]. This indicates that the assays of BNP in combination with other biomarkers could be used for the differential diagnosis of certain heart diseases.

4.3. Assessing the Severity and Prognosis of HF

BNP and NT-proBNP do not only have a great significance in the diagnosis of HF, but they also have an assistance value for assessing the severity and prognosis of HF. BNP and NT-proBNP were the strongest independent predictors for HFpEF, as determined by Doppler-echocardiography [76]. A designed trial based on the New York Heart Association (NYHA) classification system, in which patients considered to have NYHA classes I–IV were observed to have gradually increasing plasma BNP concentrations, suggesting that plasma BNP concentration increases with the severity of HF [77]. Plasma BNP and NT-proBNP levels have prognostic values in patients with cardiovascular diseases, and the reduction of BNP and NT-proBNP level predicts an improvement in clinical symptoms. There is a positive correlation between the risk of death and evaluated BNP or NT-proBNP [64]. A study of 521 AMI patients found that BNP and NT-proBNP predicted sudden cardiac death and were the strongest predictors, even after adjusting for clinical variables, including EF [78]. Plasma BNP and NT-proBNP are also used clinically to guide the management of patients with HF and cardiac dysfunction, and they are also used as prognostic indicators which can help clinicians adjust their therapy strategy and determine therapy effectiveness to improve a patient's survival [40,79].

4.4. Therapeutic Role in Cardiac Dysfunction

Recombinant human brain natriuretic peptide (rhBNP) is a synthetic endogenous hormone with the same amino acid sequence as BNP. It can directly dilate blood vessels and effectively reduce cardiac preload and afterload. Nesiritide, approved by the FDA for the therapy of acute decompensated HF in 2001, is a successful rhBNP that has several biological functions that are similar to endogenous BNP, including facilitating natriuresis, diuresis, inhibiting RAAS, increasing output of the heart, decreasing wedge pressure in pulmonary capillaries, and improving cardiac diastolic and systolic function [80–83]. As of today, rhBNP has been widely used for the therapy of HF from various causes.

5. BNP and NT-proBNP as Postmortem Biomarkers to Evaluate Cardiac Function in Forensic Medicine

5.1. Forensic Significance of Functional Biomarkers

Different from clinicians, forensic pathologists only focus on the diagnostic value of BNP and NT-proBNP. The diagnosis of HF or evaluation of cardiac dysfunction at autopsy is based predominantly on morphological and pathological findings. This includes the venous congestion of multiple organs, such as the lungs and liver, or a systemic low output state with ischemic arterioles and capillaries [84]. Acute cardiac dysfunction caused by early acute ischemic heart disease and fatal arrhythmia has become a difficult problem in the field of forensic science and pathology due to its high incidence and the lack of typical pathological changes [7,85,86]. The visible morphological changes of the myocardial structure caused by acute heart diseases, such as acute myocardial ischemic injury, are quite limited [87]. Objective evidence for the diagnosis of HF or cardiac dysfunction is extremely necessary in forensic medicine. Compared with morphological indicators, functional indicators or biomarkers, such as BNP and NT-proBNP, could reflect the cardiac function and pathophysiological processes during death and may better clarify the mechanism of death in forensic medicine [88]. The functional biomarkers BNP and NT-proBNP played quite an important role in postmortem biochemistry, and could help solve forensic problems in many routine natural deaths [88–91].

5.2. Pericardial Fluid in Postmortem Biochemistry

Being essentially the ultrafiltration of plasma, pericardial fluid is the pale yellow, clear, and transparent liquid present in the pericardial cavity which acts to lubricate and prevent adhesions. The normal pericardial fluid volume in the physiological condition is 20–30 mL [92–94]. Compared with the fact that blood and other bodily fluids are susceptible to postmortem changes, such as autolysis and spoilage, pericardial fluid is present in a closed serosa cavity and is not susceptible to contamination and postmortem changes [92,93]. It is easy to obtain during forensic autopsy and acts not only as a clinically important sample but also has wide application prospects in forensic identification. It is currently used as a substitute for serum in postmortem biochemical assays [95,96]. Forensic studies have also reported that ions and protein components in pericardial fluid could be used for forensic identification of sudden cardiac death, mechanical asphyxia, hypothermia, hyperthermia, and death inference [8,10,88,93,97–100]. Several studies have also reported the postmortem biochemical investigations of BNP and NT-proBNP in pericardial fluid, which were associated with different causes of death [7–10].

5.3. Postmortem BNP and NT-proBNP

As acute or subacute HF may occur in many acute diseases or traumatic deaths, objective evaluation of end-stage cardiac function status has great significance for forensic diagnosis [101]. Unlike other cardiac biomarkers, such as cTnT and cTnI existing in physiological cardiomyocytes, BNP is not stored in normal myocardial tissue under physiological conditions. However, the transcription of BNP mRNA and the synthesis of its protein can occur and accelerate sensitively and rapidly in a very short time under pathological conditions [30,32]. This means that BNP and NT-proBNP do not fluctuate greatly after death and might be more objective biomarkers of cardiac function [102,103]. In the past decade, a few research teams have conducted postmortem BNP and NT-proBNP studies. To investigate BNP and NT-proBNP concentrations in bodily fluids and myocardial tissue, and the expression of BNP mRNA in myocardium may objectively reflect the end-stage cardiac function status of the deceased before death, which are mainly described as below [88].

Studies in postmortem individuals have demonstrated that BNP and NT-proBNP concentrations were significantly elevated in the blood and pericardial fluid of the deceased who died from acute ischemic heart disease (with or without myocardial necrosis), chronic congestive heart disease, arrhythmogenic right ventricular cardiomyopathy, and more. BNP mRNA was also elevated in the myocardium of individuals with these diseases [7,8,90]. The concentration of BNP in pericardial fluid was closely related to the cause of death, and compared with non-cardiac death, the BNP levels were significantly increased in sudden cardiac death cases, such as acute ischemic heart disease and recurrent myocardial infarction. This further confirms that BNP is important for evaluating the cardiac function of the deceased with ischemic heart disease [9]. High levels of BNP and BNP/ANP ratios in pericardial fluid after death are hallmarks of the duration of cardiac dysfunction before death, which may be due to subacute and chronic ventricular dilatation [9]. Patients with arrhythmogenic right ventricular cardiomyopathy have elevated BNP levels in the pericardial fluid but, interestingly, BNP mRNA levels in the right ventricular myocardium are higher than those in the left ventricular myocardium [7,8]. BNP protein and mRNA were also demonstrated to be elevated in acute cardiac dysfunction caused by acute ventricular arrhythmias, indicating that BNP may be of great forensic significance in the diagnosis of acute cardiac dysfunction without any morphological changes [74]. In some forensic cases of death which are difficult to distinguish from sudden cardiac death, such as hemopericardium and pulmonary thromboembolism, neither BNP levels in the pericardial fluid nor BNP mRNA levels in the myocardial tissue increased, indicating that BNP and BNP mRNA can also be used for distinguishing different diagnoses [8].

Furthermore, in forensic medicine, NT-proBNP is expected to be a more reliable postmortem biomarker compared with BNP due to its greater stability and longer half-life of 90–120 min as mentioned above, and it is not susceptible to temperature, storage time, and storage conditions [104–108]. Several studies have focused on the postmortem investigation of NT-proBNP in different bodily fluids. Postmortem concentration of NT-proBNP in serum from femoral blood within 24 h after death has no difference with the antemortem serum NT-proBNP concentration, and it was stable within 48 h. Cardiopulmonary resuscitation before death has been found to have no effect on NT-proBNP results [19]. Serial assays of NT-proBNP in blood and pericardial fluid, which were gathered from corpses with a postmortem interval of up to 24 h, showed that NT-proBNP was stable over 24 days and, particularly, the concentration of NT-proBNP in pericardial fluid decreased by no more than 16% after storage at −20 °C for 24 days [18]. NT-proBNP concentrations in different samples, such as serum and pericardial fluid, reveal good correlations, and NT-proBNP was demonstrated to be much higher in pericardial fluid than other fluids, such as serum, which indicates that the investigation of NT-proBNP in pericardial fluid might be a much better choice in postmortem biochemical assay [18,19].

5.4. Limitation of BNP and NT-proBNP in Forensic Medicine

While BNP and NT-proBNP in pericardial fluid are not susceptible to being polluted, serious hemolysis and other postmortem changes caused by various factors, such as the preservation conditions of the corpse, may affect the postmortem biological assays of BNP and NT-proBNP. This should be taken into consideration in postmortem biochemical assays, and the affection of hemolysis can be reduced by the physical filtering of body fluids [109,110]. In addition, valuable postmortem cutoffs are still needed for the further study of both BNP and NT-proBNP in blood or pericardial fluid [89]. Currently, because of equipment and personnel, not every forensic laboratory around the world conducts postmortem biochemical assays, which is one of the reasons for the lack of postmortem cutoffs for BNP and NT-proBNP based on large amounts of data. It is worth mentioning that all diuretics, including blockers of renin angiotensin aldosterone and aldosterone receptor, could decrease BNP levels due to the amelioration of plasma volume and sodium, which should also be considered in forensic utilization of BNP or NT-proBNP.

6. Research and Application Prospects in Clinical and Forensic Medicine

BNP and NT-proBNP are currently used in the evaluation of cardiac function status in clinical and forensic practice. In recent years, many studies have confirmed that some non-coding RNAs are highly expressed in patients with cardiac dysfunction and participate in the regulation of BNP expression [111–115]. Therefore, exploring the expression patterns of BNP-related specific non-coding RNAs, such as microRNAs, in clinical and forensic samples and exploring how they regulate the expression of BNP and the expression of non-coding RNA in forensic degradation or corrupted samples may be the future research direction in this field.

In addition, exosomes, the small vesicles in different bodily fluids such as serum and urine, have been proven to be contained in different molecules such as proteins, DNA, and RNA (coding RNA and non-coding RNA). Exosomes have been expected to be a new hot issue in the field of markers for cardiovascular diseases due to their specific diagnostic value and unknown underlying mechanisms [116–118]. Exosome RNA and proteins are demonstrated to be related to cardiac dysfunction and mediate cardioprotective abilities [119–122]. Previous studies have found that exosomes containing AT1Rs were isolated from mice undergoing cardiac pressure overload. However, few studies have reported whether exosomes in bodily fluids were correlated to BNP. Thus, exosomes as a biomarker for diagnosing cardiac dysfunction in clinical and forensic medicine may also be a future research prospect [122]. As written above, pericardial fluid is an ideal biological sample for forensic pathology. Clinical research has proven that exosomes in human pericardial fluid are diagnostic and therapeutic molecules for heart disease. Whether exosomes in pericardial fluid can be used to diagnose heart disease in forensic medicine still needs to be further studied [123–126].

Postmortem biochemical assays and molecular biological methods, such as the analysis of mRNA—which should both be taken as the routine laboratory assays in forensic medicine—may be potentially useful for investigating the pathophysiology, process, and the cause of death. They may also offer powerful support by providing visible evidence for pathognomonic assessment, including cardiac function [88]. Therefore, with its advantages in assessing pathophysiological functional changes involved in the dying process, combined assays of postmortem chemistry and molecular biology of BNP and NT-proBNP may better support and reinforce morphological evidence in forensic medicine [90,127–129].

7. Conclusions

More than 30 years of research has outlined the significant contribution of BNP in cardiovascular disease, particularly in HF and cardiac dysfunction. Based on their important diagnostic, therapeutic, and prognostic roles, BNP and NT-proBNP have been used as important biomarkers in clinical and forensic medicine. With the rapid development of molecular biological technology, the accurate instigation of BNP and NT-proBNP will be better used for the assessment of clinical and forensic cardiac function status in the future.

Funding: This research was funded by National Natural Science Foundation of China (Grant No. 81273343) and Natural Science Foundation of Liaoning Province (Grant No. 20180530004).

Acknowledgments: We would like to thank Alex Schroer and MDPI English Editing Service for English language editing.

Conflicts of Interest: The authors declare no conflict of interest.

Abbreviations

ACCF	American College of Cardiology Foundation
AHA	American Heart Association
Ang	Angiotensin
ANP	Atrial natriuretic peptide
AMI	Acute myocardial infarction
AP	Activator protein
AT1R	Angiotensin II type 1 receptor
BNP	Brain natriuretic peptide
cGMP	Cyclic guanosine monophosphate
CK-MB	Creatine kinase MB
CNP	C-type natriuretic peptide
cTn	Cardiac troponin
DHF	Diastolic heart failure
DNA	Deoxyribonucleic acid
DPPIV	Dipeptidyl peptidase-IV
EF	Ejection fraction
ERK	Extracellular signal regulated kinase
ESC	European Society of Cardiology
ET	Endothelin
FDA	Food and Drug Administration
HF	Heart failure
HFnEF	Heart failure with normal ejection fraction
HFpEF	Heart failure with preserved ejection fraction
HFrEF	Heart failure with reduced ejection fraction
ICON	International Collaborative of NT-proBNP
IDE	Insulin degrading enzyme
LV	Left ventricle
LVEDD	Left ventricular end-diastolic dimension
LVEF	Left ventricular ejection fraction
NEP	Neutral endopeptidase
NF-κB	Nuclear factor kappa B
NT-proBNP	N-terminal pro-brain natriuretic peptide
NPR	Natriuretic peptide receptor
NYHA	New York Heart Association
MAPK	Mitogen-activated protein kinase
RAAS	Renin–angiotensin–aldosterone system
rhBNP	Recombinant human brain natriuretic peptide
RNA	Ribonucleic acid
SHF	Systolic heart failure
SSREs	Shear stress responsive elements

References

1. Dickstein, K.; Cohen-Solal, A.; Filippatos, G.; McMurray, J.J.; Ponikowski, P.; Poole-Wilson, P.A.; Stromberg, A.; van Veldhuisen, D.J.; Atar, D.; Hoes, A.W.; et al. ESC Guidelines for the diagnosis and treatment of acute and chronic heart failure 2008: The Task Force for the Diagnosis and Treatment of Acute and Chronic Heart Failure 2008 of the European Society of Cardiology. Developed in collaboration with the Heart Failure Association of the ESC (HFA) and endorsed by the European Society of Intensive Care Medicine (ESICM). *Eur. Heart J.* **2008**, *29*, 2388–2442. [CrossRef]
2. Huffman, M.D.; Prabhakaran, D. Heart failure: epidemiology and prevention in India. *Natl. Med. J. India* **2010**, *23*, 283–288. [PubMed]

3. Weiwei, C.; Runlin, G.; Lisheng, L.; Manlu, Z.; Wen, W.; Yongjun, W.; Zhaosu, W.; Huijun, L.; Zhe, Z.; Lixin, J.; et al. Outline of the report on cardiovascular diseases in China, 2014. *Eur. Heart J. Suppl.* **2016**, *18*, F2–F11. [CrossRef]
4. Benjamin, E.J.; Blaha, M.J.; Chiuve, S.E.; Cushman, M.; Das, S.R.; Deo, R.; de Ferranti, S.D.; Floyd, J.; Fornage, M.; Gillespie, C.; et al. Heart Disease and Stroke Statistics-2017 Update: A Report From the American Heart Association. *Circulation* **2017**, *135*, e146–e603. [CrossRef]
5. Bloom, M.W.; Greenberg, B.; Jaarsma, T.; Januzzi, J.L.; Lam, C.S.P.; Maggioni, A.P.; Trochu, J.N.; Butler, J. Heart failure with reduced ejection fraction. *Nat. Rev. Dis. Primers* **2017**, *3*, 17058. [CrossRef]
6. Savarese, G.; Lund, L.H. Global Public Health Burden of Heart Failure. *Card. Fail. Rev.* **2017**, *3*, 7–11. [CrossRef]
7. Cao, Z.P.; Xue, J.J.; Zhang, Y.; Tian, M.H.; Xiao, Y.; Jia, Y.Q.; Zhu, B.L. Differential expression of B-type natriuretic peptide between left and right ventricles, with particular regard to sudden cardiac death. *Mol. Med. Rep.* **2017**, *16*, 4763–4769. [CrossRef]
8. Chen, J.H.; Michiue, T.; Ishikawa, T.; Maeda, H. Pathophysiology of sudden cardiac death as demonstrated by molecular pathology of natriuretic peptides in the myocardium. *Forensic Sci. Int.* **2012**, *223*, 342–348. [CrossRef]
9. Zhu, B.L.; Ishikawa, T.; Michiue, T.; Li, D.R.; Zhao, D.; Tanaka, S.; Kamikodai, Y.; Tsuda, K.; Okazaki, S.; Maeda, H. Postmortem pericardial natriuretic peptides as markers of cardiac function in medico-legal autopsies. *Int. J. Legal Med.* **2007**, *121*, 28–35. [CrossRef] [PubMed]
10. Chen, J.H.; Michiue, T.; Ishikawa, T.; Maeda, H. Molecular pathology of natriuretic peptides in the myocardium with special regard to fatal intoxication, hypothermia, and hyperthermia. *Int. J. Legal Med.* **2012**, *126*, 747–756. [CrossRef] [PubMed]
11. Hijazi, Z.; Oldgren, J.; Siegbahn, A.; Granger, C.B.; Wallentin, L. Biomarkers in atrial fibrillation: A clinical review. *Eur. Heart J.* **2013**, *34*, 1475–1480. [CrossRef] [PubMed]
12. Maalouf, R.; Bailey, S. A review on B-type natriuretic peptide monitoring: assays and biosensors. *Heart Fail. Rev.* **2016**, *21*, 567–578. [CrossRef]
13. Maries, L.; Manitiu, I. Diagnostic and prognostic values of B-type natriuretic peptides (BNP) and N-terminal fragment brain natriuretic peptides (NT-pro-BNP). *Cardiovasc. J. Afr.* **2013**, *24*, 286–289. [CrossRef]
14. Troughton, R.; Michael Felker, G.; Januzzi, J.L., Jr. Natriuretic peptide-guided heart failure management. *Eur. Heart J.* **2014**, *35*, 16–24. [CrossRef]
15. Chow, S.L.; Maisel, A.S.; Anand, I.; Bozkurt, B.; de Boer, R.A.; Felker, G.M.; Fonarow, G.C.; Greenberg, B.; Januzzi, J.L., Jr.; Kiernan, M.S.; et al. Role of Biomarkers for the Prevention, Assessment, and Management of Heart Failure: A Scientific Statement From the American Heart Association. *Circulation* **2017**, *135*, e1054–e1091. [CrossRef]
16. Cocco, G.; Jerie, P. Assessing the benefits of natriuretic peptides-guided therapy in chronic heart failure. *Cardiol. J.* **2015**, *22*, 5–11. [CrossRef] [PubMed]
17. Rubattu, S.; Forte, M.; Marchitti, S.; Volpe, M. Molecular Implications of Natriuretic Peptides in the Protection from Hypertension and Target Organ Damage Development. *Int. J. Mol. Sci.* **2019**, *20*. [CrossRef]
18. Michaud, K.; Augsburger, M.; Donze, N.; Sabatasso, S.; Faouzi, M.; Bollmann, M.; Mangin, P. Evaluation of postmortem measurement of NT-proBNP as a marker for cardiac function. *Int. J. Legal Med.* **2008**, *122*, 415–420. [CrossRef]
19. Palmiere, C.; Tettamanti, C.; Bonsignore, A.; De Stefano, F.; Vanhaebost, J.; Rousseau, G.; Scarpelli, M.P.; Bardy, D. Cardiac troponins and NT-proBNP in the forensic setting: Overview of sampling site, postmortem interval, cardiopulmonary resuscitation, and review of the literature. *Forensic Sci. Int.* **2018**, *282*, 211–218. [CrossRef]
20. Sabatasso, S.; Vaucher, P.; Augsburger, M.; Donze, N.; Mangin, P.; Michaud, K. Sensitivity and specificity of NT-proBNP to detect heart failure at post mortem examination. *Int. J. Legal Med.* **2011**, *125*, 849–856. [CrossRef] [PubMed]
21. Del Ry, S.; Cabiati, M.; Clerico, A. Natriuretic peptide system and the heart. *Front. Horm. Res.* **2014**, *43*, 134–143. [CrossRef]
22. Sudoh, T.; Kangawa, K.; Minamino, N.; Matsuo, H. A new natriuretic peptide in porcine brain. *Nature* **1988**, *332*, 78–81. [CrossRef]

23. Clerico, A.; Recchia, F.A.; Passino, C.; Emdin, M. Cardiac endocrine function is an essential component of the homeostatic regulation network: physiological and clinical implications. *Am. J. Physiol. Heart Circ. Physiol.* **2006**, *290*, H17–H29. [CrossRef]
24. De Lemos, J.A.; McGuire, D.K.; Drazner, M.H. B-type natriuretic peptide in cardiovascular disease. *Lancet* **2003**, *362*, 316–322. [CrossRef]
25. Rodeheffer, R.J. Measuring plasma B-type natriuretic peptide in heart failure: good to go in 2004? *J. Am. Coll. Cardiol.* **2004**, *44*, 740–749. [CrossRef]
26. Levin, E.R.; Gardner, D.G.; Samson, W.K. Natriuretic peptides. *N. Engl. J. Med.* **1998**, *339*, 321–328. [CrossRef]
27. Grantham, J.A.; Borgeson, D.D.; Burnett, J.C., Jr. BNP: Pathophysiological and potential therapeutic roles in acute congestive heart failure. *Am. J. Physiol.* **1997**, *272*, R1077–R1083. [CrossRef]
28. Cheung, B.M.; Kumana, C.R. Natriuretic peptides–relevance in cardiovascular disease. *Jama* **1998**, *280*, 1983–1984. [CrossRef]
29. Daniels, L.B.; Maisel, A.S. Natriuretic peptides. *J. Am. Coll. Cardiol.* **2007**, *50*, 2357–2368. [CrossRef]
30. Nakagawa, O.; Ogawa, Y.; Itoh, H.; Suga, S.; Komatsu, Y.; Kishimoto, I.; Nishino, K.; Yoshimasa, T.; Nakao, K. Rapid transcriptional activation and early mRNA turnover of brain natriuretic peptide in cardiocyte hypertrophy. Evidence for brain natriuretic peptide as an "emergency" cardiac hormone against ventricular overload. *J. Clin. Invest.* **1995**, *96*, 1280–1287. [CrossRef]
31. Sudoh, T.; Maekawa, K.; Kojima, M.; Minamino, N.; Kangawa, K.; Matsuo, H. Cloning and sequence analysis of cDNA encoding a precursor for human brain natriuretic peptide. *Biochem. Biophys. Res. Commun.* **1989**, *159*, 1427–1434. [CrossRef]
32. Hama, N.; Itoh, H.; Shirakami, G.; Nakagawa, O.; Suga, S.; Ogawa, Y.; Masuda, I.; Nakanishi, K.; Yoshimasa, T.; Hashimoto, Y.; et al. Rapid ventricular induction of brain natriuretic peptide gene expression in experimental acute myocardial infarction. *Circulation* **1995**, *92*, 1558–1564. [CrossRef]
33. Kerkela, R.; Ulvila, J.; Magga, J. Natriuretic Peptides in the Regulation of Cardiovascular Physiology and Metabolic Events. *J. Am. Heart Assoc.* **2015**, *4*, e002423. [CrossRef]
34. Vanderheyden, M.; Bartunek, J.; Goethals, M. Brain and other natriuretic peptides: Molecular aspects. *Eur. J. Heart Fail.* **2004**, *6*, 261–268. [CrossRef]
35. Yamanouchi, S.; Kudo, D.; Endo, T.; Kitano, Y.; Shinozawa, Y. Blood N-terminal proBNP as a potential indicator of cardiac preload in patients with high volume load. *Tohoku J. Exp. Med.* **2010**, *221*, 175–180. [CrossRef]
36. Volpe, M.; Rubattu, S.; Burnett, J. Natriuretic peptides in cardiovascular diseases: current use and perspectives. *Eur. Heart J.* **2014**, *35*, 419–425. [CrossRef]
37. Cataliotti, A.; Boerrigter, G.; Costello-Boerrigter, L.C.; Schirger, J.A.; Tsuruda, T.; Heublein, D.M.; Chen, H.H.; Malatino, L.S.; Burnett, J.C., Jr. Brain natriuretic peptide enhances renal actions of furosemide and suppresses furosemide-induced aldosterone activation in experimental heart failure. *Circulation* **2004**, *109*, 1680–1685. [CrossRef]
38. Diez, J. Chronic heart failure as a state of reduced effectiveness of the natriuretic peptide system: implications for therapy. *Eur. J. Heart Fail.* **2017**, *19*, 167–176. [CrossRef]
39. Potter, L.R. Natriuretic peptide metabolism, clearance and degradation. *FEBS J.* **2011**, *278*, 1808–1817. [CrossRef]
40. Fu, S.; Ping, P.; Wang, F.; Luo, L. Synthesis, secretion, function, metabolism and application of natriuretic peptides in heart failure. *J. Biol. Eng.* **2018**, *12*, 2. [CrossRef]
41. Hubers, S.A.; Brown, N.J. Combined Angiotensin Receptor Antagonism and Neprilysin Inhibition. *Circulation* **2016**, *133*, 1115–1124. [CrossRef]
42. Kobalava, Z.; Kotovskaya, Y.; Averkov, O.; Pavlikova, E.; Moiseev, V.; Albrecht, D.; Chandra, P.; Ayalasomayajula, S.; Prescott, M.F.; Pal, P. Pharmacodynamic and Pharmacokinetic Profiles of Sacubitril/Valsartan (LCZ696) in Patients with Heart Failure and Reduced Ejection Fraction. *Cardiovasc. Ther.* **2016**, *34*, 191–198. [CrossRef]
43. Liang, F.; Lu, S.; Gardner, D.G. Endothelin-dependent and -independent components of strain-activated brain natriuretic peptide gene transcription require extracellular signal regulated kinase and p38 mitogen-activated protein kinase. *Hypertension* **2000**, *35*, 188–192. [CrossRef]

44. Koivisto, E.; Kaikkonen, L.; Tokola, H.; Pikkarainen, S.; Aro, J.; Pennanen, H.; Karvonen, T.; Rysa, J.; Kerkela, R.; Ruskoaho, H. Distinct regulation of B-type natriuretic peptide transcription by p38 MAPK isoforms. *Mol. Cell Endocrinol.* **2011**, *338*, 18–27. [CrossRef]
45. Piuhola, J.; Szokodi, I.; Ruskoaho, H. Endothelin-1 and angiotensin II contribute to BNP but not c-fos gene expression response to elevated load in isolated mice hearts. *Biochim. Biophys. Acta* **2007**, *1772*, 338–344. [CrossRef]
46. Pikkarainen, S.; Tokola, H.; Kerkela, R.; Majalahti-Palviainen, T.; Vuolteenaho, O.; Ruskoaho, H. Endothelin-1-specific activation of B-type natriuretic peptide gene via p38 mitogen-activated protein kinase and nuclear ETS factors. *J. Biol. Chem.* **2003**, *278*, 3969–3975. [CrossRef]
47. Majalahti, T.; Suo-Palosaari, M.; Sarman, B.; Hautala, N.; Pikkarainen, S.; Tokola, H.; Vuolteenaho, O.; Wang, J.; Paradis, P.; Nemer, M.; et al. Cardiac BNP gene activation by angiotensin II in vivo. *Mol. Cell Endocrinol.* **2007**, *273*, 59–67. [CrossRef]
48. Cheng, T.H.; Cheng, P.Y.; Shih, N.L.; Chen, I.B.; Wang, D.L.; Chen, J.J. Involvement of reactive oxygen species in angiotensin II-induced endothelin-1 gene expression in rat cardiac fibroblasts. *J. Am. Coll. Cardiol.* **2003**, *42*, 1845–1854. [CrossRef]
49. Freeman, B.D.; Machado, F.S.; Tanowitz, H.B.; Desruisseaux, M.S. Endothelin-1 and its role in the pathogenesis of infectious diseases. *Life Sci.* **2014**, *118*, 110–119. [CrossRef]
50. Hu, W.; Zhou, P.H.; Zhang, X.B.; Xu, C.G.; Wang, W. Pathophysiological functions of adrenomedullin and natriuretic peptides in patients with primary aldosteronism. *Endocrine* **2015**, *48*, 661–668. [CrossRef]
51. Liang, F.; Kapoun, A.M.; Lam, A.; Damm, D.L.; Quan, D.; O'Connell, M.; Protter, A.A. B-Type natriuretic peptide inhibited angiotensin II-stimulated cholesterol biosynthesis, cholesterol transfer, and steroidogenesis in primary human adrenocortical cells. *Endocrinology* **2007**, *148*, 3722–3729. [CrossRef] [PubMed]
52. Queisser, N.; Schupp, N. Aldosterone, oxidative stress, and NF-kappaB activation in hypertension-related cardiovascular and renal diseases. *Free Radic. Biol. Med.* **2012**, *53*, 314–327. [CrossRef]
53. Azibani, F.; Fazal, L.; Chatziantoniou, C.; Samuel, J.L.; Delcayre, C. Aldosterone mediates cardiac fibrosis in the setting of hypertension. *Curr. Hypertens. Rep.* **2013**, *15*, 395–400. [CrossRef]
54. Selvaraj, S.; Klein, I.; Danzi, S.; Akhter, N.; Bonow, R.O.; Shah, S.J. Association of serum triiodothyronine with B-type natriuretic peptide and severe left ventricular diastolic dysfunction in heart failure with preserved ejection fraction. *Am. J. Cardiol.* **2012**, *110*, 234–239. [CrossRef]
55. Liang, F.; Webb, P.; Marimuthu, A.; Zhang, S.; Gardner, D.G. Triiodothyronine increases brain natriuretic peptide (BNP) gene transcription and amplifies endothelin-dependent BNP gene transcription and hypertrophy in neonatal rat ventricular myocytes. *J. Biol. Chem.* **2003**, *278*, 15073–15083. [CrossRef]
56. Sergeeva, I.A.; Christoffels, V.M. Regulation of expression of atrial and brain natriuretic peptide, biomarkers for heart development and disease. *Biochim. Biophys. Acta* **2013**, *1832*, 2403–2413. [CrossRef]
57. Tanai, E.; Frantz, S. Pathophysiology of Heart Failure. *Compr. Physiol.* **2015**, *6*, 187–214. [CrossRef]
58. Katz, A.M.; Rolett, E.L. Heart failure: when form fails to follow function. *Eur. Heart J.* **2016**, *37*, 449–454. [CrossRef] [PubMed]
59. Ponikowski, P.; Voors, A.A.; Anker, S.D.; Bueno, H.; Cleland, J.G.; Coats, A.J.; Falk, V.; Gonzalez-Juanatey, J.R.; Harjola, V.P.; Jankowska, E.A.; et al. 2016 ESC Guidelines for the diagnosis and treatment of acute and chronic heart failure: The Task Force for the diagnosis and treatment of acute and chronic heart failure of the European Society of Cardiology (ESC). Developed with the special contribution of the Heart Failure Association (HFA) of the ESC. *Eur. J. Heart Fail.* **2016**, *18*, 891–975. [CrossRef] [PubMed]
60. Yancy, C.W.; Jessup, M.; Bozkurt, B.; Butler, J.; Casey, D.E., Jr.; Drazner, M.H.; Fonarow, G.C.; Geraci, S.A.; Horwich, T.; Januzzi, J.L.; et al. 2013 ACCF/AHA guideline for the management of heart failure: a report of the American College of Cardiology Foundation/American Heart Association Task Force on Practice Guidelines. *J. Am. Coll. Cardiol.* **2013**, *62*, e147–e239. [CrossRef]
61. Sun, Y.P.; Wei, C.P.; Ma, S.C.; Zhang, Y.F.; Qiao, L.Y.; Li, D.H.; Shan, R.B. Effect of Carvedilol on Serum Heart-type Fatty Acid-binding Protein, Brain Natriuretic Peptide, and Cardiac Function in Patients With Chronic Heart Failure. *J. Cardiovasc. Pharmacol.* **2015**, *65*, 480–484. [CrossRef]
62. Dini, F.L.; Gabutti, A.; Passino, C.; Fontanive, P.; Emdin, M.; De Tommasi, S.M. Atrial fibrillation and amino-terminal pro-brain natriuretic peptide as independent predictors of prognosis in systolic heart failure. *Int. J. Cardiol.* **2010**, *140*, 344–350. [CrossRef]

63. Shao, M.; Huang, C.; Li, Z.; Yang, H.; Feng, Q. Effects of glutamine and valsartan on the brain natriuretic peptide and N-terminal pro-B-type natriuretic peptide of patients with chronic heart failure. *Pak. J. Med. Sci.* **2015**, *31*, 82–86. [CrossRef]
64. Khanam, S.S.; Son, J.W.; Lee, J.W.; Youn, Y.J.; Yoon, J.; Lee, S.H.; Kim, J.Y.; Ahn, S.G.; Ahn, M.S.; Yoo, B.S. Prognostic value of short-term follow-up BNP in hospitalized patients with heart failure. *BMC Cardiovasc. Disord.* **2017**, *17*, 215. [CrossRef]
65. Chang, K.W.; Hsu, J.C.; Toomu, A.; Fox, S.; Maisel, A.S. Clinical Applications of Biomarkers in Atrial Fibrillation. *Am. J. Med.* **2017**, *130*, 1351–1357. [CrossRef]
66. Januzzi, J.L.; van Kimmenade, R.; Lainchbury, J.; Bayes-Genis, A.; Ordonez-Llanos, J.; Santalo-Bel, M.; Pinto, Y.M.; Richards, M. NT-proBNP testing for diagnosis and short-term prognosis in acute destabilized heart failure: an international pooled analysis of 1256 patients: the International Collaborative of NT-proBNP Study. *Eur. Heart J.* **2006**, *27*, 330–337. [CrossRef]
67. Mishra, R.K.; Beatty, A.L.; Jaganath, R.; Regan, M.; Wu, A.H.; Whooley, M.A. B-type natriuretic peptides for the prediction of cardiovascular events in patients with stable coronary heart disease: The Heart and Soul Study. *J. Am. Heart Assoc.* **2014**, *3*. [CrossRef]
68. Radwan, H.; Selem, A.; Ghazal, K. Reply to: N-terminal pro brain natriuretic peptide in coronary artery disease. *J. Saudi. Heart Assoc.* **2015**, *27*, 225. [CrossRef]
69. Radwan, H.; Selem, A.; Ghazal, K. Value of N-terminal pro brain natriuretic peptide in predicting prognosis and severity of coronary artery disease in acute coronary syndrome. *J. Saudi. Heart Assoc.* **2014**, *26*, 192–198. [CrossRef]
70. Gill, D.; Seidler, T.; Troughton, R.W.; Yandle, T.G.; Frampton, C.M.; Richards, M.; Lainchbury, J.G.; Nicholls, G. Vigorous response in plasma N-terminal pro-brain natriuretic peptide (NT-BNP) to acute myocardial infarction. *Clin. Sci.* **2004**, *106*, 135–139. [CrossRef]
71. Tesic, M.; Seferovic, J.; Trifunovic, D.; Djordjevic-Dikic, A.; Giga, V.; Jovanovic, I.; Petrovic, O.; Marinkovic, J.; Stankovic, S.; Stepanovic, J.; et al. N-terminal pro-brain natriuretic peptide is related with coronary flow velocity reserve and diastolic dysfunction in patients with asymmetric hypertrophic cardiomyopathy. *J. Cardiol.* **2017**, *70*, 323–328. [CrossRef]
72. Amorim, S.; Campelo, M.; Moura, B.; Martins, E.; Rodrigues, J.; Barroso, I.; Faria, M.; Guimaraes, T.; Macedo, F.; Silva-Cardoso, J.; et al. The role of biomarkers in dilated cardiomyopathy: Assessment of clinical severity and reverse remodeling. *Rev. Port. Cardiol.* **2017**, *36*, 709–716. [CrossRef]
73. Geske, J.B.; McKie, P.M.; Ommen, S.R.; Sorajja, P. B-type natriuretic peptide and survival in hypertrophic cardiomyopathy. *J. Am. Coll. Cardiol.* **2013**, *61*, 2456–2460. [CrossRef]
74. Cao, Z.P.; Zhang, Y.; Mi, L.; Luo, X.Y.; Tian, M.H.; Zhu, B.L. The Expression of B-Type Natriuretic Peptide After CaCl2-Induced Arrhythmias in Rats. *Am. J. Forensic Med. Pathol.* **2016**, *37*, 133–140. [CrossRef]
75. Randhawa, M.S.; Dhillon, A.S.; Taylor, H.C.; Sun, Z.; Desai, M.Y. Diagnostic utility of cardiac biomarkers in discriminating Takotsubo cardiomyopathy from acute myocardial infarction. *J. Card. Fail.* **2014**, *20*, 2–8. [CrossRef]
76. Grewal, J.; McKelvie, R.; Lonn, E.; Tait, P.; Carlsson, J.; Gianni, M.; Jarnert, C.; Persson, H. BNP and NT-proBNP predict echocardiographic severity of diastolic dysfunction. *Eur. J. Heart Fail.* **2008**, *10*, 252–259. [CrossRef]
77. Wieczorek, S.J.; Wu, A.H.; Christenson, R.; Krishnaswamy, P.; Gottlieb, S.; Rosano, T.; Hager, D.; Gardetto, N.; Chiu, A.; Bailly, K.R.; et al. A rapid B-type natriuretic peptide assay accurately diagnoses left ventricular dysfunction and heart failure: a multicenter evaluation. *Am. Heart J.* **2002**, *144*, 834–839. [CrossRef]
78. Tapanainen, J.M.; Lindgren, K.S.; Makikallio, T.H.; Vuolteenaho, O.; Leppaluoto, J.; Huikuri, H.V. Natriuretic peptides as predictors of non-sudden and sudden cardiac death after acute myocardial infarction in the beta-blocking era. *J. Am. Coll. Cardiol.* **2004**, *43*, 757–763. [CrossRef]
79. Gueant Rodriguez, R.M.; Spada, R.; Pooya, S.; Jeannesson, E.; Moreno Garcia, M.A.; Anello, G.; Bosco, P.; Elia, M.; Romano, A.; Alberto, J.M.; et al. Homocysteine predicts increased NT-pro-BNP through impaired fatty acid oxidation. *Int. J. Cardiol.* **2013**, *167*, 768–775. [CrossRef]
80. Elkayam, U.; Akhter, M.W.; Singh, H.; Khan, S.; Usman, A. Comparison of effects on left ventricular filling pressure of intravenous nesiritide and high-dose nitroglycerin in patients with decompensated heart failure. *Am. J. Cardiol.* **2004**, *93*, 237–240. [CrossRef]

81. Intravenous nesiritide vs nitroglycerin for treatment of decompensated congestive heart failure: A randomized controlled trial. *Jama* **2002**, *287*, 1531–1540.
82. Colucci, W.S.; Elkayam, U.; Horton, D.P.; Abraham, W.T.; Bourge, R.C.; Johnson, A.D.; Wagoner, L.E.; Givertz, M.M.; Liang, C.S.; Neibaur, M.; et al. Intravenous nesiritide, a natriuretic peptide, in the treatment of decompensated congestive heart failure. Nesiritide Study Group. *N. Engl. J. Med.* **2000**, *343*, 246–253. [CrossRef]
83. Zhang, S.; Wang, Z. Effect of recombinant human brain natriuretic peptide (rhBNP) versus nitroglycerin in patients with heart failure: A systematic review and meta-analysis. *Medicine* **2016**, *95*, e4757. [CrossRef]
84. Issa, V.S.; Dinardi, L.F.; Pereira, T.V.; de Almeida, L.K.; Barbosa, T.S.; Benvenutti, L.A.; Ayub-Ferreira, S.M.; Bocchi, E.A. Diagnostic discrepancies in clinical practice: An autopsy study in patients with heart failure. *Medicine* **2017**, *96*, e5978. [CrossRef]
85. Mendis, S.; Thygesen, K.; Kuulasmaa, K.; Giampaoli, S.; Mahonen, M.; Ngu Blackett, K.; Lisheng, L. World Health Organization definition of myocardial infarction: 2008–09 revision. *Int. J. Epidemiol.* **2011**, *40*, 139–146. [CrossRef]
86. Lawler, W. The negative coroner's necropsy: A personal approach and consideration of difficulties. *J. Clin. Pathol* **1990**, *43*, 977–980. [CrossRef]
87. Campuzano, O.; Allegue, C.; Partemi, S.; Iglesias, A.; Oliva, A.; Brugada, R. Negative autopsy and sudden cardiac death. *Int. J. Legal Med.* **2014**, *128*, 599–606. [CrossRef]
88. Maeda, H.; Ishikawa, T.; Michiue, T. Forensic biochemistry for functional investigation of death: Concept and practical application. *Leg. Med.* **2011**, *13*, 55–67. [CrossRef]
89. Woydt, L.; Bernhard, M.; Kirsten, H.; Burkhardt, R.; Hammer, N.; Gries, A.; Dressler, J.; Ondruschka, B. Intra-individual alterations of serum markers routinely used in forensic pathology depending on increasing post-mortem interval. *Sci. Rep.* **2018**, *8*, 12811. [CrossRef]
90. Maeda, H.; Zhu, B.L.; Ishikawa, T.; Quan, L.; Michiue, T. Significance of postmortem biochemistry in determining the cause of death. *Leg. Med.* **2009**, *11*, S46–S49. [CrossRef]
91. Madea, B.; Musshoff, F. Postmortem biochemistry. *Forensic Sci. Int.* **2007**, *165*, 165–171. [CrossRef]
92. Vogiatzidis, K.; Zarogiannis, S.G.; Aidonidis, I.; Solenov, E.I.; Molyvdas, P.A.; Gourgoulianis, K.I.; Hatzoglou, C. Physiology of pericardial fluid production and drainage. *Front. Physiol.* **2015**, *6*, 62. [CrossRef]
93. Mao, R.M.; Zheng, P.P.; Zhu, C.R.; Zhu, B.L. The analysis of pericardial fluid in forensic practice. *Fa Yi Xue Za Zhi* **2010**, *26*, 202–205.
94. Palmiere, C.; Grabherr, S. Biochemical investigations performed in pericardial fluid in forensic cases that underwent postmortem angiography. *Forensic Sci. Int.* **2019**, *297*, e11–e13. [CrossRef]
95. Comment, L.; Reggiani Bonetti, L.; Mangin, P.; Palmiere, C. Measurement of beta-tryptase in postmortem serum, pericardial fluid, urine and vitreous humor in the forensic setting. *Forensic Sci. Int.* **2014**, *240*, 29–34. [CrossRef]
96. Mizutani, T.; Yoshimoto, T.; Ishii, A. Pericardial fluid is suitable as an alternative specimen for the measurement of beta-hydroxybutyrate within 96 h after death. *Leg. Med.* **2018**, *33*, 53–54. [CrossRef]
97. Chen, J.H.; Michiue, T.; Inamori-Kawamoto, O.; Ikeda, S.; Ishikawa, T.; Maeda, H. Comprehensive investigation of postmortem glucose levels in blood and body fluids with regard to the cause of death in forensic autopsy cases. *Leg. Med.* **2015**, *17*, 475–482. [CrossRef]
98. Chen, J.H.; Inamori-Kawamoto, O.; Michiue, T.; Ikeda, S.; Ishikawa, T.; Maeda, H. Cardiac biomarkers in blood, and pericardial and cerebrospinal fluids of forensic autopsy cases: A reassessment with special regard to postmortem interval. *Leg. Med.* **2015**, *17*, 343–350. [CrossRef]
99. Ishikawa, T.; Quan, L.; Michiue, T.; Kawamoto, O.; Wang, Q.; Chen, J.H.; Zhu, B.L.; Maeda, H. Postmortem catecholamine levels in pericardial and cerebrospinal fluids with regard to the cause of death in medicolegal autopsy. *Forensic Sci. Int.* **2013**, *228*, 52–60. [CrossRef]
100. Kounis, N.G.; Koniari, I.; Soufras, G.; Koutsogiannis, N.; Hahalis, G. Specific IgE levels in pericardial and cerebrospinal fluids in forensic casework: The presence of additional molecules for sudden cardiac death diagnosis. *Forensic Sci. Int.* **2018**, *282*, 79. [CrossRef]

101. Zhu, B.L.; Ishikawa, T.; Michiue, T.; Li, D.R.; Zhao, D.; Kamikodai, Y.; Tsuda, K.; Okazaki, S.; Maeda, H. Postmortem cardiac troponin T levels in the blood and pericardial fluid. Part 2: Analysis for application in the diagnosis of sudden cardiac death with regard to pathology. *Leg. Med.* **2006**, *8*, 94–101. [CrossRef] [PubMed]
102. Semenov, A.G.; Seferian, K.R. Biochemistry of the human B-type natriuretic peptide precursor and molecular aspects of its processing. *Clin. Chim. Acta* **2011**, *412*, 850–860. [CrossRef] [PubMed]
103. Yasue, H.; Yoshimura, M.; Sumida, H.; Kikuta, K.; Kugiyama, K.; Jougasaki, M.; Ogawa, H.; Okumura, K.; Mukoyama, M.; Nakao, K. Localization and mechanism of secretion of B-type natriuretic peptide in comparison with those of A-type natriuretic peptide in normal subjects and patients with heart failure. *Circulation* **1994**, *90*, 195–203. [CrossRef]
104. Omland, T.; Aakvaag, A.; Bonarjee, V.V.; Caidahl, K.; Lie, R.T.; Nilsen, D.W.; Sundsfjord, J.A.; Dickstein, K. Plasma brain natriuretic peptide as an indicator of left ventricular systolic function and long-term survival after acute myocardial infarction. Comparison with plasma atrial natriuretic peptide and N-terminal proatrial natriuretic peptide. *Circulation* **1996**, *93*, 1963–1969. [CrossRef] [PubMed]
105. Pfister, R.; Scholz, M.; Wielckens, K.; Erdmann, E.; Schneider, C.A. Use of NT-proBNP in routine testing and comparison to BNP. *Eur. J. Heart Fail.* **2004**, *6*, 289–293. [CrossRef] [PubMed]
106. Kragelund, C.; Gronning, B.; Kober, L.; Hildebrandt, P.; Steffensen, R. N-terminal pro-B-type natriuretic peptide and long-term mortality in stable coronary heart disease. *N. Engl. J. Med.* **2005**, *352*, 666–675. [CrossRef] [PubMed]
107. Nowatzke, W.L.; Cole, T.G. Stability of N-terminal pro-brain natriuretic peptide after storage frozen for one year and after multiple freeze-thaw cycles. *Clin. Chem.* **2003**, *49*, 1560–1562. [CrossRef]
108. Wu, A.H.; Packer, M.; Smith, A.; Bijou, R.; Fink, D.; Mair, J.; Wallentin, L.; Johnston, N.; Feldcamp, C.S.; Haverstick, D.M.; et al. Analytical and clinical evaluation of the Bayer ADVIA Centaur automated B-type natriuretic peptide assay in patients with heart failure: A multisite study. *Clin. Chem.* **2004**, *50*, 867–873. [CrossRef]
109. Koseoglu, M.; Hur, A.; Atay, A.; Cuhadar, S. Effects of hemolysis interferences on routine biochemistry parameters. *Biochem. Med.* **2011**, *21*, 79–85. [CrossRef]
110. Nishiumi, S.; Shima, K.; Azuma, T.; Yoshida, M. Evaluation of a novel system for analyzing hydrophilic blood metabolites. *J. Biosci. Bioeng.* **2017**, *123*, 754–759. [CrossRef]
111. Zhao, Y.; Yan, M.; Chen, C.; Gong, W.; Yin, Z.; Li, H.; Fan, J.; Zhang, X.A.; Wang, D.W.; Zuo, H. MiR-124 aggravates failing hearts by suppressing CD151-facilitated angiogenesis in heart. *Oncotarget* **2018**, *9*, 14382–14396. [CrossRef]
112. Bao, Q.; Chen, L.; Li, J.; Zhao, M.; Wu, S.; Wu, W.; Liu, X. Role of microRNA-124 in cardiomyocyte hypertrophy inducedby angiotensin II. *Cell. Mol. Biol.* **2017**, *63*, 23–27. [CrossRef] [PubMed]
113. Murach, K.A.; McCarthy, J.J. MicroRNAs, heart failure, and aging: Potential interactions with skeletal muscle. *Heart Fail. Rev.* **2017**, *22*, 209–218. [CrossRef] [PubMed]
114. Watson, C.J.; Gupta, S.K.; O'Connell, E.; Thum, S.; Glezeva, N.; Fendrich, J.; Gallagher, J.; Ledwidge, M.; Grote-Levi, L.; McDonald, K.; et al. MicroRNA signatures differentiate preserved from reduced ejection fraction heart failure. *Eur. J. Heart Fail.* **2015**, *17*, 405–415. [CrossRef]
115. Zhang, X.; Sha, M.; Yao, Y.; Da, J.; Jing, D. Increased B-type-natriuretic peptide promotes myocardial cell apoptosis via the B-type-natriuretic peptide/long non-coding RNA LSINCT5/caspase-1/interleukin 1beta signaling pathway. *Mol. Med. Rep.* **2015**, *12*, 6761–6767. [CrossRef]
116. Bi, S.; Wang, C.; Jin, Y.; Lv, Z.; Xing, X.; Lu, Q. Correlation between serum exosome derived miR-208a and acute coronary syndrome. *Int. J. Clin. Exp. Med.* **2015**, *8*, 4275–4280.
117. Staals, R.H.; Pruijn, G.J. The human exosome and disease. *Adv. Exp. Med. Biol.* **2011**, *702*, 132–142. [CrossRef]
118. Ye, W.; Tang, X.; Yang, Z.; Liu, C.; Zhang, X.; Jin, J.; Lyu, J. Plasma-derived exosomes contribute to inflammation via the TLR9-NF-kappaB pathway in chronic heart failure patients. *Mol. Immunol.* **2017**, *87*, 114–121. [CrossRef]
119. Yang, V.K.; Loughran, K.A.; Meola, D.M.; Juhr, C.M.; Thane, K.E.; Davis, A.M.; Hoffman, A.M. Circulating exosome microRNA associated with heart failure secondary to myxomatous mitral valve disease in a naturally occurring canine model. *J. Extracell. Vesicles* **2017**, *6*, 1350088. [CrossRef]

120. Wendt, S.; Goetzenich, A.; Goettsch, C.; Stoppe, C.; Bleilevens, C.; Kraemer, S.; Benstoem, C. Evaluation of the cardioprotective potential of extracellular vesicles-a systematic review and meta-analysis. *Sci. Rep.* **2018**, *8*, 15702. [CrossRef]
121. Gartz, M.; Strande, J.L. Examining the Paracrine Effects of Exosomes in Cardiovascular Disease and Repair. *J. Am. Heart Assoc.* **2018**, *7*. [CrossRef]
122. Poe, A.J.; Knowlton, A.A. Exosomes as agents of change in the cardiovascular system. *J. Mol. Cell. Cardiol.* **2017**, *111*, 40–50. [CrossRef]
123. Sahoo, S.; Mathiyalagan, P.; Hajjar, R.J. Pericardial Fluid Exosomes: A New Material to Treat Cardiovascular Disease. *Mol. Ther.* **2017**, *25*, 568–569. [CrossRef] [PubMed]
124. Beltrami, C.; Besnier, M.; Shantikumar, S.; Shearn, A.I.; Rajakaruna, C.; Laftah, A.; Sessa, F.; Spinetti, G.; Petretto, E.; Angelini, G.D.; et al. Human pericardial fluid contains exosomes enriched with cardiovascular-expressed microRNAs and Promotes therapeutic angiogenesis. *Mol. Ther.* **2017**, *25*, 679–693. [CrossRef] [PubMed]
125. Kuosmanen, S.M.; Hartikainen, J.; Hippelainen, M.; Kokki, H.; Levonen, A.L.; Tavi, P. MicroRNA profiling of pericardial fluid samples from patients with heart failure. *PLoS ONE* **2015**, *10*, e0119646. [CrossRef] [PubMed]
126. Foglio, E.; Puddighinu, G.; Fasanaro, P.; D'Arcangelo, D.; Perrone, G.A.; Mocini, D.; Campanella, C.; Coppola, L.; Logozzi, M.; Azzarito, T.; et al. Exosomal clusterin, identified in the pericardial fluid, improves myocardial performance following MI through epicardial activation, enhanced arteriogenesis and reduced apoptosis. *Int. J. Cardiol.* **2015**, *197*, 333–347. [CrossRef] [PubMed]
127. Maeda, H.; Ishikawa, T.; Michiue, T. Forensic molecular pathology: Its impacts on routine work, education and training. *Leg. Med.* **2014**, *16*, 61–69. [CrossRef]
128. Maeda, H.; Zhu, B.L.; Ishikawa, T.; Michiue, T. Forensic molecular pathology of violent deaths. *Forensic Sci. Int.* **2010**, *203*, 83–92. [CrossRef] [PubMed]
129. Zhao, D.; Ishikawa, T.; Quan, L.; Michiue, T.; Zhu, B.L.; Maeda, H. Postmortem quantitative mRNA analyses of death investigation in forensic pathology: An overview and prospects. *Leg. Med.* **2009**, *11*, S43–S45. [CrossRef]

© 2019 by the authors. Licensee MDPI, Basel, Switzerland. This article is an open access article distributed under the terms and conditions of the Creative Commons Attribution (CC BY) license (http://creativecommons.org/licenses/by/4.0/).

Review

C-Type Natriuretic Peptide: A Multifaceted Paracrine Regulator in the Heart and Vasculature

Amie J. Moyes *,† and Adrian J. Hobbs †

William Harvey Research Institute, Barts and The London School of Medicine & Dentistry, Queen Mary University of London, Charterhouse Square, London EC1M 6BQ, UK; a.j.hobbs@qmul.ac.uk
* Correspondence: a.j.moyes@qmul.ac.uk; Tel.: +44-(0)207-882-5780
† These authors equally contributed to the work.

Received: 15 April 2019; Accepted: 2 May 2019; Published: 8 May 2019

Abstract: C-type natriuretic peptide (CNP) is an autocrine and paracrine mediator released by endothelial cells, cardiomyocytes and fibroblasts that regulates vital physiological functions in the cardiovascular system. These roles are conveyed via two cognate receptors, natriuretic peptide receptor B (NPR-B) and natriuretic peptide receptor C (NPR-C), which activate different signalling pathways that mediate complementary yet distinct cellular responses. Traditionally, CNP has been deemed the endothelial component of the natriuretic peptide system, while its sibling peptides, atrial natriuretic peptide (ANP) and brain natriuretic peptide (BNP), are considered the endocrine guardians of cardiac function and blood volume. However, accumulating evidence indicates that CNP not only modulates vascular tone and blood pressure, but also governs a wide range of cardiovascular effects including the control of inflammation, angiogenesis, smooth muscle and endothelial cell proliferation, atherosclerosis, cardiomyocyte contractility, hypertrophy, fibrosis, and cardiac electrophysiology. This review will focus on the novel physiological functions ascribed to CNP, the receptors/signalling mechanisms involved in mediating its cardioprotective effects, and the development of therapeutics targeting CNP signalling pathways in different disease pathologies.

Keywords: natriuretic peptide; vascular; endothelial cell; cardiomyocyte; fibroblast; inflammation; heart failure; hypertension; angiogenesis

---

1. Introduction

The natriuretic peptides are a family of three structurally related hormones that play unique and distinctive roles within the cardiovascular system. The physiological functions of atrial natriuretic peptide (ANP) and brain natriuretic peptide (BNP) have been intensively investigated over the past few decades, however there has been considerably less focus on C-type natriuretic peptide (CNP). ANP and BNP are expressed in the heart [1–4] and are released in response to a volume-induced stretch of the atria and ventricles, respectively [5,6]. These peptides act as endocrine hormones and contribute to the regulation of cardiac structure, blood pressure and blood volume [7]. In contrast, the tissue distribution and mode of action of CNP is different, with recent studies revealing diverse endogenous roles of CNP including the control of vascular tone, leukocyte activation, angiogenesis, smooth muscle and endothelial cell proliferation, vascular integrity, coronary blood flow, cardiac fibrosis, cardiac hypertrophy, and electrophysiology. These aspects of CNP physiology and pathology will be detailed herein.

2. CNP Expression, Release & Degradation

CNP is a 22 amino acid peptide that is produced following the processing of preproCNP by a signal peptidase and subsequent cleavage of proCNP by the endoprotease furin [8]. Two forms of CNP exist in tissue and plasma, CNP-53 and CNP-22 [8], although the protease responsible for processing the elongated peptide into its shorter, more prevalent form is not known. CNP-22 was initially discovered in extracts from porcine brain [9]. In addition to its abundant expression in the CNS, high levels of CNP are found in chondrocytes [10] and endothelial cells [11,12], which constitutively release the peptide. Other cells within the cardiovascular system, including cardiomyocytes [13,14] and fibroblasts [15], also produce CNP, however, tissue expression and plasma levels are relatively low in healthy individuals, suggesting that CNP most likely acts as a local paracrine/autocrine mediator in the heart and blood vessels.

The half-life of CNP in plasma is short (2.6 min) [16], therefore, degradation is rapid which may account for the low concentrations (fmol–pmol range) of the peptide measured in the circulation [11,17]. There are two main pathways by which CNP is inactivated, cleavage by neutral endopeptidase (NEP) [18], or internalisation by natriuretic peptide receptor C (NPR-C) followed by endocytosis and lysosomal degradation [19,20]. Overall, the contribution of each pathway to the degradation of natriuretic peptides appears to be equal [21] in healthy subjects but there is evidence to suggest that during pathophysiological conditions where natriuretic peptide levels are raised and NPR-C may be saturated, NEP may play a major role in clearance [22]. Furthermore, the tissue distribution of NEP and/or NPR-C may affect CNP inactivation in different organs, for example, CNP is internalised more readily by NPR-C in the kidney compared to the lungs [23].

Most of the stimuli that are known to increase gene expression and/or trigger the release of CNP are pertinent to cardiovascular health including shear stress [24,25], inflammatory cytokines such as tumour necrosis factor (TNF)-α [26], interleukin (IL)1β [26,27], transforming growth factor (TGF)-β [12,28], and bacterial lipopolysaccharide [26,29]. In accordance with these findings are studies showing that plasma levels of CNP are elevated in patients with heart failure (HF) [30] and sepsis [31]. In contrast, CNP release is attenuated by oxidised low-density lipoprotein [32] and vascular endothelial growth factor [33].

3. Natriuretic Peptide Receptors

CNP exerts its biological effects via the activation of two cell surface receptors, natriuretic peptide receptor B (NPR-B, also termed guanylyl cyclase-B, GC-B) and natriuretic peptide receptor C (NPR-C) [34,35]. The peptide has a very low binding affinity for natriuretic peptide receptor A (NPR-A) [36], which is the endogenous receptor for the ligands ANP and BNP. Both NPR-B and NPR-C are widely expressed and are found on endothelial cells [37,38], smooth muscle cells [37,39], cardiomyocytes [14,40], and fibroblasts [15,41]. NPR-C is the most abundant natriuretic peptide receptor and accounts for ~95% of the total natriuretic peptide receptor population in endothelial cells [42].

CNP has a similar binding affinity for NPR-B and NPR-C [36] but the signalling pathways activated by each receptor are markedly different. NPR-B is a particulate guanylyl cyclase receptor that catalyses the conversion of guanosine-5′-triphosphate to cyclic guanosine-3′,5′-monophosphate (cGMP), a second messenger that activates protein kinase G I and II [43–45], which in turn alters cellular functions by phosphorylating specific target proteins. NPR-C was originally considered to be a clearance receptor [46] devoid of signalling activity but later it was shown to contain pertussis toxin sensitive Gi binding domains within the intracellular C-terminal tail that couple to adenylyl cyclase inhibition (by G_i α subunit) and phospholipase C-β activation (by G_i $\beta\gamma$ subunits) [47–50]. Two subtypes of NPR-C have been reported with different molecular masses, a 67-kDa protein and a 77-kDa protein, but it is not known if their capacity to signal and clear natriuretic peptides in vivo is distinct [51,52]. A study in isolated rat glomerular membranes showed that the 67-kDa NPR-C has a high affinity for CNP and activation of this subtype reduces cAMP synthesis via G_i signalling, whereas the 77-kDa receptor has a very low affinity for CNP and is involved in ligand internalization [48,53].

4. CNP Regulates Vascular Tone and Blood Pressure

Pharmacological experiments on isolated blood vessel preparations have shown that CNP is a potent vasodilator of conduit and resistance arteries throughout the vascular tree [54–66]. In the microvasculature, CNP is more efficacious than ANP and BNP, suggesting it may play a role in regulating peripheral vascular resistance [64,67]. Numerous studies have shown that CNP infusion reduces systemic blood pressure in both humans and animals [68–72]. Despite this, the physiological role of endogenous CNP in the cardiovascular system had not been elucidated until recently. Early studies of global CNP knockout (KO) mice were confounded by the effects of CNP deletion on bone development [10]. These animals exhibit skeletal abnormalities, dwarfism, and a high mortality rate, thus, investigations utilising the Cre/Lox recombination system to generate animals with cell-restricted deletion of CNP have been key to gaining a fuller understanding of the function of this peptide in vivo. Endothelial-specific deletion of CNP in mice results in elevated blood pressure and impaired responses to endothelium-dependent vasodilators, providing definitive evidence that the constitutive release of CNP contributes to the regulation of vascular tone [73–75]. The (patho)physiological relevance of these experimental findings is exemplified by the discovery of polymorphisms in the CNP and furin genes that are associated with hypertension in humans [76,77].

CNP-mediated vasodilation occurs via different mechanisms depending on the species, vessel studied, and/or natriuretic peptide receptor activated. In conduit arteries, CNP-induced relaxation is blocked by the dual NPR-A/B antagonist HS-142-1, suggesting NPR-B activation and subsequent production of cGMP mediates the dilatory effects of CNP in large vessels [57,60,63]. However, in the resistance vasculature the importance of NPR-C in the vasoreactivity of CNP increases. In both rodents and humans, a similar pathway exists involving activation of NPR-C and smooth muscle cell hyperpolarisation [66,71,78–80]. In the rat mesenteric artery, the release of CNP accounts for a major component of the endothelium-derived hyperpolarisation (EDH) induced by acetylcholine, a response that can be inhibited by NPR-C antagonists, blockade of small and intermediate conductance calcium-sensitive potassium channels (SK_{Ca} and IK_{Ca}), and G-protein inwardly rectifying potassium channels (GIRK) [65]. It is proposed that the opening of SK_{Ca} and IK_{Ca} on the endothelial cell triggers the release of CNP that binds to NPR-C on the smooth muscle cell resulting in the $G_{i/o}$-mediated activation of GIRK, potassium efflux, and hyperpolarisation. In human arteries, both GIRK and large conductance calcium-activated K^+ channels (BK_{Ca} channels) have been implicated in CNP-evoked vasodilation [79]. Alternatively, studies in rats have shown that NPR-C can also couple to eNOS, resulting in the production of nitric oxide, although, this mechanism has only been reported in larger diameter vessels [81,82].

The receptor that mediates the endogenous regulatory effects of endothelial-derived CNP on vascular tone in vivo is still under debate. Global and smooth muscle-specific NPR-B KO mice are normotensive and their vascular function is normal despite the vasodilator responses to exogenous CNP being impaired [74,83]. It has been proposed that CNP maintains endothelial function independently of smooth muscle NPR-B and that alterations in the production of the vasoconstrictor endothelin-1 account for the elevations in blood pressure observed in ecCNP KO animals, however, the mechanism(s) involved has not been elucidated. A recent study proposed that NPR-B signalling in pericytes may be more important than vascular smooth muscle. This latest research shows that the disruption of NPR-B under the control of the PDGFRβ promotor in mice results in a hypertensive phenotype, indicating CNP may participate in paracrine communication between endothelial cells and pericytes to regulate peripheral vascular resistance [75].

In contrast, accumulating evidence suggests that NPR-C mediates a large proportion of the vasodilator effects of CNP in the vasculature. NPR-C KO mice exhibit impaired endothelial function and a diminished hypotensive response to CNP in vitro and in vivo [73]. The original publication describing global deletion of NPR-C in mice reports a lower blood pressure in these animals (males only). However, more recently it was shown that female NPR-C KO exhibit elevated blood pressure and diminished vascular endothelial function [73,84]. This discrepancy may be due to sex differences in the clearance versus signalling functions of NPR-C in mice, however, data from human genome wide association studies linking variants of the NPR-C gene with hypertension did not find a disparity between sexes, suggesting the NPR-C signalling pathway is equally important in men and women [85]. Patients with the blood pressure elevating NPR-C genotype have lower levels of receptor mRNA and protein in their vascular smooth muscle cells, supporting the theory that diminished CNP activation of NPR-C may underlie this association (as opposed to altered clearance) [86]. Further evidence in favour of a role of NPR-C in the pathogenesis of hypertension was published in a study investigating the effects of the endogenous secretory peptide musclin (also known as osteocrin). Musclin is a competitive ligand at NPR-C and the infusion of musclin increases systolic blood pressure in vivo [87]. Gene expression levels of musclin are elevated in spontaneously hypertensive rats (SHR) and the administration of anti-musclin antibodies reduces blood pressure in these animals, intimating that interference with NP binding at NPR-C by musclin may contribute to the hypertensive state of these animals [87]. However, contrary to this research is a study showing that the infusion of musclin (osteocrin) lowers blood pressure in mice, an effect that is absent in NPR-A KO animals, suggesting that an increase in circulating levels of ANP and/or BNP due to the blockade of peptide clearance by NPR-C accounts for this response [88]. It is possible that musclin interferes with both the signalling and clearance functions of NPR-C and plays a different role in the regulation of blood pressure in healthy and diseased animals. Indeed, the vasoconstrictor and blood pressure elevating effects of musclin are significantly enhanced in SHR compared to normotensive controls, intimating that a change in receptor expression or the signalling/clearance function of NPR-C occurs in hypertension that alters the response to musclin in this model of disease [87]. This is further supported by data demonstrating that NPR-C agonism in SHR attenuates the development of high blood pressure, an effect that is not observed in control Wistar-Kyoto rats [89].

5. CNP Influences Vascular Remodelling and Promotes Angiogenesis

CNP has direct effects on the mitogenesis of endothelial and smooth muscle cells and it promotes wound healing and vascular repair by stimulating endothelial growth, whilst concomitantly inhibiting smooth muscle cell proliferation. This dual protective role of CNP was first described in animal models of vein graft and balloon angioplasty, clearly showing that CNP treatment accelerates re-endothelialisation and reduces deleterious neointimal hyperplasia [90–92]. A similar response to CNP has been observed in carotid arteries subjected to physical damage [93]. Many of these studies report an increase in cGMP production following treatment with CNP [91,93–95], intimating the involvement of NPR-B, however, others have shown that CNP influences the growth of endothelial and smooth muscle cells via NPR-C in a cGMP-independent manner. These experiments revealed that the pro- and anti-mitogenic effects of CNP are mediated by the extracellular signal-related kinase (ERK) 1/2 and can be blocked by the NPR-C antagonist, M372049, and by the $G_{i/o}$ inhibitor, *Pertussis* toxin, despite significant increases in cGMP production by both cell types [37,96]. Activation of ERK 1/2 by CNP results in the enhanced expression of cell cycle promotors (cyclin D1) in endothelial cells and inhibitory cell cycle proteins in smooth muscle cells (p21 and p27). This is further supported by the observation that primary microvascular lung endothelial cells, isolated from NPR-C KO mice, proliferate more slowly than wildtype (WT) cells, whilst aortic smooth muscle cells, isolated from KO animals, grow at a faster rate [37]. Indeed, in vivo studies show that mice lacking endothelial-derived CNP and NPR-C exhibit slower wound healing and greater intimal hyperplasia following vascular injury, indicating that vascular CNP release is a vital step in tissue repair [97].

The ability of CNP to influence endothelial cell growth led researchers to question the role of this peptide in angiogenesis. The potential angiogenic effects of CNP were initially tested in classical assays of endothelial tube formation in vitro and revealed that CNP-induced increases in capillary network formation are of a similar magnitude to the potent pro-angiogenic mediator, VEGF [98]. In addition to this, the gene transfer of CNP directly into ischaemic muscle has been reported to enhance blood flow recovery and increase capillary density following ligation and excision of the femoral artery in mice [98]. Research concurs that these angiogenic responses are dependent on the activation of ERK 1/2, however, there are opposing data published regarding the receptor involved.

A comprehensive study performed in KO animals suggests that the endogenous effects of endothelial-derived CNP on angiogenesis are mediated by NPR-C, whereas both receptors are implicated when CNP is administered pharmacologically. For example, branching angiogenesis in human umbilical vein endothelial cells (HUVEC) has been shown to be blocked by an inhibitor of cGMP-dependent protein kinase, suggesting the involvement of NPR-B signalling [98]. In contrast, tube formation in murine pulmonary endothelial cells is inhibited by *Pertussis* toxin and NPR-C antagonism [97]. Experiments performed in transgenic mice show that basal endothelial tubule formation, de novo aortic sprouting, and restoration of blood flow following hindlimb ischaemia is diminished in ecCNP KO and NPR-C KO tissues/animals, whilst NPR-B KO display a similar angiogenic capacity to WT mice [97]. In addition to this, the same study reported that patients with critical limb ischaemia have lower levels of CNP and NPR-C in biopsies of the gastrocnemius muscle, suggesting that diminished signalling via this pathway may contribute to the insufficient angiogenic response to hypoxia associated with peripheral arterial disease.

While the majority of studies indicate that CNP promotes angiogenesis, there is also evidence demonstrating that the NPR-C agonist cANF$^{4\text{-}23}$ reduces neovascularization in murine sponge implants [99]. This finding was accompanied by reduced levels of VEGF which corroborates with other studies showing that CNP and cANF$^{4\text{-}23}$ inhibit VEGF expression and signalling in vascular smooth muscle and endothelial cells [100]. Contrary to this, VEGF has also been shown to reduce CNP secretion from cultured endothelial cells [33], suggesting there may be a reciprocal relationship between the two vascular mediators, however, it is not known if an interplay between the two factors modulates angiogenesis.

6. CNP Inhibits Inflammation and Slows the Development of Atherosclerosis

The first indication that CNP may influence the inflammatory response to infection and disease comes from research showing that the cytokines IL-1α, IL-1β, and tumour necrosis factor (TNF)α stimulate the release of CNP from endothelial cells [26,27]. The most potent of these cytokines (at inducing CNP secretion) is TNFα, which is released by macrophages during the acute phase of inflammation. Another strong stimulus for triggering CNP release from the endothelium is the endotoxin bacterial lipopolysaccharide (LPS) [26,29]. Indeed, CNP levels are markedly increased in patients with septic shock, a 5–10-fold increase in plasma CNP concentrations has been reported in several studies [31,101–103]. Furthermore, plasma concentrations of NT-proCNP are strongly associated with inflammation-induced organ dysfunction and are predictive of a detrimental outcome [101,104]. It has also been suggested that measurement of NT-proCNP in the early phase of septic shock might help to predict the emergence of sepsis-induced encephalopathy [103]. Together, these data suggest that the acute release of CNP may modulate the progression of sepsis and other inflammatory disorders.

Endothelial activation by inflammatory mediators is a key event in the pathogenesis of sepsis and cardiovascular diseases such as atherosclerosis. Changes in the expression of cell adhesion molecules, such as integrins and selectins, facilitate the recruitment and adherence of leukocytes during the initial phase of the immune response [105,106]. CNP has been shown to dampen endothelial activation induced by range of inflammatory stimuli both in vitro and in vivo. IL-1β and histamine-induced leukocyte rolling in murine post-capillary venules is inhibited by CNP and cANF$^{4\text{-}23}$ via the suppression of P-selectin expression [107]. CNP infusion via mini-pump inhibits LPS-stimulated leukocyte infiltration into the lungs, attenuates E-selectin gene expression, and reduces the levels of the inflammatory mediators TNFα, macrophage inflammatory protein-2, monocyte chemoattractant protein-1 (MCP-1), and interleukin-6 (IL-6) [108]. In addition, CNP inhibits elevations of intercellular adhesion molecule-1, vascular cell adhesion molecule-1, E-selectin, and P-selectin expression in HUVECs stimulated with LPS [109]. The mechanism by which CNP attenuates this response at least in vitro is via the inhibition of pro-inflammatory NF-κB and p38 signalling pathways and the activation of the pro-survival PI3K/Akt pathway.

Endothelial-derived CNP also appears to maintain a resting anti-inflammatory influence on the vascular wall as ecCNP KO mice exhibit greater leukocyte rolling at baseline prior to stimulation with an inflammogen [73]. In addition, the response to LPS and TNFα-induced peritonitis is significantly exacerbated in these animals. The anti-leukocyte effects of the endogenous peptide may involve similar mechanisms to those induced by an exogenous application of the peptide, as higher vascular P-selectin expression is observed in mice lacking endothelial CNP. Given that a similar increase in leukocyte recruitment was observed in NPR-C KO mice following treatment with LPS, it is hypothesised that NPR-C-driven suppression of cell adhesion molecule expression may underpin the immune dampening effect of CNP [73]. This facet of CNP biology is clearly important in the context of atherosclerosis as genetic ablation of CNP leads to an increase in the development of atherosclerotic lesions, greater infiltration of macrophages, and the formation of aortic and abdominal aneurysms in ecCNP/ApoE double-KO mice [73]. Indeed, this finding fits with previous work demonstrating reduced CNP immunoreactivity in diseased human coronary arteries and the inverse relationship discovered between the expression of CNP and lesion severity [110]. Moreover, CNP inhibits the proliferative and pro-migratory effects of oxidised LDL on smooth muscle cells which may affect the growth and stability of atherosclerotic plaques [111]. The spontaneous development of aneurysms in ApoE KO mice is rare, however, ecCNP/ApoE double-KO mice are more susceptible to this phenomenon, suggesting that CNP may help to maintain the structural integrity of the vessel wall [73]. Interestingly, aneurysms were only observed in male double-KO mice, this observation aligns with the human condition that predominantly affects the elderly male population [112]. It is possible that CNP regulates the expression and release of matrix metalloproteinases (MMPs) which are implicated in the development of aneurysms. In support of this thesis are data showing that CNP modulates the expression of MMP-2 and MMP-9 in chondrocytes and the kidney [113,114].

There are a number of other inflammatory disorders where CNP has proven beneficial in experimental models of disease. CNP reduces the number of macrophages, neutrophils, and lymphocytes accumulating in the lungs of mice exposed to bleomycin in a model of pulmonary hypertension [115]. In a rat model of haemorrhagic shock, CNP reduces markers of oxidative stress and the expression of tumour necrosis factor (TNF)-α, interleukin (IL)-6, and IL-1β in the kidney, suggesting it may improve symptoms of acute renal injury associated with this condition [116]. Furthermore, studies performed using transgenic mice overexpressing CNP in endothelial cells suggest that CNP regulates inflammation associated with obesity. Overexpression of endothelial-derived CNP improves glucose tolerance, decreases insulin resistance, and inhibits adipose macrophage infiltration in mice that are fed a high-fat diet [117]. Using the same animals, these authors also demonstrate that CNP inhibits expression of inflammatory markers IL-6, MCP-1, and CD68 in the liver of mice fed high fat diets in a model of non-alcoholic steatohepatitis [118]. Thus, the anti-inflammatory benefits of endothelial-derived CNP are not entirely limited to cardiovascular disease.

7. CNP is a Novel Regulator of Cardiac Structure and Function

7.1. CNP and HF

For many years the role of CNP in the heart was largely ignored as the majority of research focused on the cardiac hormones ANP and BNP. In addition to this, the expression of CNP in cardiomyocytes is much lower than that of ANP and BNP, suggesting that under basal conditions it does not play a major role in regulating cardiac function [14,119]. However, it has been widely reported that cardiac gene expression and plasma levels of CNP are increased in patients with HF [13,30,120–123]. Elevated circulating levels of CNP are associated with a high-risk phenotype in patients with cardiovascular comorbidities and left ventricular dysfunction [17]. Furthermore, plasma NT-pro CNP levels in patients with HF are correlated with disease severity and are a strong predictor of all cause mortality and hospitalization in patients with HF with preserved ejection fraction (HFpEF) [124]. Yet these studies do not tell us if CNP is produced by the heart or by other organs during HF. A key experiment comparing plasma CNP levels in the coronary sinus and aortic root of failing hearts discovered that concentrations of CNP are significantly higher in the coronary bloodstream than those measured in the systemic circulation, providing the first direct evidence that CNP is released by the heart in HF [125].

In non-failing hearts, levels of CNP are higher in the atria than in the ventricles but studies in mini-pigs have shown that cardiac pacing induces a 15-fold increase in CNP expression in the ventricles along with elevated levels of CNP protein, demonstrating that acute cardiac stress elicits immediate upregulation of the gene and an increase in CNP release [13]. There also appears to be a switch between natriuretic peptide signalling in the failing ventricle. In sham hearts, ANP induces a greater increase in guanylyl cyclase activity than CNP, however, in pressure overload-induced HF, CNP elicits twice as much cGMP production than ANP. This might be due to a reduction in NPR-A expression, suggesting that CNP signalling via NPRB may be more important during HF [126]. However, other studies have reported that NPR-B expression decreases in the ventricles of the failing heart [13,127], whereas NPR-C increases are the most pronounced of the three NP receptors in end-stage disease [128].

7.2. CNP Directly Modulates Cardiomyocyte Contractility, Fibrosis, And Hypertrophy

CNP has been shown to exert direct effects on cardiac contractility, although both positive and negative inotropic responses have been reported. For example, CNP increases myocyte contractile force in canine isolated atrial and ventricular preparations [129,130], whereas positive lusitropic and negative inotropic effects have been observed in rat heart muscle preparations [131,132]. A number of studies have shown that CNP induces phosphorylation of the sarcoplasmic reticulum calcium pump (SERCA) 2 regulator, phospholamban (PLB), and cardiac troponin I (cTnI), a regulatory protein that controls the calcium-mediated interaction between actin and myosin [131–134]. The positive lusitropic and inotropic effects of CNP reported in the failing rat heart are associated with phosphorylation of both of these regulatory proteins in addition to an increase in sarcoplasmic reticulum (SR) calcium load [132]. Further investigations demonstrated that this negative inotropic effect of CNP is sensitive to PKG inhibition (i.e., NPR-B-dependent), SERCA2 inhibition, and is absent in SERCA2 KO mice [135]. It is proposed that an increase in SERCA2 activation via phosphorylation of PLB by CNP causes a higher fraction of the cytosolic Ca^{2+} to be sequestered back into the SR, therefore reducing Ca^{2+} activation of the myofilaments resulting in a negative inotropic effect. In contrast, others have demonstrated that CNP induces a positive inotropic response in the heart and that PLB phosphorylation results in a greater uptake of Ca^{2+} into the SR, creating a larger pool of Ca^{2+} available for contraction [134].

Biphasic responses to CNP have also been reported by a number of studies, where an initial, transient positive inotropic response is followed by a slow developing reduction in contractility [133,135,136]. The nature of this biphasic response was investigated to elucidate if the two opposing effects were due to activation of different receptors, however it appears that both phases of the response are mediated by NPR-B as a cGMP analogue mimicked both the immediate and delayed phase of the contractile response. The NPR-C agonist cANF$^{4-23}$ did not affect contractility and no changes in cardiac cAMP were observed [133]. The reason why there is such ambiguity in the contractile responses elicited by CNP could be attributed to cross-talk between the cGMP and cAMP signalling systems in the heart. It has been shown that the negative inotropic response to CNP is the dominating effect when the cAMP signalling is reduced (e.g., during β-adrenoceptor blockade), whilst the effect is completely lost in the presence of maximal β-adrenoceptor stimulation by isoprenaline [135]. In both failing and non-failing hearts, CNP increases the positive inotropic effect of β-adrenoceptor stimulation due to cGMP inhibition of phosphodiesterase (PDE)3, an enzyme responsible for the breakdown of cAMP produced during β-adrenoceptor stimulation [137,138]. Thus, the contractile effect of CNP observed in different experimental models is likely to be influenced by intracellular cAMP levels and β-adrenoceptor stimulation. NPR-C signalling may also influence cardiac contractility, although this has not been directly investigated, receptor stimulation would likely result in a reduction in cAMP via the inhibition of adenylyl cyclase [50,139]. cANF$^{4-23}$ has also been shown to inhibit L-type calcium currents in atrial myocytes [140], if a similar pathway is present in ventricular myocytes NPR-C activation may induce a negative inotropic response.

Cardiac remodelling during HF is characterised by fibroblast proliferation, myofibroblast transformation, and collagen deposition resulting in the development of cardiac fibrosis. This leads to ventricular distortion and myocardial stiffness, which has significant consequences for heart function [141]. CNP exerts anti-fibrotic effects in the heart and is significantly more potent at reducing fibroblast growth and extracellular matrix production than other natriuretic peptides [15]. CNP is expressed and released by cardiac fibroblasts in response to the basic fibroblast growth factor (BFGF), TGFβ, and endothelin 1. It induces a greater increase in cGMP and more suppression of collagen synthesis than ANP and BNP [15]. In addition, fibroblast differentiation, migration, and the production of the pathologic mediators, MCP-1 and PAI-1, are attenuated by CNP in vitro [128,142,143].

In addition to attenuating fibrosis, CNP also exerts anti-hypertrophic effects in the heart. Experiments performed in isolated rat cardiomyocytes have shown that CNP inhibits basal and endothelin-1-induced protein expression, ANP secretion, and the expression of the hypertrophic genes GATA-4 and MEF-2. Endothelin-1-induced increases in calcium/calmodulin-dependent kinase and ERK activities are also attenuated by CNP. These effects are recapitulated using a cGMP analogue, suggesting that the mechanism involves activation of NPR-B [144]. Similarly, CNP has been shown to reduce angiotensin II-induced increases in murine cardiomyocyte size, indicating that CNP directly supresses hypertrophic signalling cascades [145].

8. Endogenous CNP Is Cardioprotective in Animal Models of Heart Failure

The generation of tissue-specific knockouts has facilitated a greater understanding of the cell types responsible for CNP release in the heart and how each source of CNP impacts cardiac structure and function in disease. At baseline, mice lacking cardiomyocyte- (cmCNP KO) and fibroblast-derived CNP (fbCNP KO) exhibit no overt changes in cardiac contractility, structure, or fibrosis, confirming previous speculation that CNP plays a minimal role in healthy hearts [145]. However, following aortic banding (pressure overload-induced HF) both cmCNP KO and fbCNP KO mice display a greater decline in ejection fraction, increased ventricular dilation, greater cardiac hypertrophy (cmCNP KO only), and more collagen deposition compared to littermate controls. In contrast, endothelial-derived CNP does not appear to contribute to cardioprotection, at least in this model. Thus, endogenous CNP secreted from cardiomyocytes and fibroblasts reduces the deleterious pathological changes that occur during heart failure. Comparable cardiac dysfunction, hypertrophy, and fibrosis is observed in NPR-C KO animals subjected to aortic banding, suggesting that NPR-C mediates the effects of CNP in myocytes and fibroblasts. Indeed, CNP infusion via mini-pump reverses cardiac dysfunction and fibrosis during HF in WT animals but not NPR-C KO mice. cmCNP KO animals fared worse than WT animals upon stimulation with isoprenaline (i.e., sympathetic hyperactivation models of HF), whilst the loss of NPR-B did not adversely affect the hypertrophic or fibrotic response [145]. This contrasts with previous studies that show transgenic rats expressing a dominant negative form of NPR-B exhibit cardiac hypertrophy at baseline [146]. However, these mutants do not exhibit cardiac fibrosis, nor changes in contractile function before or after chronic volume overload, so perhaps, in the longer-term, NPR-B plays a predominant role in regulating compensatory hypertrophy, whereas NPR-C regulates the maladaptive hypertrophy and anti-fibrotic effects of CNP. In support of this are data showing that cardiomyocyte-specific NPR-B deletion does not alter the response to pressure overload-induced HF in mice [147], intimating that endogenous NPR-B signalling is either not vital in pathologic remodelling in the heart, or another system compensates for the loss of NPR-B. However, NPR-B heterozygote mice are susceptible to aortic stenosis [148], suggesting that the importance of this NPR subtype might sit outside the cardiomyocyte. This does not mean that NPR-B cannot be targeted pharmacologically, as recent studies suggest that novel designer peptides that bind to NPR-B can reduce fibroblast proliferation and collagen secretion in vitro and in vivo [128,149,150]. This role of endogenous NPR-C signalling in regulating fibrosis is supported by other studies that have observed greater cardiac dysfunction, atrial collagen deposition, and higher levels of TGFβ and TIMP1 in NPR-C KO mice subjected to angiotensin II-induced pressure overload [151,152]. Furthermore, a functional genetic variant in NPR-C has been discovered in humans that is associated with diastolic dysfunction. This single nucleotide polymorphism (SNP) does not affect the protein expression of NPR-C or the circulating plasma levels of natriuretic peptides, suggesting that downstream signalling is affected. It is postulated that this SNP leads to dysfunction of the catalytic domain of NPR-C, and that aberrant signalling in fibroblasts contributes to cardiac fibrosis and impaired diastolic function [153]. The precise mechanism by which NPR-C signalling inhibits fibroblast proliferation/collagen synthesis is unknown, but it has been shown that CNP- and cANF[4-23]-mediated stimulation of NPR-C can activate a non-selective cation current that is partly carried by transient receptor potential C channels, and the authors tentatively suggest that this may affect the secretory state of the cell [154].

However, it appears that NPR-C may play a Janus-faced role in HF as other work suggests that the removal/blockade of the clearance function of NPR-C is beneficial in cardiac disease. NPR-C KO mice cross-bred with animals that spontaneously develop atrial fibrosis (TGFβ1 overexpression) display significantly less fibrosis and collagen deposition than controls. Also, NPR-C knockdown in cultured fibroblasts stimulated with TGFβ1 results in a lower expression of pro-fibrotic markers pSmad and collagen. These effects are reversed by NPR-A knockdown, intimating that the reduced clearance of ANP and the subsequent increase in ANP signalling underlies this effect [88,155]. In addition, transgenic mice overexpressing osteocrin (OSTN-Tg) have an improved prognosis and higher survival rates after myocardial infarction (MI) [88]. ANP and CNP levels are elevated in OSTN-Tg mice, thus the cardioprotective effects of osteocrin in this model have been ascribed to the inhibition of NPR-C-mediated natriuretic peptide clearance. Clearly, further investigation is required to understand more about the switch between NPR-C signalling and clearance and if the balance changes in different pathological conditions contributing to cardiac disease (e.g., pressure overload or MI).

9. Coronary Vasodilator Effects of CNP

The first studies of the vascular actions of CNP in the heart were performed in porcine coronary arteries. These early experiments provided the first evidence that CNP exerts coronary vascular relaxation via hyperpolarisation. CNP responses could be inhibited by the potassium channel blockers charybdotoxin and glibenclamide [55]. The same authors also discovered that HS-142-1 inhibits reductions in coronary flow induced by CNP in dogs, suggesting there is a NPR-B/cGMP component of CNP relaxation in the heart [156]. In contrast, studies in rodent Langendorff-perfused hearts showed that CNP and cANF[4-23] induce reductions in coronary perfusion pressure via the activation of NPR-C and the opening of GIRK channels, in a mechanism analogous to the mesenteric vasculature [157]. Furthermore, increases in CNP peptide could be measured in coronary effluent following stimulation with ACh, suggesting CNP is released as an endothelium-derived hyperpolarising factor (EDHF) by coronary vessels. Intriguingly, in ecCNP KO and NPR-C KO mice, the response to endothelium-dependent vasodilators and flow-mediated dilatation (a shear stress response) are diminished [145]. Thus, CNP may be released during cardiac stress in response to changes in flow. Interestingly, NT-pro CNP levels predict mortality and cardiac readmission in patients with unstable angina, a condition characterised by high wall shear and altered coronary vascular flow [158].

10. Role of CNP in Ischemia Reperfusion Injury and MI

Microvascular obstruction is a pathological feature of acute MI and frequently occurs despite the restoration of flow to ischaemic tissue following coronary interventions. Two major contributing factors are impaired vasodilation and neutrophil plugging, which lead to mechanical obstruction of the vessels and the release of oxidants and pro-inflammatory mediators [159]. Given the coronary vasodilator capacity of CNP and its release in response to shear stress, it is hypothesised that this vasoactive mediator may improve coronary flow and reduce tissue damage by inhibiting inflammatory cell accumulation and obstruction of the coronary vessels. Data from human studies show that CNP gene expression is elevated in failing ischaemic hearts, suggesting it may play a role in the physiological protective response during MI [160]. The acute effects of CNP during myocardial ischaemia reperfusion (I/R) injury have been studied in isolated hearts, an experimental system devoid of circulating inflammatory cells. In this setting, infusion of CNP attenuates the increase in coronary perfusion pressure during reperfusion, reduces infarct size, and improves left ventricular contractility [157]. The protective effect of CNP in this model is abolished by the NPR-C antagonist M372049 and recapitulated by the infusion of cANF$^{4-23}$. Furthermore, a larger infarct size and poorer functional recovery of the heart has been observed in NPR-C KO animals subjected to the same experimental protocol, intimating that NPR-C activation by CNP is beneficial during I/R [145]. However, it should be noted that the coronary vasodilator responses to CNP are not completely abolished in NPR-C KO mice, suggesting that NPR-B activation may, in part, mediate some of the vasorelaxant effects of CNP in the heart. It is also likely that NPR-B-mediated increases in cGMP/PKG I-signalling contribute to the cardioprotective effects of CNP following ischaemia [161,162].

Patients with pre-existing microvascular dysfunction are more vulnerable to myocardial injury following percutaneous coronary intervention, therefore one might expect that a loss of endothelial-derived CNP would make the heart more susceptible to damage following I/R. However, genetic ablation of CNP from cardiomyocytes results in poorer recovery from I/R injury, whilst deletion of endothelial CNP does not worsen the phenotype [145]. It is possible that cardiomyocyte-derived CNP has a direct effect on contractility following I/R in the isolated heart, however the mechanism involved has not been investigated. It has been postulated that NPR-C coupling to K_{ATP} channels may confer the beneficial effects of CNP during I/R injury as CNP can induce the opening of K_{ATP} which is known to reduce cardiac and metabolic stress during ischaemic injury [162].

The effect of CNP overexpression and knockdown has also been explored in chronic models of MI with the aim of understanding its role in mediating inflammation and cardiac remodelling in the long term. The latest research employing CNP gene silencing in rats demonstrates that abrogation of the endogenous production of CNP by cardiomyocytes results in a larger infarct size following I/R, greater cardiac fibrosis, and an increase in the inflammatory markers TNFα and IL-6 [163]. In contrast, others have reported that cardiomyocyte overexpression of CNP does not affect infarct size but does reduce cardiac hypertrophy and the number of mononuclear infiltrates observed in the myocardium [164]. Similarly, chronic infusion of CNP in a model of permanent coronary artery ligation reduces left ventricular enlargement, collagen deposition, and increases cardiac output [165]. In addition to this, an increase in CNP expression has been reported in the infarct border zone in swine hearts, where it is believed it may contribute to myocardial restoration by increasing capillary density (dovetailing well with the pro-angiogenic actions of the peptide) [166]. Together, these findings suggest that CNP could be a therapeutic target in MI as it is effective at reducing infarct size, cardiac inflammation, and the adverse ventricular remodelling that occurs following MI which may slow the progression of HF.

11. CNP Regulates Heart Rate and Electrical Conduction in the Sinoatrial Node (SAN)

CNP affects heart rate via two mechanisms, the alteration of ionic currents in the SAN, and the modulation of sympathetic drive. It has been reported to induce both positive and negative chronotropic effects in the heart via the modulation of L-type Ca^{2+} currents in the SAN [129,140,167,168]. Under basal conditions or mild stimulation with a β-adrenoceptor agonist, CNP elicits an increase in heart rate and electrical conduction through the SAN. This response is attenuated by the PDE3 inhibitor milrinone, suggesting that the mechanism involves NPR-B/cGMP-mediated inhibition of PDE3 and an increase in cAMP, akin to the positive inotropic effects of CNP observed in myocytes [169]. In contrast, when heart rate is elevated, CNP induces a negative chronotropic effect and decreases conduction velocity within the SAN. NPR-C is believed to mediate this response as cANF$^{4-23}$ reduces the chronotropic effect of isoprenaline but has no effect under basal conditions [170]. The importance of NPR-C signalling in the SAN has been demonstrated in studies using KO mice. Deletion of NPR-C results in SAN dysfunction, prolongation of SAN recovery time, and increased susceptibility to atrial fibrillation [171]. These mice also exhibit atrial fibrosis at baseline which is thought to contribute to aberrant SAN conduction. This is exacerbated in models of heart failure, although treatment with cANF$^{4-23}$ reduces the number of arrhythmias and the changes in electrophysiology [151].

The second mechanism by which CNP modulates heart rate is via the inhibition of cardiacsympathetic neurotransmission in the heart. CNP treatment reduces tachycardia during right stellate (sympathetic) ganglion stimulation in rats and inhibits the release of norepinephrine from isolated atria [172]. Evidence for an endogenous CNP/NPR-B pathway regulating sympathetic activity is demonstrated in transgenic rats with neuron-specific overexpression of a dominant negative form of NPR-B. These animals exhibit elevated heart rates, greater heart rate variability, and frequency domain analyses reveal a higher low-frequency (LF)/high-frequency (HF) ratio, indicative of a shift towards sympathoexcitation [172]. Similar findings have also been reported in mice lacking NPR-C. These animals display a reduction in circadian changes of heart rate, a loss of dynamic changes due to alterations in activity, and a greater LF/HF ratio, suggesting that sympathetic activity is enhanced [173]. Nevertheless, regardless of the receptor that mediates this sympatho-inhibitory effect of CNP, the ability to dampen sympathetic activity in the heart may be an important protective mechanism in diseases characterised by autonomic dysregulation.

12. Current and Future Therapeutics

The past decade has yielded a vast amount of evidence supporting a broad homeostatic role for CNP in maintaining vascular and cardiac function. Moreover, the latest research employing transgenic models has enabled a greater depth of understanding of the key physiological functions of the peptide in the cardiovascular system. Therapeutics designed to bind to the cognate receptors for CNP could have wide-ranging clinical applications in diseases such as hypertension, atherosclerosis, restenosis, critical limb ischaemia, peripheral arterial disease, I/R injury, MI, HF, and heart rhythm disorders. Currently, there are two therapies that target the natriuretic peptide system that have been tested in clinical trials, NEP inhibitors (inhibit the breakdown of natriuretic peptides) and cenderitide (a chimeric NPR-A/NPR-B agonist). NEP inhibitors, used in combination with angiotensin converting enzyme inhibitors (ACEi), have been trialled in patients with hypertension and HF. Initial results were promising, however there was a higher occurrence of angioedema reported in patients on dual treatment compared to ACE inhibitor alone, therefore development was halted [174]. However, the NEP inhibitor sacubitril, given in combination with the angiotensin receptor blocker (ARB) valsartan (LCZ696), has been used with more success. This drug appears to be more efficacious at reducing blood pressure than the currently available ACEi and ARBs, with a similar safety and tolerability profile [175,176]. Furthermore, LCZ696 had impressive results in the PARADIGM-HF trial for the treatment of patients with HF and reduced ejection fraction (EF). The results from the trial showed significantly greater benefits of this combination therapy compared to standard therapy (ACEi treatment alone) [177]. A significant reduction in cardiovascular mortality and heart failure related hospitalization (20%) was reported and the trial was terminated early due to the overwhelming benefit with regard to the primary endpoint. LCZ696 is now licensed as Entresto and is currently being used in a clinical trial (PARAGON-HF) for patients with HFpEF [178]. Given that hypertension is common in this group of patients and the disease is associated with reduced cGMP availability [179], boosting natriuretic levels may be advantageous. Theoretically, NEP inhibition could increase the levels of all natriuretic peptides and enhance their beneficial effects, however, it should be noted that NEP also cleaves other vasoactive peptides, such as bradykinin, so the outcome of this treatment may not be solely down to a reduction of natriuretic peptide degradation. Although, higher levels of cGMP and BNP have been reported in patients receiving LCZ696, indicating that elevated natriuretic peptide levels likely contribute to the protective effect of NEP inhibition. Moreover, CNP is more susceptible than ANP and BNP to NEP degradation, so it may be an important contributor to LCZ696 efficacy [18].

Cenderitide (CD-NP) is a novel 'designer' natriuretic peptide that consists of CNP plus the C terminus of *Dendroaspis* natriuretic peptide (isolated from the green mamba snake). It is a dual

NPR-A/NPR-B agonist that has been engineered to harness the anti-fibrotic, anti-proliferative, and vascular regenerating properties of CNP and the beneficial renal effects of NPR-A activation [180]. A key benefit of this drug is the fact that it is more resistant to NEP degradation than the native natriuretic peptides [181]. The first clinical target of this drug is HF as it has proven to be efficacious in a rat model of early stage disease in which CD-NP reduces fibrosis and diastolic dysfunction [149]. In addition, CD-NP causes a greater reduction in collagen production by human cardiac fibroblasts than BNP or CNP alone [128,182]. It has been suggested that targeting NPR-B in the heart could potentially be detrimental if NPR-B/cGMP signalling increases adrenergic drive (via PDE3 inhibition) in vivo as it has been shown to do in vitro [138]. Indeed, clinical studies of the PDE3 inhibitor milrinone, demonstrate increased mortality, sudden death, and arrhythmias in HF patients, so the effects of NPR-B agonists on PDE3 activity would need to be investigated thoroughly. However, the first trial in man has shown that Cenderitide is safe, well-tolerated, and causes increases in plasma and urinary cGMP in patients with HF, suggesting that this could prove a promising therapeutic agent in the future [182]. More recently, Burnett et al. have developed other designer natriuretic peptides, such as C53, a long-acting NPR-B activator that is resistant to NEP and has limited interaction with NPR-C which elicits potent anti-fibrotic effects in renal and cardiac fibroblasts [150]. The newest compound in this drug development pipeline, CRRL269, a non-hypotensive activator of NPR-A, is being considered for use in acute kidney injury [183].

The rationale for the development of NPR-C agonists came from studies indicating that this receptor mediates a large proportion of the effects of CNP on vascular tone [65,73], in addition, NPR-C mutations are linked to hypertension in GWAS [85]. NPR-C has also been shown to mediate, at least in part, the endogenous effects of CNP in failing hearts, vascular regeneration/angiogenesis, and inflammation. Furthermore, NPR-C is the receptor that is upregulated the most in HF. Targeting NPR-C could also potentially avoid effects on bone development, which are mediated primarily by NPR-B. Thus, small molecule agonists of NPR-C have been designed according to the crystal structure of the receptor bound with CNP [184] and the selective antagonist M372049 [185]. The lead compound 118 has been shown to reduce blood pressure in vivo and relaxes mesenteric arteries in vitro [73]. Furthermore, 118 has high affinity and slow dissociation characteristics at the receptor so it could compete for the clearance function of NPR-C. Accordingly, NPR-C agonists may have the additional benefit of being able to reduce the degradation of all natriuretic peptides and could have broader therapeutic effects than those conferred by NPR-C signalling alone. Further development and optimisation are ongoing.

13. Summary

CNP drives a multitude of cardiac and vascular protective effects via its two cognate receptors, NPR-B and NPR-C. These beneficial actions are mediated by a number of distinct molecular pathways (Figure 1). Pharmacological targeting of NPR-B and/or NPR-C harnesses these salutary functions and holds wide-reaching therapeutic promise for cardiovascular disease.

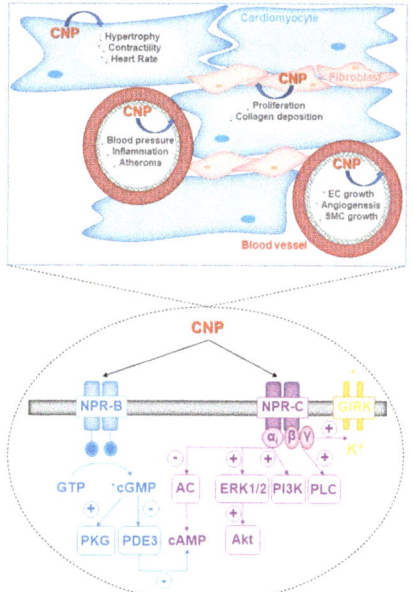

Figure 1. (Patho)physiological functions (upper panel) and signalling pathways (lower panel) activated by CNP in endothelial cells, cardiomyocytes, and fibroblasts. In the heart, CNP reduces cardiac hypertrophy, inhibits fibrosis, and modulates myocyte inotropy and chronotropy. In the vasculature, CNP lowers blood pressure, inhibits inflammation, reduces atherosclerotic plaque deposition, modulates endothelial cell (EC) and smooth muscle cell (SMC) growth, and stimulates angiogenesis. The cellular effects of CNP are mediated via two cognate receptors, NPR-B and NPR-C. NPR-B is a particulate guanylyl cyclase receptor and stimulation results in the production of cGMP and the activation of protein kinase G (PKG) I. NPR-C is G_i protein-linked receptor that modulates various intracellular enzymes including adenylyl cyclase (AC), phospholipase C (PLC), extracellular signal-related kinase (ERK) 1/2, phosphoinositide-3-kinase (PI3K), and protein kinase B (Akt). NPR-C activation also triggers the opening of G-protein gated inwardly rectifying potassium (GIRK) channels. Cross-talk occurs between the two receptor signalling pathways via cGMP-mediated inhibition of phosphodiesterase (PDE) 3, the enzyme responsible for the hydrolysis of cAMP in cardiomyocytes.

Conflicts of Interest: Adrian J. Hobbs is a scientific advisory board member for Palatin Technologies Inc and is a named inventor on a patent describing NPR-C agonists. The other authors report no conflicts.

References

1. De Bold, A.J. Atrial natriuretic factor: A hormone produced by the heart. *Science* **1985**, *230*, 767–770. [CrossRef] [PubMed]
2. Mukoyama, M.; Nakao, K.; Saito, Y.; Ogawa, Y.; Hosoda, K.; Suga, S.; Shirakami, G.; Jougasaki, M.; Imura, H. Human brain natriuretic peptide, a novel cardiac hormone. *Lancet* **1990**, *335*, 801–802. [CrossRef]
3. Mukoyama, M.; Nakao, K.; Hosoda, K.; Suga, S.; Saito, Y.; Ogawa, Y.; Shirakami, G.; Jougasaki, M.; Obata, K.; Yasue, H.; et al. Brain natriuretic peptide as a novel cardiac hormone in humans. Evidence for an exquisite dual natriuretic peptide system, atrial natriuretic peptide and brain natriuretic peptide. *J. Clin. Investig.* **1991**, *87*, 1402–1412. [CrossRef] [PubMed]
4. Oikawa, S.; Imai, M.; Ueno, A.; Tanaka, S.; Noguchi, T.; Nakazato, H.; Kangawa, K.; Fukuda, A.; Matsuo, H. Cloning and sequence analysis of cDNA encoding a precursor for human atrial natriuretic polypeptide. *Nature* **1984**, *309*, 724–726. [CrossRef]

5. Edwards, B.S.; Zimmerman, R.S.; Schwab, T.R.; Heublein, D.M.; Burnett, J.C., Jr. Atrial stretch, not pressure, is the principal determinant controlling the acute release of atrial natriuretic factor. *Circ. Res.* **1988**, *62*, 191–195. [CrossRef]
6. Kinnunen, P.; Vuolteenaho, O.; Ruskoaho, H. Mechanisms of atrial and brain natriuretic peptide release from rat ventricular myocardium: Effect of stretching. *Endocrinology* **1993**, *132*, 1961–1970. [CrossRef] [PubMed]
7. Potter, L.R.; Yoder, A.R.; Flora, D.R.; Antos, L.K.; Dickey, D.M. Natriuretic peptides: Their structures, receptors, physiologic functions and therapeutic applications. *Handb. Exp. Pharmacol.* **2009**, 341–366. [CrossRef]
8. Wu, C.; Wu, F.; Pan, J.; Morser, J.; Wu, Q. Furin-mediated processing of Pro-C-type natriuretic peptide. *J. Biol. Chem.* **2003**, *278*, 25847–25852. [CrossRef] [PubMed]
9. Sudoh, T.; Minamino, N.; Kangawa, K.; Matsuo, H. C-type natriuretic peptide (CNP): A new member of natriuretic peptide family identified in porcine brain. *Biochem. Biophys. Res. Commun.* **1990**, *168*, 863–870. [CrossRef]
10. Chusho, H.; Tamura, N.; Ogawa, Y.; Yasoda, A.; Suda, M.; Miyazawa, T.; Nakamura, K.; Nakao, K.; Kurihara, T.; Komatsu, Y.; et al. Dwarfism and early death in mice lacking C-type natriuretic peptide. *Proc. Natl. Acad. Sci. USA* **2001**, *98*, 4016–4021. [CrossRef] [PubMed]
11. Stingo, A.J.; Clavell, A.L.; Heublein, D.M.; Wei, C.M.; Pittelkow, M.R.; Burnett, J.C., Jr. Presence of C-type natriuretic peptide in cultured human endothelial cells and plasma. *Am. J. Physiol.* **1992**, *263* Pt 2, H1318–H1321. [CrossRef]
12. Suga, S.; Nakao, K.; Itoh, H.; Komatsu, Y.; Ogawa, Y.; Hama, N.; Imura, H. Endothelial production of C-type natriuretic peptide and its marked augmentation by transforming growth factor-beta. Possible existence of "vascular natriuretic peptide system". *J. Clin. Investig.* **1992**, *90*, 1145–1149. [CrossRef] [PubMed]
13. Del Ry, S.; Cabiati, M.; Lionetti, V.; Emdin, M.; Recchia, F.A.; Giannessi, D. Expression of C-type natriuretic peptide and of its receptor NPR-B in normal and failing heart. *Peptides* **2008**, *29*, 2208–2215. [CrossRef]
14. Del Ry, S.; Cabiati, M.; Vozzi, F.; Battolla, B.; Caselli, C.; Forini, F.; Segnani, C.; Prescimone, T.; Giannessi, D.; Mattii, L. Expression of C-type natriuretic peptide and its receptor NPR-B in cardiomyocytes. *Peptides* **2011**, *32*, 1713–1718. [CrossRef]
15. Horio, T.; Tokudome, T.; Maki, T.; Yoshihara, F.; Suga, S.; Nishikimi, T.; Kojima, M.; Kawano, Y.; Kangawa, K. Gene expression, secretion, and autocrine action of C-type natriuretic peptide in cultured adult rat cardiac fibroblasts. *Endocrinology* **2003**, *144*, 2279–2284. [CrossRef]
16. Hunt, P.J.; Richards, A.M.; Espiner, E.A.; Nicholls, M.G.; Yandle, T.G. Bioactivity and metabolism of C-type natriuretic peptide in normal man. *J. Clin. Endocrinol. Metab.* **1994**, *78*, 1428–1435.
17. Sangaralingham, S.J.; McKie, P.M.; Ichiki, T.; Scott, C.G.; Heublein, D.M.; Chen, H.H.; Bailey, K.R.; Redfield, M.M.; Rodeheffer, R.J.; Burnett, J.C., Jr. Circulating C-type natriuretic peptide and its relationship to cardiovascular disease in the general population. *Hypertension* **2015**, *65*, 1187–1194. [CrossRef] [PubMed]
18. Kenny, A.J.; Bourne, A.; Ingram, J. Hydrolysis of human and pig brain natriuretic peptides, urodilatin, C-type natriuretic peptide and some C-receptor ligands by endopeptidase-24.11. *Biochem. J.* **1993**, *291* Pt 1, 83–88. [CrossRef]
19. Maack, T.; Suzuki, M.; Almeida, F.A.; Nussenzveig, D.; Scarborough, R.M.; McEnroe, G.A.; Lewicki, J.A. Physiological role of silent receptors of atrial natriuretic factor. *Science* **1987**, *238*, 675–678. [CrossRef] [PubMed]
20. Cohen, D.; Koh, G.Y.; Nikonova, L.N.; Porter, J.G.; Maack, T. Molecular determinants of the clearance function of type C receptors of natriuretic peptides. *J. Biol. Chem.* **1996**, *271*, 9863–9869. [CrossRef]
21. Charles, C.J.; Espiner, E.A.; Nicholls, M.G.; Richards, A.M.; Yandle, T.G.; Protter, A.; Kosoglou, T. Clearance receptors and endopeptidase 24.11: Equal role in natriuretic peptide metabolism in conscious sheep. *Am. J. Physiol.* **1996**, *271* Pt 2, R373–R380. [CrossRef] [PubMed]
22. Hashimoto, Y.; Nakao, K.; Hama, N.; Imura, H.; Mori, S.; Yamaguchi, M.; Yasuhara, M.; Hori, R. Clearance mechanisms of atrial and brain natriuretic peptides in rats. *Pharm. Res.* **1994**, *11*, 60–64. [CrossRef]
23. Brandt, R.R.; Heublein, D.M.; Aarhus, L.L.; Lewicki, J.A.; Burnett, J.C., Jr. Role of natriuretic peptide clearance receptor in in vivo control of C-type natriuretic peptide. *Am. J. Physiol.* **1995**, *269* Pt 2, H326–H331. [CrossRef]
24. Okahara, K.; Kambayashi, J.; Ohnishi, T.; Fujiwara, Y.; Kawasaki, T.; Monden, M. Shear stress induces expression of CNP gene in human endothelial cells. *FEBS Lett.* **1995**, *373*, 108–110. [CrossRef]
25. Zhang, Z.; Xiao, Z.; Diamond, S.L. Shear stress induction of C-type natriuretic peptide (CNP) in endothelial cells is independent of NO autocrine signaling. *Ann. Biomed. Eng.* **1999**, *27*, 419–426. [CrossRef]

26. Suga, S.; Itoh, H.; Komatsu, Y.; Ogawa, Y.; Hama, N.; Yoshimasa, T.; Nakao, K. Cytokine-induced C-type natriuretic peptide (CNP) secretion from vascular endothelial cells—Evidence for CNP as a novel autocrine/paracrine regulator from endothelial cells. *Endocrinology* **1993**, *133*, 3038–3041. [CrossRef]
27. Osterbur, K.; Yu, D.H.; DeClue, A.E. Interleukin-1beta, tumour necrosis factor-alpha and lipopolysaccharide induce C-type natriuretic peptide from canine aortic endothelial cells. *Res. Vet. Sci.* **2013**, *94*, 478–483. [CrossRef]
28. Mendonca, M.C.; Koles, N.; Doi, S.Q.; Sellitti, D.F. Transforming growth factor-beta1 regulation of C-type natriuretic peptide expression in human vascular smooth muscle cells: dependence on TSC22D1. *Am. J. Physiol. Heart Circ. Physiol.* **2010**, *299*, H2018–H2027. [CrossRef]
29. Vollmar, A.M.; Schulz, R. Expression and differential regulation of natriuretic peptides in mouse macrophages. *J. Clin. Investig.* **1995**, *95*, 2442–2450. [CrossRef] [PubMed]
30. Del Ry, S.; Passino, C.; Maltinti, M.; Emdin, M.; Giannessi, D. C-type natriuretic peptide plasma levels increase in patients with chronic heart failure as a function of clinical severity. *Eur. J. Heart Fail.* **2005**, *7*, 1145–1148. [CrossRef] [PubMed]
31. Hama, N.; Itoh, H.; Shirakami, G.; Suga, S.; Komatsu, Y.; Yoshimasa, T.; Tanaka, I.; Mori, K.; Nakao, K. Detection of C-type natriuretic peptide in human circulation and marked increase of plasma CNP level in septic shock patients. *Biochem. Biophys. Res. Commun.* **1994**, *198*, 1177–1182. [CrossRef] [PubMed]
32. Sugiyama, S.; Kugiyama, K.; Matsumura, T.; Suga, S.; Itoh, H.; Nakao, K.; Yasue, H. Lipoproteins regulate C-type natriuretic peptide secretion from cultured vascular endothelial cells. *Arterioscler. Thromb. Vasc. Biol.* **1995**, *15*, 1968–1974. [CrossRef] [PubMed]
33. Doi, K.; Itoh, H.; Komatsu, Y.; Igaki, T.; Chun, T.H.; Takaya, K.; Yamashita, J.; Inoue, M.; Yoshimasa, T.; Nakao, K. Vascular endothelial growth factor suppresses C-type natriuretic peptide secretion. *Hypertension* **1996**, *27 Pt 2*, 811–815. [CrossRef]
34. Levin, E.R.; Gardner, D.G.; Samson, W.K. Natriuretic peptides. *N. Engl. J. Med.* **1998**, *339*, 321–328.
35. Potter, L.R.; Abbey-Hosch, S.; Dickey, D.M. Natriuretic peptides, their receptors, and cyclic guanosine monophosphate-dependent signaling functions. *Endocr. Rev.* **2006**, *27*, 47–72. [CrossRef]
36. He, X.L.; Dukkipati, A.; Garcia, K.C. Structural determinants of natriuretic peptide receptor specificity and degeneracy. *J. Mol. Biol.* **2006**, *361*, 698–714. [CrossRef]
37. Khambata, R.S.; Panayiotou, C.M.; Hobbs, A.J. Natriuretic peptide receptor-3 underpins the disparate regulation of endothelial and vascular smooth muscle cell proliferation by C-type natriuretic peptide. *Br. J. Pharmacol.* **2011**, *164*, 584–597. [CrossRef]
38. Suga, S.; Nakao, K.; Mukoyama, M.; Arai, H.; Hosoda, K.; Ogawa, Y.; Imura, H. Characterization of natriuretic peptide receptors in cultured cells. *Hypertension* **1992**, *19*, 762–765. [CrossRef]
39. Fujio, N.; Gossard, F.; Bayard, F.; Tremblay, J. Regulation of natriuretic peptide receptor A and B expression by transforming growth factor-beta 1 in cultured aortic smooth muscle cells. *Hypertension* **1994**, *23 Pt 2*, 908–913. [CrossRef]
40. Lin, X.; Hanze, J.; Heese, F.; Sodmann, R.; Lang, R.E. Gene expression of natriuretic peptide receptors in myocardial cells. *Circ. Res.* **1995**, *77*, 750–758. [CrossRef] [PubMed]
41. Huntley, B.K.; Sandberg, S.M.; Noser, J.A.; Cataliotti, A.; Redfield, M.M.; Matsuda, Y.; Burnett, J.C., Jr. BNP-induced activation of cGMP in human cardiac fibroblasts: Interactions with fibronectin and natriuretic peptide receptors. *J. Cell Physiol.* **2006**, *209*, 943–949. [CrossRef] [PubMed]
42. Leitman, D.C.; Andresen, J.W.; Kuno, T.; Kamisaki, Y.; Chang, J.K.; Murad, F. Identification of multiple binding sites for atrial natriuretic factor by affinity cross-linking in cultured endothelial cells. *J. Biol. Chem.* **1986**, *261*, 11650–11655. [PubMed]
43. Koller, K.J.; Lowe, D.G.; Bennett, G.L.; Minamino, N.; Kangawa, K.; Matsuo, H.; Goeddel, D.V. Selective activation of the B natriuretic peptide receptor by C-type natriuretic peptide (CNP). *Science* **1991**, *252*, 120–123. [CrossRef]
44. Schlossmann, J.; Feil, R.; Hofmann, F. Insights into cGMP signalling derived from cGMP kinase knockout mice. *Front. Biosci.* **2005**, *10*, 1279–1289. [CrossRef]
45. Miyazawa, T.; Ogawa, Y.; Chusho, H.; Yasoda, A.; Tamura, N.; Komatsu, Y.; Pfeifer, A.; Hofmann, F.; Nakao, K. Cyclic GMP-dependent protein kinase II plays a critical role in C-type natriuretic peptide-mediated endochondral ossification. *Endocrinology* **2002**, *143*, 3604–3610. [CrossRef] [PubMed]

46. Nussenzveig, D.R.; Lewicki, J.A.; Maack, T. Cellular mechanisms of the clearance function of type C receptors of atrial natriuretic factor. *J. Biol. Chem.* **1990**, *265*, 20952–20958. [PubMed]
47. Murthy, K.S.; Makhlouf, G.M. Identification of the G protein-activating domain of the natriuretic peptide clearance receptor (NPR-C). *J. Biol. Chem.* **1999**, *274*, 17587–17592. [CrossRef]
48. Anand-Srivastava, M.B.; Sairam, M.R.; Cantin, M. Ring-deleted analogs of atrial natriuretic factor inhibit adenylate cyclase/cAMP system. Possible coupling of clearance atrial natriuretic factor receptors to adenylate cyclase/cAMP signal transduction system. *J. Biol. Chem.* **1990**, *265*, 8566–8572. [PubMed]
49. Murthy, K.S.; Teng, B.Q.; Zhou, H.; Jin, J.G.; Grider, J.R.; Makhlouf, G.M. G(i-1)/G(i-2)-dependent signaling by single-transmembrane natriuretic peptide clearance receptor. *Am. J. Physiol. Gastrointest. Liver Physiol.* **2000**, *278*, G974–G980. [CrossRef]
50. Pagano, M.; Anand-Srivastava, M.B. Cytoplasmic domain of natriuretic peptide receptor C constitutes Gi activator sequences that inhibit adenylyl cyclase activity. *J. Biol. Chem.* **2001**, *276*, 22064–22070. [CrossRef] [PubMed]
51. Trachte, G.J.; Kanwal, S.; Elmquist, B.J.; Ziegler, R.J. C-type natriuretic peptide neuromodulates via "clearance" receptors. *Am. J. Physiol.* **1995**, *268 Pt 1*, C978–C984. [CrossRef]
52. Savoie, P.; de Champlain, J.; Anand-Srivastava, M.B. C-type natriuretic peptide and brain natriuretic peptide inhibit adenylyl cyclase activity: interaction with ANF-R2/ANP-C receptors. *FEBS Lett.* **1995**, *370*, 6–10. [CrossRef]
53. Brown, J.; Zuo, Z. Receptor proteins and biological effects of C-type natriuretic peptides in the renal glomerulus of the rat. *Am. J. Physiol.* **1994**, *266 Pt 2*, R1383–R1394. [CrossRef]
54. Wei, C.M.; Aarhus, L.L.; Miller, V.M.; Burnett, J.C., Jr. Action of C-type natriuretic peptide in isolated canine arteries and veins. *Am. J. Physiol.* **1993**, *264 Pt 2*, H71–H73. [CrossRef]
55. Wei, C.M.; Hu, S.; Miller, V.M.; Burnett, J.C., Jr. Vascular actions of C-type natriuretic peptide in isolated porcine coronary arteries and coronary vascular smooth muscle cells. *Biochem. Biophys. Res. Commun.* **1994**, *205*, 765–771. [CrossRef] [PubMed]
56. Klinger, J.R.; Siddiq, F.M.; Swift, R.A.; Jackson, C.; Pietras, L.; Warburton, R.R.; Alia, C.; Hill, N.S. C-type natriuretic peptide expression and pulmonary vasodilation in hypoxia-adapted rats. *Am. J. Physiol.* **1998**, *275*, L645–L652. [CrossRef]
57. Drewett, J.G.; Fendly, B.M.; Garbers, D.L.; Lowe, D.G. Natriuretic peptide receptor-B (guanylyl cyclase-B) mediates C-type natriuretic peptide relaxation of precontracted rat aorta. *J. Biol. Chem.* **1995**, *270*, 4668–4674. [CrossRef]
58. Mori, Y.; Takayasu, M.; Suzuki, Y.; Shibuya, M.; Yoshida, J.; Hidaka, H. Vasodilator effects of C-type natriuretic peptide on cerebral arterioles in rats. *Eur. J. Pharmacol.* **1997**, *320*, 183–186. [CrossRef]
59. Barber, D.A.; Burnett, J.C., Jr.; Fitzpatrick, L.A.; Sieck, G.C.; Miller, V.M. Gender and relaxation to C-type natriuretic peptide in porcine coronary arteries. *J. Cardiovasc. Pharmacol.* **1998**, *32*, 5–11. [CrossRef] [PubMed]
60. Wennberg, P.W.; Miller, V.M.; Rabelink, T.; Burnett, J.C., Jr. Further attenuation of endothelium-dependent relaxation imparted by natriuretic peptide receptor antagonism. *Am. J. Physiol.* **1999**, *277*, H1618–H1621. [CrossRef]
61. Brunner, F.; Wolkart, G. Relaxant effect of C-type natriuretic peptide involves endothelium and nitric oxide-cGMP system in rat coronary microvasculature. *Cardiovasc. Res.* **2001**, *51*, 577–584. [CrossRef]
62. Otsuka, K.; Tanaka, H.; Horinouchi, T.; Koike, K.; Shigenobu, K.; Tanaka, Y. Functional contribution of voltage-dependent and Ca2+ activated K+ (BK(Ca)) channels to the relaxation of guinea-pig aorta in response to natriuretic peptides. *J. Smooth Muscle Res.* **2002**, *38*, 117–129. [CrossRef] [PubMed]
63. Madhani, M.; Scotland, R.S.; MacAllister, R.J.; Hobbs, A.J. Vascular natriuretic peptide receptor-linked particulate guanylate cyclases are modulated by nitric oxide-cyclic GMP signalling. *Br. J. Pharmacol.* **2003**, *139*, 1289–1296. [CrossRef]
64. Garcha, R.S.; Hughes, A.D. CNP, but not ANP or BNP, relax human isolated subcutaneous resistance arteries by an action involving cyclic GMP and BKCa channels. *J. Renin Angiotensin Aldosterone Syst.* **2006**, *7*, 87–91. [CrossRef]
65. Villar, I.C.; Panayiotou, C.M.; Sheraz, A.; Madhani, M.; Scotland, R.S.; Nobles, M.; Kemp-Harper, B.; Ahluwalia, A.; Hobbs, A.J. Definitive role for natriuretic peptide receptor-C in mediating the vasorelaxant activity of C-type natriuretic peptide and endothelium-derived hyperpolarising factor. *Cardiovasc. Res.* **2007**, *74*, 515–525. [CrossRef] [PubMed]

66. Leuranguer, V.; Vanhoutte, P.M.; Verbeuren, T.; Feletou, M. C-type natriuretic peptide and endothelium-dependent hyperpolarization in the guinea-pig carotid artery. *Br. J. Pharmacol.* **2008**, *153*, 57–65. [CrossRef] [PubMed]
67. Edvinsson, M.L.; Ahnstedt, H.; Edvinsson, L.; Andersson, S.E. Characterization of Relaxant Responses to Natriuretic Peptides in the Human Microcirculation In Vitro and In Vivo. *Microcirculation* **2016**, *23*, 438–446. [CrossRef] [PubMed]
68. Clavell, A.L.; Stingo, A.J.; Wei, C.M.; Heublein, D.M.; Burnett, J.C., Jr. C-type natriuretic peptide: A selective cardiovascular peptide. *Am. J. Physiol.* **1993**, *264 Pt 2*, R290–R295. [CrossRef]
69. Nakamura, M.; Arakawa, N.; Yoshida, H.; Makita, S.; Hiramori, K. Vasodilatory effects of C-type natriuretic peptide on forearm resistance vessels are distinct from those of atrial natriuretic peptide in chronic heart failure. *Circulation* **1994**, *90*, 1210–1214. [CrossRef] [PubMed]
70. Igaki, T.; Itoh, H.; Suga, S.; Hama, N.; Ogawa, Y.; Komatsu, Y.; Mukoyama, M.; Sugawara, A.; Yoshimasa, T.; Tanaka, I.; et al. C-type natriuretic peptide in chronic renal failure and its action in humans. *Kidney Int. Suppl.* **1996**, *55*, S144–S147.
71. Honing, M.L.; Smits, P.; Morrison, P.J.; Burnett, J.C., Jr.; Rabelink, T.J. C-type natriuretic peptide-induced vasodilation is dependent on hyperpolarization in human forearm resistance vessels. *Hypertension* **2001**, *37*, 1179–1183. [CrossRef]
72. Aizawa, N.; Ishizuka, O.; Ogawa, T.; Mizusawa, H.; Igawa, Y.; Nishizawa, O.; Andersson, K.E. Effects of natriuretic peptides on intracavernous pressure and blood pressure in conscious rats. *J. Sex. Med.* **2008**, *5*, 2312–2317. [CrossRef] [PubMed]
73. Moyes, A.J.; Khambata, R.S.; Villar, I.; Bubb, K.J.; Baliga, R.S.; Lumsden, N.G.; Xiao, F.; Gane, P.J.; Rebstock, A.S.; Worthington, R.J.; et al. Endothelial C-type natriuretic peptide maintains vascular homeostasis. *J. Clin. Investig.* **2014**, *124*, 4039–4051. [CrossRef]
74. Nakao, K.; Kuwahara, K.; Nishikimi, T.; Nakagawa, Y.; Kinoshita, H.; Minami, T.; Kuwabara, Y.; Yamada, C.; Yamada, Y.; Tokudome, T.; et al. Endothelium-Derived C-Type Natriuretic Peptide Contributes to Blood Pressure Regulation by Maintaining Endothelial Integrity. *Hypertension* **2017**, *69*, 286–296. [CrossRef]
75. Spiranec, K.; Chen, W.; Werner, F.; Nikolaev, V.O.; Naruke, T.; Koch, F.; Werner, A.; Eder-Negrin, P.; Dieguez-Hurtado, R.; Adams, R.H.; et al. Endothelial C-Type Natriuretic Peptide Acts on Pericytes to Regulate Microcirculatory Flow and Blood Pressure. *Circulation* **2018**, *138*, 494–508. [CrossRef]
76. Ono, K.; Mannami, T.; Baba, S.; Tomoike, H.; Suga, S.; Iwai, N. A single-nucleotide polymorphism in C-type natriuretic peptide gene may be associated with hypertension. *Hypertens. Res.* **2002**, *25*, 727–730. [CrossRef]
77. Li, N.; Luo, W.; Juhong, Z.; Yang, J.; Wang, H.; Zhou, L.; Chang, J. Associations between genetic variations in the FURIN gene and hypertension. *BMC Med. Genet.* **2010**, *11*, 124. [CrossRef] [PubMed]
78. Chauhan, S.D.; Nilsson, H.; Ahluwalia, A.; Hobbs, A.J. Release of C-type natriuretic peptide accounts for the biological activity of endothelium-derived hyperpolarizing factor. *Proc. Natl. Acad. Sci. USA* **2003**, *100*, 1426–1431. [CrossRef]
79. Kun, A.; Kiraly, I.; Pataricza, J.; Marton, Z.; Krassoi, I.; Varro, A.; Simonsen, U.; Papp, J.G.; Pajor, L. C-type natriuretic peptide hyperpolarizes and relaxes human penile resistance arteries. *J. Sex. Med.* **2008**, *5*, 1114–1125. [CrossRef]
80. Simon, A.; Harrington, E.O.; Liu, G.X.; Koren, G.; Choudhary, G. Mechanism of C-type natriuretic peptide-induced endothelial cell hyperpolarization. *Am. J. Physiol. Lung Cell Mol. Physiol.* **2009**, *296*, L248–L256. [CrossRef] [PubMed]
81. Caniffi, C.; Elesgaray, R.; Gironacci, M.; Arranz, C.; Costa, M.A. C-type natriuretic peptide effects on cardiovascular nitric oxide system in spontaneously hypertensive rats. *Peptides* **2010**, *31*, 1309–1318. [CrossRef]
82. Caniffi, C.; Cerniello, F.M.; Gobetto, M.N.; Sueiro, M.L.; Costa, M.A.; Arranz, C. Vascular Tone Regulation Induced by C-Type Natriuretic Peptide: Differences in Endothelium-Dependent and -Independent Mechanisms Involved in Normotensive and Spontaneously Hypertensive Rats. *PLoS ONE* **2016**, *11*, e0167817. [CrossRef]
83. Tamura, N.; Doolittle, L.K.; Hammer, R.E.; Shelton, J.M.; Richardson, J.A.; Garbers, D.L. Critical roles of the guanylyl cyclase B receptor in endochondral ossification and development of female reproductive organs. *Proc. Natl. Acad. Sci. USA* **2004**, *101*, 17300–17305. [CrossRef]
84. Matsukawa, N.; Grzesik, W.J.; Takahashi, N.; Pandey, K.N.; Pang, S.; Yamauchi, M.; Smithies, O. The natriuretic peptide clearance receptor locally modulates the physiological effects of the natriuretic peptide system. *Proc. Natl. Acad. Sci. USA* **1999**, *96*, 7403–7408. [CrossRef] [PubMed]

85. Ehret, G.B.; Munroe, P.B.; Rice, K.M.; Bochud, M.; Johnson, A.D.; Chasman, D.I.; Smith, A.V.; Tobin, M.D.; Verwoert, G.C.; Hwang, S.J.; et al. Genetic variants in novel pathways influence blood pressure and cardiovascular disease risk. *Nature* **2011**, *478*, 103–109. [CrossRef] [PubMed]
86. Ren, M.; Ng, F.L.; Warren, H.R.; Witkowska, K.; Baron, M.; Jia, Z.; Cabrera, C.; Zhang, R.; Mifsud, B.; Munroe, P.B.; et al. The biological impact of blood pressure-associated genetic variants in the natriuretic peptide receptor C gene on human vascular smooth muscle. *Hum. Mol. Genet.* **2018**, *27*, 199–210. [CrossRef] [PubMed]
87. Li, Y.X.; Cheng, K.C.; Asakawa, A.; Kato, I.; Sato, Y.; Amitani, H.; Kawamura, N.; Cheng, J.T.; Inui, A. Role of musclin in the pathogenesis of hypertension in rat. *PLoS ONE* **2013**, *8*, e72004. [CrossRef] [PubMed]
88. Miyazaki, T.; Otani, K.; Chiba, A.; Nishimura, H.; Tokudome, T.; Takano-Watanabe, H.; Matsuo, A.; Ishikawa, H.; Shimamoto, K.; Fukui, H.; et al. A New Secretory Peptide of Natriuretic Peptide Family, Osteocrin, Suppresses the Progression of Congestive Heart Failure After Myocardial Infarction. *Circ. Res.* **2018**, *122*, 742–751. [CrossRef]
89. Li, Y.; Sarkar, O.; Brochu, M.; Anand-Srivastava, M.B. Natriuretic peptide receptor-C attenuates hypertension in spontaneously hypertensive rats: role of nitroxidative stress and Gi proteins. *Hypertension* **2014**, *63*, 846–855. [CrossRef] [PubMed]
90. Schachner, T.; Zou, Y.; Oberhuber, A.; Mairinger, T.; Tzankov, A.; Laufer, G.; Ott, H.; Bonatti, J. Perivascular application of C-type natriuretic peptide attenuates neointimal hyperplasia in experimental vein grafts. *Eur. J. Cardiothorac. Surg.* **2004**, *25*, 585–590. [CrossRef]
91. Doi, K.; Ikeda, T.; Itoh, H.; Ueyama, K.; Hosoda, K.; Ogawa, Y.; Yamashita, J.; Chun, T.H.; Inoue, M.; Masatsugu, K.; et al. C-type natriuretic peptide induces redifferentiation of vascular smooth muscle cells with accelerated reendothelialization. *Arterioscler. Thromb. Vasc. Biol.* **2001**, *21*, 930–936. [CrossRef]
92. Ohno, N.; Itoh, H.; Ikeda, T.; Ueyama, K.; Yamahara, K.; Doi, K.; Yamashita, J.; Inoue, M.; Masatsugu, K.; Sawada, N.; et al. Accelerated reendothelialization with suppressed thrombogenic property and neointimal hyperplasia of rabbit jugular vein grafts by adenovirus-mediated gene transfer of C-type natriuretic peptide. *Circulation* **2002**, *105*, 1623–1626. [CrossRef]
93. Furuya, M.; Aisaka, K.; Miyazaki, T.; Honbou, N.; Kawashima, K.; Ohno, T.; Tanaka, S.; Minamino, N.; Kangawa, K.; Matsuo, H. C-type natriuretic peptide inhibits intimal thickening after vascular injury. *Biochem. Biophys. Res. Commun.* **1993**, *193*, 248–253. [CrossRef]
94. Furuya, M.; Yoshida, M.; Hayashi, Y.; Ohnuma, N.; Minamino, N.; Kangawa, K.; Matsuo, H. C-type natriuretic peptide is a growth inhibitor of rat vascular smooth muscle cells. *Biochem. Biophys. Res. Commun.* **1991**, *177*, 927–931. [CrossRef]
95. Hutchinson, H.G.; Trindade, P.T.; Cunanan, D.B.; Wu, C.F.; Pratt, R.E. Mechanisms of natriuretic-peptide-induced growth inhibition of vascular smooth muscle cells. *Cardiovasc. Res.* **1997**, *35*, 158–167. [CrossRef]
96. Cahill, P.A.; Hassid, A. ANF-C-receptor-mediated inhibition of aortic smooth muscle cell proliferation and thymidine kinase activity. *Am. J. Physiol.* **1994**, *266 Pt 2*, R194–R1203. [CrossRef]
97. Bubb, K.J.; Aubdool, A.A.; Moyes, A.J.; Lewis, S.; Drayton, J.P.; Tang, O.; Mehta, V.; Zachary, I.C.; Abraham, D.J.; Tsui, J.; et al. Endothelial C-Type Natriuretic Peptide Is a Critical Regulator of Angiogenesis and Vascular Remodeling. *Circulation* **2019**, *139*, 1612–1628. [CrossRef]
98. Yamahara, K.; Itoh, H.; Chun, T.H.; Ogawa, Y.; Yamashita, J.; Sawada, N.; Fukunaga, Y.; Sone, M.; Yurugi-Kobayashi, T.; Miyashita, K.; et al. Significance and therapeutic potential of the natriuretic peptides/cGMP/cGMP-dependent protein kinase pathway in vascular regeneration. *Proc. Natl. Acad. Sci. USA* **2003**, *100*, 3404–3409. [CrossRef] [PubMed]
99. Almeida, S.A.; Cardoso, C.C.; Orellano, L.A.; Reis, A.M.; Barcelos, L.S.; Andrade, S.P. Natriuretic peptide clearance receptor ligand (C-ANP4-23) attenuates angiogenesis in a murine sponge implant model. *Clin. Exp. Pharmacol. Physiol.* **2014**, *41*, 691–697. [CrossRef]
100. Pedram, A.; Razandi, M.; Hu, R.M.; Levin, E.R. Vasoactive peptides modulate vascular endothelial cell growth factor production and endothelial cell proliferation and invasion. *J. Biol. Chem.* **1997**, *272*, 17097–17103. [CrossRef]
101. Koch, A.; Voigt, S.; Sanson, E.; Duckers, H.; Horn, A.; Zimmermann, H.W.; Trautwein, C.; Tacke, F. Prognostic value of circulating amino-terminal pro-C-type natriuretic peptide in critically ill patients. *Crit. Care* **2011**, *15*, R45. [CrossRef]

102. Bahrami, S.; Pelinka, L.; Khadem, A.; Maitzen, S.; Hawa, G.; van Griensven, M.; Redl, H. Circulating NT-proCNP predicts sepsis in multiple-traumatized patients without traumatic brain injury. *Crit. Care Med.* **2010**, *38*, 161–166. [CrossRef]
103. Ehler, J.; Saller, T.; Wittstock, M.; Rommer, P.S.; Chappell, D.; Zwissler, B.; Grossmann, A.; Richter, G.; Reuter, D.A.; Noldge-Schomburg, G.; et al. Diagnostic value of NT-proCNP compared to NSE and S100B in cerebrospinal fluid and plasma of patients with sepsis-associated encephalopathy. *Neurosci. Lett.* **2019**, *692*, 167–173. [CrossRef]
104. Tomasiuk, R.; Mikaszewska-Sokolewicz, M.; Szlufik, S.; Rzepecki, P.; Lazowski, T. The prognostic value of concomitant assessment of NT-proCNP, C-reactive protein, procalcitonin and inflammatory cytokines in septic patients. *Crit. Care* **2014**, *18*, 440. [CrossRef]
105. Ince, C.; Mayeux, P.R.; Nguyen, T.; Gomez, H.; Kellum, J.A.; Ospina-Tascon, G.A.; Hernandez, G.; Murray, P.; De Backer, D. The endothelium in sepsis. *Shock* **2016**, *45*, 259–270. [CrossRef]
106. Ross, R. Atherosclerosis—An inflammatory disease. *N. Engl. J. Med.* **1999**, *340*, 115–126. [CrossRef]
107. Scotland, R.S.; Cohen, M.; Foster, P.; Lovell, M.; Mathur, A.; Ahluwalia, A.; Hobbs, A.J. C-type natriuretic peptide inhibits leukocyte recruitment and platelet-leukocyte interactions via suppression of P-selectin expression. *Proc. Natl. Acad. Sci. USA* **2005**, *102*, 14452–14457. [CrossRef]
108. Kimura, T.; Nojiri, T.; Hosoda, H.; Ishikane, S.; Shintani, Y.; Inoue, M.; Miyazato, M.; Okumura, M.; Kangawa, K. C-type natriuretic peptide attenuates lipopolysaccharide-induced acute lung injury in mice. *J. Surg. Res.* **2015**, *194*, 631–637. [CrossRef] [PubMed]
109. Chen, G.; Zhao, J.; Yin, Y.; Wang, B.; Liu, Q.; Li, P.; Zhao, L.; Zhou, H. C-type natriuretic peptide attenuates LPS-induced endothelial activation: Involvement of p38, Akt, and NF-kappaB pathways. *Amino Acids* **2014**, *46*, 2653–2663. [CrossRef] [PubMed]
110. Casco, V.H.; Veinot, J.P.; Kuroski de Bold, M.L.; Masters, R.G.; Stevenson, M.M.; de Bold, A.J. Natriuretic peptide system gene expression in human coronary arteries. *J. Histochem. Cytochem.* **2002**, *50*, 799–809. [CrossRef]
111. Kohno, M.; Yokokawa, K.; Yasunari, K.; Kano, H.; Minami, M.; Ueda, M.; Yoshikawa, J. Effect of natriuretic peptide family on the oxidized LDL-induced migration of human coronary artery smooth muscle cells. *Circ. Res.* **1997**, *81*, 585–590. [CrossRef]
112. Villard, C.; Hultgren, R. Abdominal aortic aneurysm: Sex differences. *Maturitas* **2018**, *109*, 63–69. [CrossRef]
113. Hu, P.; Wang, J.; Zhao, X.Q.; Hu, B.; Lu, L.; Qin, Y.H. Overexpressed C-type natriuretic peptide serves as an early compensatory response to counteract extracellular matrix remodeling in unilateral ureteral obstruction rats. *Mol. Biol. Rep.* **2013**, *40*, 1429–1441. [CrossRef] [PubMed]
114. Krejci, P.; Masri, B.; Fontaine, V.; Mekikian, P.B.; Weis, M.; Prats, H.; Wilcox, W.R. Interaction of fibroblast growth factor and C-natriuretic peptide signaling in regulation of chondrocyte proliferation and extracellular matrix homeostasis. *J. Cell Sci.* **2005**, *118 Pt 21*, 5089–5100. [CrossRef]
115. Murakami, S.; Nagaya, N.; Itoh, T.; Fujii, T.; Iwase, T.; Hamada, K.; Kimura, H.; Kangawa, K. C-type natriuretic peptide attenuates bleomycin-induced pulmonary fibrosis in mice. *Am. J. Physiol. Lung Cell. Mol. Physiol.* **2004**, *287*, L1172–L1177. [CrossRef]
116. Chen, G.; Song, X.; Yin, Y.; Xia, S.; Liu, Q.; You, G.; Zhao, L.; Zhou, H. C-type natriuretic peptide prevents kidney injury and attenuates oxidative and inflammatory responses in hemorrhagic shock. *Amino Acids* **2017**, *49*, 347–354. [CrossRef] [PubMed]
117. Bae, C.R.; Hino, J.; Hosoda, H.; Arai, Y.; Son, C.; Makino, H.; Tokudome, T.; Tomita, T.; Kimura, T.; Nojiri, T.; et al. Overexpression of C-type Natriuretic Peptide in Endothelial Cells Protects against Insulin Resistance and Inflammation during Diet-induced Obesity. *Sci. Rep.* **2017**, *7*, 9807. [CrossRef] [PubMed]
118. Bae, C.R.; Hino, J.; Hosoda, H.; Miyazato, M.; Kangawa, K. C-type natriuretic peptide (CNP) in endothelial cells attenuates hepatic fibrosis and inflammation in non-alcoholic steatohepatitis. *Life Sci.* **2018**, *209*, 349–356. [CrossRef]
119. Del Ry, S. C-type natriuretic peptide: a new cardiac mediator. *Peptides* **2013**, *40*, 93–98. [CrossRef]
120. Wei, C.M.; Heublein, D.M.; Perrella, M.A.; Lerman, A.; Rodeheffer, R.J.; McGregor, C.G.; Edwards, W.D.; Schaff, H.V.; Burnett, J.C., Jr. Natriuretic peptide system in human heart failure. *Circulation* **1993**, *88*, 1004–1009. [CrossRef]

121. Palmer, S.C.; Prickett, T.C.; Espiner, E.A.; Yandle, T.G.; Richards, A.M. Regional release and clearance of C-type natriuretic peptides in the human circulation and relation to cardiac function. *Hypertension* **2009**, *54*, 612–618. [CrossRef]
122. Prickett, T.C.; Yandle, T.G.; Nicholls, M.G.; Espiner, E.A.; Richards, A.M. Identification of amino-terminal pro-C-type natriuretic peptide in human plasma. *Biochem. Biophys. Res. Commun.* **2001**, *286*, 513–517. [CrossRef]
123. Wright, S.P.; Prickett, T.C.; Doughty, R.N.; Frampton, C.; Gamble, G.D.; Yandle, T.G.; Sharpe, N.; Richards, M. Amino-terminal pro-C-type natriuretic peptide in heart failure. *Hypertension* **2004**, *43*, 94–100. [CrossRef] [PubMed]
124. Lok, D.J.; Klip, I.T.; Voors, A.A.; Lok, S.I.; Bruggink-Andre de la Porte, P.W.; Hillege, H.L.; Jaarsma, T.; van Veldhuisen, D.J.; van der Meer, P. Prognostic value of N-terminal pro C-type natriuretic peptide in heart failure patients with preserved and reduced ejection fraction. *Eur. J. Heart Fail.* **2014**, *16*, 958–966. [CrossRef]
125. Del Ry, S.; Maltinti, M.; Piacenti, M.; Passino, C.; Emdin, M.; Giannessi, D. Cardiac production of C-type natriuretic peptide in heart failure. *J. Cardiovasc. Med. (Hagerstown)* **2006**, *7*, 397–399. [CrossRef] [PubMed]
126. Dickey, D.M.; Flora, D.R.; Bryan, P.M.; Xu, X.; Chen, Y.; Potter, L.R. Differential regulation of membrane guanylyl cyclases in congestive heart failure: Natriuretic peptide receptor (NPR)-B, Not NPR-A, is the predominant natriuretic peptide receptor in the failing heart. *Endocrinology* **2007**, *148*, 3518–3522. [CrossRef] [PubMed]
127. Dickey, D.M.; Dries, D.L.; Margulies, K.B.; Potter, L.R. Guanylyl cyclase (GC)-A and GC-B activities in ventricles and cardiomyocytes from failed and non-failed human hearts: GC-A is inactive in the failed cardiomyocyte. *J. Mol. Cell. Cardiol.* **2012**, *52*, 727–732. [CrossRef] [PubMed]
128. Ichiki, T.; Schirger, J.A.; Huntley, B.K.; Brozovich, F.V.; Maleszewski, J.J.; Sandberg, S.M.; Sangaralingham, S.J.; Park, S.J.; Burnett, J.C., Jr. Cardiac fibrosis in end-stage human heart failure and the cardiac natriuretic peptide guanylyl cyclase system: regulation and therapeutic implications. *J. Mol. Cell. Cardiol.* **2014**, *75*, 199–205. [CrossRef]
129. Beaulieu, P.; Cardinal, R.; Page, P.; Francoeur, F.; Tremblay, J.; Lambert, C. Positive chronotropic and inotropic effects of C-type natriuretic peptide in dogs. *Am. J. Physiol.* **1997**, *273 Pt 2*, H1933–H1940. [CrossRef]
130. Hirose, M.; Furukawa, Y.; Kurogouchi, F.; Nakajima, K.; Miyashita, Y.; Chiba, S. C-type natriuretic peptide increases myocardial contractility and sinus rate mediated by guanylyl cyclase-linked natriuretic peptide receptors in isolated, blood-perfused dog heart preparations. *J. Pharmacol. Exp. Ther.* **1998**, *286*, 70–76. [PubMed]
131. Brusq, J.M.; Mayoux, E.; Guigui, L.; Kirilovsky, J. Effects of C-type natriuretic peptide on rat cardiac contractility. *Br. J. Pharmacol.* **1999**, *128*, 206–212. [CrossRef] [PubMed]
132. Moltzau, L.R.; Aronsen, J.M.; Meier, S.; Skogestad, J.; Orstavik, O.; Lothe, G.B.; Sjaastad, I.; Skomedal, T.; Osnes, J.B.; Levy, F.O.; et al. Different compartmentation of responses to brain natriuretic peptide and C-type natriuretic peptide in failing rat ventricle. *J. Pharmacol. Exp. Ther.* **2014**, *350*, 681–690. [CrossRef]
133. Pierkes, M.; Gambaryan, S.; Boknik, P.; Lohmann, S.M.; Schmitz, W.; Potthast, R.; Holtwick, R.; Kuhn, M. Increased effects of C-type natriuretic peptide on cardiac ventricular contractility and relaxation in guanylyl cyclase A-deficient mice. *Cardiovasc. Res.* **2002**, *53*, 852–861. [CrossRef]
134. Wollert, K.C.; Yurukova, S.; Kilic, A.; Begrow, F.; Fiedler, B.; Gambaryan, S.; Walter, U.; Lohmann, S.M.; Kuhn, M. Increased effects of C-type natriuretic peptide on contractility and calcium regulation in murine hearts overexpressing cyclic GMP-dependent protein kinase I. *Br. J. Pharmacol.* **2003**, *140*, 1227–1236. [CrossRef]
135. Moltzau, L.R.; Aronsen, J.M.; Meier, S.; Nguyen, C.H.; Hougen, K.; Orstavik, O.; Sjaastad, I.; Christensen, G.; Skomedal, T.; Osnes, J.B.; et al. SERCA2 activity is involved in the CNP-mediated functional responses in failing rat myocardium. *Br. J. Pharmacol.* **2013**, *170*, 366–379. [CrossRef] [PubMed]
136. Frantz, S.; Klaiber, M.; Baba, H.A.; Oberwinkler, H.; Volker, K.; Gabetaner, B.; Bayer, B.; Abebetaer, M.; Schuh, K.; Feil, R.; et al. Stress-dependent dilated cardiomyopathy in mice with cardiomyocyte-restricted inactivation of cyclic GMP-dependent protein kinase I. *Eur. Heart J.* **2013**, *34*, 1233–1244. [CrossRef] [PubMed]
137. Qvigstad, E.; Moltzau, L.R.; Aronsen, J.M.; Nguyen, C.H.; Hougen, K.; Sjaastad, I.; Levy, F.O.; Skomedal, T.; Osnes, J.B. Natriuretic peptides increase beta1-adrenoceptor signalling in failing hearts through phosphodiesterase 3 inhibition. *Cardiovasc. Res.* **2010**, *85*, 763–772. [CrossRef]

138. Meier, S.; Andressen, K.W.; Aronsen, J.M.; Sjaastad, I.; Hougen, K.; Skomedal, T.; Osnes, J.B.; Qvigstad, E.; Levy, F.O.; Moltzau, L.R. PDE3 inhibition by C-type natriuretic peptide-induced cGMP enhances cAMP-mediated signaling in both non-failing and failing hearts. *Eur. J. Pharmacol.* **2017**, *812*, 174–183. [CrossRef] [PubMed]
139. Anand-Srivastava, M.B. Natriuretic peptide receptor-C signaling and regulation. *Peptides* **2005**, *26*, 1044–1059. [CrossRef]
140. Rose, R.A.; Lomax, A.E.; Kondo, C.S.; Anand-Srivastava, M.B.; Giles, W.R. Effects of C-type natriuretic peptide on ionic currents in mouse sinoatrial node: A role for the NPR-C receptor. *Am. J. Physiol. Heart Circ. Physiol.* **2004**, *286*, H1970–H1977. [CrossRef] [PubMed]
141. Travers, J.G.; Kamal, F.A.; Robbins, J.; Yutzey, K.E.; Blaxall, B.C. Cardiac Fibrosis: The Fibroblast Awakens. *Circ. Res.* **2016**, *118*, 1021–1040. [CrossRef]
142. Li, Z.Q.; Liu, Y.L.; Li, G.; Li, B.; Liu, Y.; Li, X.F.; Liu, A.J. Inhibitory effects of C-type natriuretic peptide on the differentiation of cardiac fibroblasts, and secretion of monocyte chemoattractant protein-1 and plasminogen activator inhibitor-1. *Mol. Med. Rep.* **2015**, *11*, 159–165. [CrossRef] [PubMed]
143. Sangaralingham, S.J.; Huntley, B.K.; Martin, F.L.; McKie, P.M.; Bellavia, D.; Ichiki, T.; Harders, G.E.; Chen, H.H.; Burnett, J.C., Jr. The aging heart, myocardial fibrosis, and its relationship to circulating C-type natriuretic Peptide. *Hypertension* **2011**, *57*, 201–207. [CrossRef]
144. Tokudome, T.; Horio, T.; Soeki, T.; Mori, K.; Kishimoto, I.; Suga, S.; Yoshihara, F.; Kawano, Y.; Kohno, M.; Kangawa, K. Inhibitory effect of C-type natriuretic peptide (CNP) on cultured cardiac myocyte hypertrophy: Interference between CNP and endothelin-1 signaling pathways. *Endocrinology* **2004**, *145*, 2131–2140. [CrossRef]
145. Moyes, A.J.; Chu, S.M.; Aubdool, A.A.; Dukinfield, M.S.; Margulies, K.B.; Bedi, K.C.; Hodivala-Dilke, K.; Baliga, R.S.; Hobbs, A.J. C-type natriuretic peptide co-ordinates cardiac structure and function. *Eur. Heart J.* **2019**. [CrossRef] [PubMed]
146. Langenickel, T.H.; Buttgereit, J.; Pagel-Langenickel, I.; Lindner, M.; Monti, J.; Beuerlein, K.; Al-Saadi, N.; Plehm, R.; Popova, E.; Tank, J.; et al. Cardiac hypertrophy in transgenic rats expressing a dominant-negative mutant of the natriuretic peptide receptor B. *Proc. Natl. Acad. Sci. USA* **2006**, *103*, 4735–4740. [CrossRef] [PubMed]
147. Michel, K.; Werner, F.; Prentki, E.; Abesser, M.; Voelker, K.; Baba, H.A.; Skryabin, B.V.; Schuh, K.; Herwig, M.; Hamdani, N.; et al. Blood pressure independent actions of C-type natriuretic peptide in hypertensive heart disease. *Clin. Res. Cardiol.* **2018**, *107* (Suppl. 1).
148. Blaser, M.C.; Wei, K.; Adams, R.L.E.; Zhou, Y.Q.; Caruso, L.L.; Mirzaei, Z.; Lam, A.Y.; Tam, R.K.K.; Zhang, H.; Heximer, S.P.; et al. Deficiency of Natriuretic Peptide Receptor 2 Promotes Bicuspid Aortic Valves, Aortic Valve Disease, Left Ventricular Dysfunction, and Ascending Aortic Dilatations in Mice. *Circ. Res.* **2018**, *122*, 405–416. [CrossRef]
149. Martin, F.L.; Sangaralingham, S.J.; Huntley, B.K.; McKie, P.M.; Ichiki, T.; Chen, H.H.; Korinek, J.; Harders, G.E.; Burnett, J.C., Jr. CD-NP: A novel engineered dual guanylyl cyclase activator with anti-fibrotic actions in the heart. *PLoS ONE* **2012**, *7*, e52422. [CrossRef]
150. Chen, Y.; Zheng, Y.; Iyer, S.R.; Harders, G.E.; Pan, S.; Chen, H.H.; Ichiki, T.; Burnett, J.C., Jr.; Sangaralingham, S.J. C53: A novel particulate guanylyl cyclase B receptor activator that has sustained activity in vivo with anti-fibrotic actions in human cardiac and renal fibroblasts. *J. Mol. Cell. Cardiol.* **2019**, *130*, 140–150. [CrossRef]
151. Jansen, H.J.; Mackasey, M.; Moghtadaei, M.; Liu, Y.; Kaur, J.; Egom, E.E.; Tuomi, J.M.; Rafferty, S.A.; Kirkby, A.W.; Rose, R.A. NPR-C (Natriuretic Peptide Receptor-C) Modulates the Progression of Angiotensin II-Mediated Atrial Fibrillation and Atrial Remodeling in Mice. *Circ. Arrhythm. Electrophysiol.* **2019**, *12*, e006863. [CrossRef] [PubMed]
152. Mackasey, M.; Egom, E.E.; Jansen, H.J.; Hua, R.; Moghtadaei, M.; Liu, Y.; Kaur, J.; McRae, M.D.; Bogachev, O.; Rafferty, S.A.; et al. Natriuretic Peptide Receptor-C Protects Against Angiotensin II-Mediated Sinoatrial Node Disease in Mice. *JACC Basic Transl. Sci.* **2018**, *3*, 824–843. [CrossRef] [PubMed]
153. Pereira, N.L.; Redfield, M.M.; Scott, C.; Tosakulwong, N.; Olson, T.M.; Bailey, K.R.; Rodeheffer, R.J.; Burnett, J.C., Jr. A functional genetic variant (N521D) in natriuretic peptide receptor 3 is associated with diastolic dysfunction: The prevalence of asymptomatic ventricular dysfunction study. *PLoS ONE* **2014**, *9*, e85708. [CrossRef]

154. Rose, R.A.; Hatano, N.; Ohya, S.; Imaizumi, Y.; Giles, W.R. C-type natriuretic peptide activates a non-selective cation current in acutely isolated rat cardiac fibroblasts via natriuretic peptide C receptor-mediated signalling. *J. Physiol.* **2007**, *580 Pt 1*, 255–274. [CrossRef]
155. Rahmutula, D.; Zhang, H.; Wilson, E.E.; Olgin, J.E. Absence of natriuretic peptide clearance receptor attenuates TGF-beta1-induced selective atrial fibrosis and atrial fibrillation. *Cardiovasc. Res.* **2019**, *115*, 357–372. [CrossRef]
156. Wright, R.S.; Wei, C.M.; Kim, C.H.; Kinoshita, M.; Matsuda, Y.; Aarhus, L.L.; Burnett, J.C., Jr.; Miller, W.L. C-type natriuretic peptide-mediated coronary vasodilation: role of the coronary nitric oxide and particulate guanylate cyclase systems. *J. Am. Coll. Cardiol.* **1996**, *28*, 1031–1038. [CrossRef]
157. Hobbs, A.; Foster, P.; Prescott, C.; Scotland, R.; Ahluwalia, A. Natriuretic peptide receptor-C regulates coronary blood flow and prevents myocardial ischemia/reperfusion injury: Novel cardioprotective role for endothelium-derived C-type natriuretic peptide. *Circulation* **2004**, *110*, 1231–1235. [CrossRef]
158. Prickett, T.C.; Doughty, R.N.; Troughton, R.W.; Frampton, C.M.; Whalley, G.A.; Ellis, C.J.; Espiner, E.A.; Richards, A.M. C-Type Natriuretic Peptides in Coronary Disease. *Clin. Chem.* **2017**, *63*, 316–324. [CrossRef]
159. Hausenloy, D.J.; Yellon, D.M. Myocardial ischemia-reperfusion injury: A neglected therapeutic target. *J. Clin. Investig.* **2013**, *123*, 92–100. [CrossRef]
160. Tarazon, E.; Rosello-Lleti, E.; Ortega, A.; Molina-Navarro, M.M.; Sanchez-Lazaro, I.; Lago, F.; Gonzalez-Juanatey, J.R.; Rivera, M.; Portoles, M. Differential gene expression of C-type natriuretic peptide and its related molecules in dilated and ischemic cardiomyopathy. A new option for the management of heart failure. *Int. J. Cardiol.* **2014**, *174*, e84–e86. [CrossRef]
161. Gorbe, A.; Giricz, Z.; Szunyog, A.; Csont, T.; Burley, D.S.; Baxter, G.F.; Ferdinandy, P. Role of cGMP-PKG signaling in the protection of neonatal rat cardiac myocytes subjected to simulated ischemia/reoxygenation. *Basic Res. Cardiol.* **2010**, *105*, 643–650. [CrossRef] [PubMed]
162. Burley, D.S.; Cox, C.D.; Zhang, J.; Wann, K.T.; Baxter, G.F. Natriuretic peptides modulate ATP-sensitive K(+) channels in rat ventricular cardiomyocytes. *Basic Res. Cardiol.* **2014**, *109*, 402. [CrossRef]
163. Wu, L.H.; Zhang, Q.; Zhang, S.; Meng, L.Y.; Wang, Y.C.; Sheng, C.J. Effects of gene knockdown of CNP on ventricular remodeling after myocardial ischemia-reperfusion injury through NPRB/Cgmp signaling pathway in rats. *J. Cell. Biochem.* **2018**, *119*, 1804–1818. [CrossRef]
164. Wang, Y.; de Waard, M.C.; Sterner-Kock, A.; Stepan, H.; Schultheiss, H.P.; Duncker, D.J.; Walther, T. Cardiomyocyte-restricted over-expression of C-type natriuretic peptide prevents cardiac hypertrophy induced by myocardial infarction in mice. *Eur. J. Heart Fail.* **2007**, *9*, 548–557. [CrossRef] [PubMed]
165. Soeki, T.; Kishimoto, I.; Okumura, H.; Tokudome, T.; Horio, T.; Mori, K.; Kangawa, K. C-type natriuretic peptide, a novel antifibrotic and antihypertrophic agent, prevents cardiac remodeling after myocardial infarction. *J. Am. Coll. Cardiol.* **2005**, *45*, 608–616. [CrossRef] [PubMed]
166. Del Ry, S.; Cabiati, M.; Martino, A.; Cavallini, C.; Caselli, C.; Aquaro, G.D.; Battolla, B.; Prescimone, T.; Giannessi, D.; Mattii, L.; et al. High concentration of C-type natriuretic peptide promotes VEGF-dependent vasculogenesis in the remodeled region of infarcted swine heart with preserved left ventricular ejection fraction. *Int. J. Cardiol.* **2013**, *168*, 2426–2434. [CrossRef] [PubMed]
167. Rose, R.A.; Lomax, A.E.; Giles, W.R. Inhibition of L-type Ca2+ current by C-type natriuretic peptide in bullfrog atrial myocytes: An NPR-C-mediated effect. *Am. J. Physiol. Heart Circ. Physiol.* **2003**, *285*, H2454–H2462. [CrossRef] [PubMed]
168. Rose, R.A.; Giles, W.R. Natriuretic peptide C receptor signalling in the heart and vasculature. *J. Physiol.* **2008**, *586*, 353–366. [CrossRef]
169. Springer, J.; Azer, J.; Hua, R.; Robbins, C.; Adamczyk, A.; McBoyle, S.; Bissell, M.B.; Rose, R.A. The natriuretic peptides BNP and CNP increase heart rate and electrical conduction by stimulating ionic currents in the sinoatrial node and atrial myocardium following activation of guanylyl cyclase-linked natriuretic peptide receptors. *J. Mol. Cell. Cardiol.* **2012**, *52*, 1122–1134. [CrossRef] [PubMed]
170. Azer, J.; Hua, R.; Vella, K.; Rose, R.A. Natriuretic peptides regulate heart rate and sinoatrial node function by activating multiple natriuretic peptide receptors. *J. Mol. Cell. Cardiol.* **2012**, *53*, 715–724. [CrossRef]
171. Egom, E.E.; Vella, K.; Hua, R.; Jansen, H.J.; Moghtadaei, M.; Polina, I.; Bogachev, O.; Hurnik, R.; Mackasey, M.; Rafferty, S.; et al. Impaired sinoatrial node function and increased susceptibility to atrial fibrillation in mice lacking natriuretic peptide receptor C. *J. Physiol.* **2015**, *593*, 1127–1146. [CrossRef]

172. Buttgereit, J.; Shanks, J.; Li, D.; Hao, G.; Athwal, A.; Langenickel, T.H.; Wright, H.; da Costa Goncalves, A.C.; Monti, J.; Plehm, R.; et al. C-type natriuretic peptide and natriuretic peptide receptor B signalling inhibits cardiac sympathetic neurotransmission and autonomic function. *Cardiovasc. Res.* **2016**, *112*, 637–644. [CrossRef]
173. Moghtadaei, M.; Langille, E.; Rafferty, S.A.; Bogachev, O.; Rose, R.A. Altered heart rate regulation by the autonomic nervous system in mice lacking natriuretic peptide receptor C (NPR-C). *Sci. Rep.* **2017**, *7*, 17564. [CrossRef]
174. Zanchi, A.; Maillard, M.; Burnier, M. Recent clinical trials with omapatrilat: new developments. *Curr. Hypertens. Rep.* **2003**, *5*, 346–352. [CrossRef]
175. Ruilope, L.M.; Dukat, A.; Bohm, M.; Lacourciere, Y.; Gong, J.; Lefkowitz, M.P. Blood-pressure reduction with LCZ696, a novel dual-acting inhibitor of the angiotensin II receptor and neprilysin: A randomised, double-blind, placebo-controlled, active comparator study. *Lancet* **2010**, *375*, 1255–1266. [CrossRef]
176. Kario, K.; Sun, N.; Chiang, F.T.; Supasyndh, O.; Baek, S.H.; Inubushi-Molessa, A.; Zhang, Y.; Gotou, H.; Lefkowitz, M.; Zhang, J. Efficacy and safety of LCZ696, a first-in-class angiotensin receptor neprilysin inhibitor, in Asian patients with hypertension: A randomized, double-blind, placebo-controlled study. *Hypertension* **2014**, *63*, 698–705. [CrossRef] [PubMed]
177. McMurray, J.J.; Packer, M.; Desai, A.S.; Gong, J.; Lefkowitz, M.P.; Rizkala, A.R.; Rouleau, J.L.; Shi, V.C.; Solomon, S.D.; Swedberg, K.; et al. Angiotensin-neprilysin inhibition versus enalapril in heart failure. *N. Engl. J. Med.* **2014**, *371*, 993–1004. [CrossRef]
178. Solomon, S.D.; Rizkala, A.R.; Gong, J.; Wang, W.; Anand, I.S.; Ge, J.; Lam, C.S.P.; Maggioni, A.P.; Martinez, F.; et al. Angiotensin Receptor Neprilysin Inhibition in Heart Failure With Preserved Ejection Fraction: Rationale and Design of the PARAGON-HF Trial. *JACC. Heart Fail.* **2017**, *5*, 471–482. [CrossRef] [PubMed]
179. Kovacs, A.; Alogna, A.; Post, H.; Hamdani, N. Is enhancing cGMP-PKG signalling a promising therapeutic target for heart failure with preserved ejection fraction? *Neth. Heart J.* **2016**, *24*, 268–274. [CrossRef]
180. Ichiki, T.; Dzhoyashvili, N.; Burnett, J.C., Jr. Natriuretic peptide based therapeutics for heart failure: Cenderitide: A novel first-in-class designer natriuretic peptide. *Int. J. Cardiol.* **2019**, *281*, 166–171. [CrossRef] [PubMed]
181. Dickey, D.M.; Potter, L.R. Dendroaspis natriuretic peptide and the designer natriuretic peptide, CD-NP, are resistant to proteolytic inactivation. *J. Mol. Cell. Cardiol.* **2011**, *51*, 67–71. [CrossRef]
182. Lee, C.Y.; Chen, H.H.; Lisy, O.; Swan, S.; Cannon, C.; Lieu, H.D.; Burnett, J.C., Jr. Pharmacodynamics of a novel designer natriuretic peptide, CD-NP, in a first-in-human clinical trial in healthy subjects. *J. Clin. Pharmacol.* **2009**, *49*, 668–673. [CrossRef] [PubMed]
183. Chen, Y.; Harty, G.J.; Zheng, Y.; Iyer, S.R.; Sugihara, S.; Sangaralingham, S.J.; Ichiki, T.; Grande, J.P.; Lee, H.C.; Wang, X.L.; et al. CRRL269: A Novel Particulate Guanylyl Cyclase A Receptor Peptide Activator For Acute Kidney Injury. *Circ. Res.* **2019**. [CrossRef] [PubMed]
184. He, X.; Chow, D.; Martick, M.M.; Garcia, K.C. Allosteric activation of a spring-loaded natriuretic peptide receptor dimer by hormone. *Science* **2001**, *293*, 1657–1662. [CrossRef]
185. Veale, C.A.; Alford, V.C.; Aharony, D.; Banville, D.L.; Bialecki, R.A.; Brown, F.J.; Damewood, J.R., Jr.; Dantzman, C.L.; Edwards, P.D.; Jacobs, R.T.; et al. The discovery of non-basic atrial natriuretic peptide clearance receptor antagonists. Part 1. *Bioorg. Med. Chem. Lett.* **2000**, *10*, 1949–1952. [CrossRef]

© 2019 by the authors. Licensee MDPI, Basel, Switzerland. This article is an open access article distributed under the terms and conditions of the Creative Commons Attribution (CC BY) license (http://creativecommons.org/licenses/by/4.0/).

Review

Pulmonary Arterial Hypertension Due to NPR-C Mutation: A Novel Paradigm for Normal and Pathologic Remodeling?

Emmanuel Eroume-A Egom

St Martha's Regional Hospital, Dalhousie University, Antigonish, B2G 2G7 NS, Canada; egomemmanuel@gmail.com

Received: 24 May 2019; Accepted: 21 June 2019; Published: 22 June 2019

Abstract: Idiopathic Pulmonary Arterial Hypertension (IPAH) is a deadly and disabling disease characterized by severe vascular remodeling of small pulmonary vessels by fibroblasts, myofibroblasts and vascular smooth muscle cell proliferation. Recent studies suggest that the Natriuretic Peptide Clearance Receptor (NPR-C) signaling pathways may play a crucial role in the development of IPAH. Reduced expression or function of NPR-C signaling in pulmonary artery smooth muscle cells may contribute to the pulmonary vascular remodeling, which is characteristic of this disease. The likely mechanisms may involve an impaired interaction between NPR-C, specific growth factors and other signal transduction pathways including but not limited to Gqα/mitogen-activated protein kinase (MAPK)/PI3K and AKT signaling. The resulting failure of growth suppression in pulmonary artery smooth muscle cells provides critical clues to the cellular pathobiology of IPAH. The reciprocal regulation of NPR-C signaling in models of tissue remodeling may thus provide new insights to our understanding of IPAH.

Keywords: Idiopathic Pulmonary Arterial Hypertension (IPAH); Natriuretic Peptide Clearance Receptor (NPR-C) signaling

1. Introduction

Pulmonary arterial hypertension (PAH) is a devastating disease, which if not interrupted, leads to progressive right-sided heart failure and death within 2 to 3 years after diagnosis [1–3]. Although schistosomiasis may be the most common cause of PAH worldwide, evidence suggests that over half of cases of PAH in regions of the world without endemic schistosomiasis are idiopathic (IPAH) [4–6]. Pathobiologically, IPAH is a proliferative vasculopathy, characterized by vasoconstriction, vascular muscle cell proliferation, fibrosis, and microthrombosis [2,3]. Histologic findings may include hyperplasia and hypertrophy of all three layers of the pulmonary vascular wall (intima, media, adventitia) as well as fibrosis and in situ thrombi of the small pulmonary arteries and arterioles (plexiform lesions) [2,7,8]. It is during the transition from normal to remodeled pulmonary vascular cells that occur critical pathophysiological processes including, but not limited to, changes in key signal transduction pathways, from which the nitric oxide (NO) signaling is believed to be one of, if not, the major contributor to its homeostasis [2,3,9,10]. Several clinical studies have thus focused on targeting the NO pathway in patients with IPAH; however, they all fall short as to re-establishment of structural as well as functional pulmonary vascular integrity, as a basis for handicap-free long-term survival [11,12]. These findings could have been anticipated as individuals with IPAH are well known to have endothelial dysfunction and therefore reduced, if not, loss of NO signaling [12]. Interestingly, evidence also suggests that in the presence of a reduced or loss of NO signal transduction, there may be an enhanced natriuretic peptide clearance receptor (NPR-C)-mediated vasorelaxant effect [12,13]. They may thus be synergistic as well as complementary cardioprotective actions for NPR-C signal transduction and NO-mediated pathway in the vasculature [12,13]. The inhibition of one signaling pathway may therefore be compensated for by the upregulation of the other [12,13]. These striking observations raise the question of whether NPR-C signal transduction pathway may play a crucial role in IPAH pathobiology.

2. Normal NPR-C Signaling

Natriuretic peptides (NPs) constitute a family of at least four structurally related peptide hormones named Atrial Natriuretic Peptide (ANP), Brain Natriuretic Peptide (BNP), C-type Natriuretic Peptide (CNP), and Urodilatin (URO), which may regulate several biological processes including plasma volume and blood pressure control [10,14,15]. Most of the biological actions of NPs appear to be mediated via binding to three specific cell membranes receptors known as natriuretic peptide receptors-A, B, and C (NPR-A, NPR-B, and NPR-C) [10,14,15]. NPR-C is a disulfide-linked homodimer of a single transmembrane domain, an extracellular domain of ~440 amino acids; and a short 37 amino acid cytoplasmic domain with several inhibitory guanine nucleotide regulatory protein (Gi) activator peptide sequences that bind and activate the Gi-dependent signal transduction [16]. NPR-C has been found to lack the seven transmembrane domains of the typical G-protein-coupled receptors (GPCRs) and may thus be considered as an atypical GPCR [17]. NPR-C is widely distributed in several tissues and cells including but not limited to cardiac fibroblasts and myocytes, endothelial cells (EC) and vascular smooth muscles cells (VSMC) [16,17]. The binding affinity of the NPs for NPR-C is as follows: ANP > BNP > CNP [17,18]. Although originally classified as a clearance receptor with no signaling function, evidence suggests that NPR-C may be coupled to different intracellular signaling pathways including the adenylyl cyclase (AC)/cAMP signal transduction [16], the phospholipase C (PLC) signaling pathway, the nitric oxide (NO) pathway and Gqα/mitogen-activated protein kinase (MAPK)/PI3K and AKT pathways (As illustrated in Figure 1) [17,19–21].

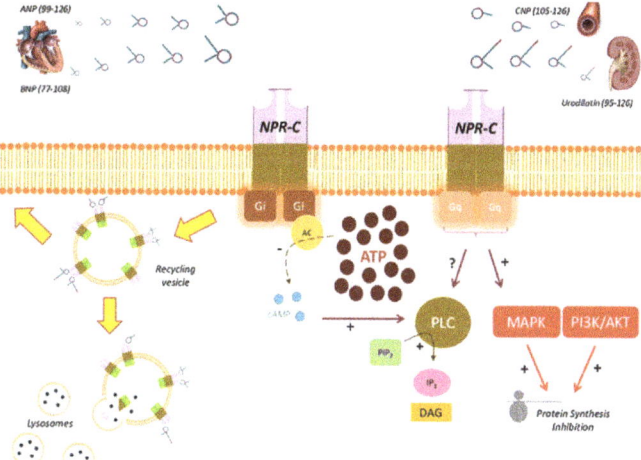

Figure 1. NPR-C signaling pathway. AC: adenylyl cyclase; ANP: Atrial Natriuretic Peptide; AKT: protein kinase B; ATP: adenosine triphosphate; BNP: Brain Natriuretic Peptide; cAMP: cyclic adenosine monophospate; CNP: C-type Natriuretic Peptide; DAG: diacylglycerol; Gi: G inhibitory protein; Gq: Gq protein; IP3: inositol triphosphate; MAPK: mitogen-activated protein kinase; PI3K: phosphatidylinositol 3 kinase; PIP2: phosphatidylinositol 4,5-bisphosphate; PLC: phospholipase C. +: stimulation; -: inhibition; ?: unknow. Reprint from [17] with permission from Springer.

Various studies have demonstrated that NPR-C can inhibit the AC/cAMP signal transduction and elicit physiological functions [22]. Natriuretic peptides such as ANP, CNP and BNP have been reported to inhibit AC and decrease cAMP levels in a variety of tissues and cells by interacting with NPR-C receptor [22,23]. In addition, activation of NPR-C signaling may decrease L-type calcium current in single sinuatrial node cells by AC inhibition and decreased cAMP levels [24]. Furthermore, the NPR-C's agonist C-ANP$_{4-23}$ may decrease the cAMP levels in VSMCs [23]. NPR-C has also been involved in the modulation of PLC signaling pathway. Phosphatidyl inositol turnover signaling is a major signal transduction pathway involved in intracellular calcium mobilization and protein kinase C (PKC) activation [16]. Evidence suggests that activation of NPR-C signaling may stimulate PI turnover in cultured bovine aortic smooth muscle cells [20]. Murthy and colleagues demonstrated that small peptide fragments of cytoplasmic domain of NPR-C may stimulate PLC-β activity in guinea pig tenia coli smooth muscle cells [23]. In addition, the NPR-C's agonist C-ANP$_{4-23}$ may trigger inositol triphosphate (IP3) formation in A10 vascular smooth muscle cells [25]. Whether the activation of PI turnover by C-ANP$_{4-23}$ is a secondary event mediated through the AC/cAMP system coupled to NPR-C or a primary event is still not well understood. There may definitely be a cross-talk between NPR-C-mediated AC/cAMP and PLC signaling pathways as the inhibition of AC and decreased levels of cAMP triggered by NPRC activation may contribute to stimulation of PI turnover [16,25]. The modulation of other NPR-C-induced signal transduction pathways have also been suggested. As discussed in detail later, evidence suggests that NPR-C signaling may inhibit the platelet-derived growth factor (PDGF) and endothelin-3-induced mitogen-activated protein kinase (MAPK) through the inhibition of upstream kinases including MAPK kinase [26]. NPs antiproliferative effects in cardiac fibroblasts may be a result of NPR-C activation [27]. In addition, NPs-induced NPR-C signal transduction has also been implicated in modulating endothelial and VSMC proliferation, endothelial permeability in coronary endothelial cells as well as L-type calcium influx [17]. These observations suggest that the NPR-C signaling pathway may play a critical role in cell proliferation via ihinhibition of MAPK signaling pathway [16]. Finally, evidence also suggests that the NPR-C's agonist C-ANP$_{4-23}$ as well as other vasoactive peptides may trigger the activation of constitutive nitric oxide synthase (NOS) smooth muscle cells [28].

3. Consequences of Npr3 Mutation and/or Alterations in NPR-C Signaling

3.1. Studies in Transgenic and Knockout Mice

Both heterozygous and homozygous knockout mice for *Npr3* exhibit alterations in cardiovascular system, and several evidences suggest that mutations in *Npr3* may lead to cardiovascular diseases [17]. Recent evidence indicates a critical role of NPR-C signaling in the pathobiology of PH. We recently described the cardiac structure and function of mice lacking NPR-C (NPR-C$^{-/-}$) by echocardiography [1,10]. NPR-C$^{-/-}$ mice exhibit important structural features including right atrial and ventricular enlargement, hypertrophy of the right ventricular free wall, tricuspid regurgitation as well as echocardiographic findings suggestive of right ventricular pressure overload manifested as abnormal motion of the interventricular septum, which are all findings typically seen in humans with PAH [1]. Moreover, Doppler Echocardiography assessment demonstrated a significantly higher right ventricular systolic pressure compared with wild-type littermates [1]. These novel findings were also confirmed by resting right heart catheterization [1]. The above results suggest that NPR-C signal transduction may play a crucial role in the patho-biology of PAH.

3.2. NPR-C signaling in Hypoxia-Induced Pulmonary Hypertension

Group 3 Pulmonary hypertension (PH) includes PH due to lung diseases or hypoxia. Among the etiologies of group 3 PH, the strongest evidence favors hypoxic pulmonary vasoconstriction with remodeling of the pulmonary vascular bed. The pulmonary vascular remodeling is characterized by the proliferation, hypertrophy and extension of smooth muscle cells to previously unmuscularized pulmonary arterioles [29]. The mechanisms underlying hypoxia-induced PH and remodeling are poorly understood. One candidate that may play a critical role in the pathobiology of the condition is

NPR-C signaling pathway. Evidence suggests NPR-C gene expression is selectively downregulated in the setting of hypoxia, which in turn may, at least in part, contribute to the development of hypoxic PH. Li and colleagues demonstrated that steady-state mRNA levels of NPR-C may be decreased to 20–30% of air control levels in lungs of hypoxia-adapted experimental models [30]. In addition, nuclear-runoff analysis revealed a huge and significant decreased transcription of the NPR-C gene in lung of hypoxia-exposed experimental models compared with air control, suggesting that the hypoxia-induced reduction of NPR-C steady-state mRNA levels is due to the downregulated gene transcription. As NPR-C in the lungs is located in pulmonary vascular smooth muscle cells (VSMCs), Sun and colleagues used pulmonary arterial smooth muscle cells (PASMCs) cultured in vitro to investigate mechanisms underlying the downregulation of NPR-C gene expression in lung of hypoxia-adapted experimental models [31]. The authors found that the diminished expression of NPR-C mRNA observed under hypoxic conditions in lung might be mediated through the tyrosine kinase receptor-associated fibroblast growth factor (FGF) and platelet-derived growth factor (PDGF), the gene expression of which is enhanced in lung of hypoxia-adapted animals [31]. Various FGFs and PDGFs are expressed in lung and may play critical roles in diverse aspects of pulmonary vascular remodeling, including but not limited to lung epithelial cells, VSMCs, and myofibroblast proliferation, differentiation, and angiogenesis, as well as adaptation to environmental hypoxia [31,32]. Evidence suggests that these growth factors may, at least in part, contribute to hyperproliferation of PASMCs and muscularization of pulmonary vasculature in hypoxia-induced PH [33–36]. The important relationship between the overexpression of these hypoxia-responsive growth factors and the downregulation of NPR-C has also been demonstrated in several studies performed in other experimental models [16,31]. Multiple signaling pathways may mediate the mitogenic activity of FGF and PDGF [31,37,38]. In fact, Sun and colleagues also demonstrated that the FGF- and PDGF-mediated downregulation of NPR-C gene expression is dependent on the activation of Ras-Raf-MEK-MAP kinase in PASMCs [31]. As activation of NPR-C signaling may have anti-proliferative effects, hypoxia-induced down regulation of NPR-C expression and associated impaired NPR-C pathways may lead to failure of the NPRC-related antiproliferative effects in the pulmonary vasculature, which may then ultimately lead to PAH [1].

3.3. Studies implicating NPR-C Signaling in Other Cardiovascular Diseases

Several evidence also suggests a potential role of NPR-C signaling pathways in the pathophysiology of other cardiovascular diseases [17,18]. Alterations in NPR-C signaling have been found in various experimental hypertensive models [17,27]. Consistently, a study in 200,000 European descents demonstrated an association between rs1173771 polymorphism in NPR-C with hypertension [39]. NPR-C signal transduction may also play an important role in the pathogenesis of atherosclerosis [17]. Recently, several novel single nucleotide polymorphisms (SNPs) of NPR-C were identified to be associated with atherosclerotic cardiovascular disease (ASCVD) in Han Chinese population [40]. Interestingly, multivariate logistical regression analysis demonstrated that the association between these SNPs and ASCVD remained significant even after adjustment for all the conventional risk factors of ASCVD including hypertension [40]. Although the underlying molecular mechanisms of the association between NPR-C gene polymorphism and ASCVD are unclear, some studies have shown that percutaneous coronary intervention-induced injury to neointima may lead to an increase in NPR-C expression in neointimal smooth muscle cells [17]. In addition, NPR-C is over-expressed near the luminal surface of atherosclerotic plaques and in VSMC [17]. After acute myocardial infarction, NPR-C expression may be up-regulated in infarcted and non-infarcted areas of the left ventricular wall [17]. NPR-C expression appears also upregulated in patients with heart failure [17,27]. Fox and colleagues identified four Npr3 SNPs that may be associated to left ventricular dysfunction, further suggesting an established effects of NPR-C variant on ASCVD [41].

4. NPR-C Signaling as a Therapeutic Target in Tissue Remodeling

Excessive vascular smooth muscle cell (VSMC) proliferation contributes to tissue remodeling that occurs in several vasculopathies including but not limited to PH [42,43]. Evidence suggests that activation of NPR-C signaling may attenuate vasoactive peptide-induced hyperproliferation of VSMC via MAP kinase and phosphatidylinositol 3-kinase (PI3K) pathways [44]. Consistently, Andalousi and colleagues have demonstrated that the NPR-C agonist, C-ANP$_{4-23}$ may attenuate the enhanced proliferation of VSMC by decreasing the expression of cell cycle proteins [42]. The authors showed that the antiproliferative effect of C-ANP$_{4-23}$ in VSMC was, again, mediated through the inhibition of MAP kinase/PI3-kinase/AKT signaling pathways [42]. Cahill and colleagues demonstrated that NPR-C may inhibit aortic smooth muscle cell mitogenesis and proliferation via inhibition of thymidine kinase activity [23]. In addition, NPR-C may mediate, at least partially, the anti-proliferative actions of BNP in human cardiac fibroblasts [45]. Sangaralingham and colleagues also demonstrated that NPR-C may mediate the antifibrotic and antiproliferative peptide CNP in cultured adult human cardiac fibroblasts [46]. Furthermore, we recently showed that mice lacking NPR-C exhibit enhanced collagen expression and deposition in the atria [47,48], whereas selective NPR-C activation may prevent pathological collagen deposition [49]. The activation of NPR-C signaling may inhibit the hypoxia-induced vascular endothelial cell growth factor (VEGF) transcription and protein production, suggesting that this receptor may have both direct [50], and indirect effects as antiproliferation factors for endothelial cells (EC), the latter potentially mediated via modulating VEGF synthesis [51]. Pedram and colleagues also demonstrated that NPR-C activation may inhibit the vasoactive peptides endothelin (ET)-stimulated secretion of VEGF and the subsequent EC proliferation and invasion [51]. These observations further support the potential therapeutic functions of NPR-C in vascular remodeling and angiogenesis [51]. Overall, the ability of the NPR-C signaling to impede cardiac or vascular remodeling may limit the cellular response to various chronic pulmonary vascular insults, making this pathway an attractive therapeutic target to prevent or reverse PH.

5. Conclusions

For several decades, NPR-C has been considered a NP clearance receptor responsible for receptor-mediated NPs degradation [17]. More recent data showing evidence of NPR-C multiple actions on different cells, organs and systems; and the existence of a specific intracellular signaling pathway have overcome this view. The role of NPR-C signaling in pulmonary diseases remains at an early stage. Although clearly of direct and critical relevance to PH, NPR-C signaling pathway is very likely to contribute to other pulmonary pathologies characterized by tissue remodeling, such as pulmonary fibrosis and chronic obstructive pulmonary disease. As activation of NPR-C signaling may have anti-proliferative effects, any alterations in this pathway may thus lead to failure of its antiproliferative effect within the pulmonary vasculature, which in turn may ultimately lead to pulmonary vascular remodeling. Large-scale and carefully designed genetic studies are needed to investigate alterations of NPR-C function and structure as well as the existence of gene polymorphisms in patients with PH, particularly patients with FPAH. Further studies are also required to completely elucidate the pathobiological role played by NPR-C signaling and to better understand how NPR-C alterations at gene and protein levels could contribute to tissue remodeling, particularly pulmonary vascular remodeling.

Funding: Egom Clinical & Translational Research Services Ltd.

Conflicts of Interest: The author declares no conflict of interest.

References

1. Egom, E.E.-A.; Feridooni, T.; Pharithi, R.B.; Khan, B.; Shiwani, H.A.; Maher, V.; El Hiani, Y.; Rose, R.A.; Pasumarthi, K.B.S.; Ribama, H.A. New insights and new hope for pulmonary arterial hypertension: natriuretic peptides clearance receptor as a novel therapeutic target for a complex disease. *Int. J. Physiol. Pathophysiol. Pharmacol.* **2017**, *9*, 112–118. [CrossRef]
2. Humbert, M.; Guignabert, C.; Bonnet, S.; Dorfmuller, P.; Klinger, J.R.; Nicolls, M.R.; Olschewski, A.J.; Pullamsetti, S.S.; Schermuly, R.T.; Stenmark, K.R.; et al. Pathology and pathobiology of pulmonary hypertension: state of the art and research perspectives. *Eur. Respir. J.* **2019**, *53*. [CrossRef] [PubMed]
3. Simonneau, G.; Montani, D.; Celermajer, D.S.; Denton, C.P.; Gatzoulis, M.A.; Krowka, M.; Williams, P.G.; Souza, R. Haemodynamic definitions and updated clinical classification of pulmonary hypertension. *Eur. Respir. J.* **2019**, *53*. [CrossRef] [PubMed]
4. Humbert, M.; Sitbon, O.; Chaouat, A.; Bertocchi, M.; Habib, G.; Gressin, V.; Yaici, A.; Weitzenblum, E.; Cordier, J.-F.; Chabot, F. Pulmonary arterial hypertension in France: results from a national registry. *Am. J. Respir. Crit. Care Med.* **2006**, *173*, 1023–1030. [CrossRef] [PubMed]
5. Badesch, D.B.; Raskob, G.E.; Elliott, C.G.; Krichman, A.M.; Farber, H.W.; Frost, A.E.; Barst, R.J.; Benza, R.L.; Liou, T.G.; Turner, M. Pulmonary arterial hypertension: baseline characteristics from the REVEAL Registry. *Chest* **2010**, *137*, 376–387. [CrossRef] [PubMed]
6. Peacock, A.; Murphy, N.; McMurray, J.; Caballero, L.; Stewart, S. An epidemiological study of pulmonary arterial hypertension. *Eur. Respir. J.* **2007**, *30*, 104–109. [CrossRef] [PubMed]
7. Humbert, M.; Morrell, N.W.; Archer, S.L.; Stenmark, K.R.; MacLean, M.R.; Lang, I.M.; Christman, B.W.; Weir, E.K.; Eickelberg, O.; Voelkel, N.F. Cellular and molecular pathobiology of pulmonary arterial hypertension. *J. Am. Coll. Cardiol.* **2004**, *43*, S13–S24. [CrossRef] [PubMed]
8. Pietra, G.G.; Capron, F.; Stewart, S.; Leone, O.; Humbert, M.; Robbins, I.M.; Reid, L.M.; Tuder, R. Pathologic assessment of vasculopathies in pulmonary hypertension. *J. Am. Coll. Cardiol.* **2004**, *43*, S25–S32. [CrossRef]
9. Dupont, M.; Tang, W.W. Right ventricular afterload and the role of nitric oxide metabolism in left-sided heart failure. *J. Card. Fail.* **2013**, *19*, 712–721. [CrossRef]
10. Egom, E.E.; Maher, V.; El Hiani, Y. Evolving use of natriuretic peptide receptor type-C as part of strategies for the treatment of pulmonary hypertension due to left ventricle heart failure. *Int. J. Cardiol.* **2019**, *281*, 172–178. [CrossRef]
11. Rosenkranz, S.; Gibbs, J.S.; Wachter, R.; De Marco, T.; Vonk-Noordegraaf, A.; Vachiery, J.L. Left ventricular heart failure and pulmonary hypertension. *Eur. Heart J.* **2016**, *37*, 942–954. [CrossRef] [PubMed]
12. Egom, E.; Feridooni, T.; Pharithi, R.; Maher, V.; El Hiani, Y.; Pasumarthi, K.; Ribama, H. New Insights and New Hope for Pulmonary Arterial Hypertension: Natriuretic Peptides Clearance Receptor as a Novel Therapeutic Target for a Complex Disease. *J. Am. Coll. Cardiol.* **2017**, *69* (Suppl. 11), 1902. [CrossRef]
13. Hobbs, A.; Foster, P.; Prescott, C.; Scotland, R.; Ahluwalia, A. Natriuretic peptide receptor-C regulates coronary blood flow and prevents myocardial ischemia/reperfusion injury: novel cardioprotective role for endothelium-derived C-type natriuretic peptide. *Circulation* **2004**, *110*, 1231–1235. [CrossRef] [PubMed]
14. Egom, E.E. BNP and Heart Failure: Preclinical and Clinical Trial Data. *J. Cardiovasc. Transl. Res.* **2015**, *8*, 149–157. [CrossRef] [PubMed]
15. Egom, E.E.; Feridooni, T.; Hotchkiss, A.; Kruzliak, P.; Pasumarthi, K.B. Mechanisms of renal hyporesponsiveness to BNP in heart failure. *Can. J. Physiol. Pharmacol.* **2015**, *93*, 399–403. [CrossRef] [PubMed]
16. Anand-Srivastava, M.B. Natriuretic peptide receptor-C signaling and regulation. *Peptides* **2005**, *26*, 1044–1059. [CrossRef] [PubMed]
17. Kouyoumdzian, N.M.; Mikusic, N.L.R.; Lee, H.J.; Fernández, B.E.; Choi, M.R. Natriuretic Peptide Receptor Type C (NPRC). In *Encyclopedia of Signaling Molecules*; Choi, S., Ed.; Ed. Springer New York: New York, NY, USA, 2017; pp. 1–7.
18. Cantú, S.M.; Donoso, A.S.; Kouyoumdzian, N.M.; Mikusic, R.; Lucía, N.; Puyó, A.M.; Choi, M.R. Clinical aspects of c-type natriuretic peptide on the cardiovascular system. *Int. J. Clin. Endocrinol. Metab.* **2015**, *1*, 3–36.
19. Li, Y.; Hashim, S.; Anand-Srivastava, M.B. Intracellular peptides of natriuretic peptide receptor-C inhibit vascular hypertrophy via Gqα/MAP kinase signaling pathways. *Cardiovasc. Res.* **2006**, *72*, 464–472. [CrossRef]

20. Jain, A.; Anand-Srivastava, M.B. Natriuretic peptide receptor-C-mediated attenuation of vascular smooth muscle cell hypertrophy involves Gqalpha/PLCbeta1 proteins and ROS-associated signaling. *Pharmacol. Res. Perspect.* **2018**, *6*. [CrossRef]
21. Ventimiglia, M.S.; Najenson, A.C.; Perazzo, J.C.; Carozzo, A.; Vatta, M.S.; Davio, C.A.; Bianciotti, L.G. Blockade of Multidrug Resistance-Associated Proteins Aggravates Acute Pancreatitis and Blunts Atrial Natriuretic Factor's Beneficial Effect in Rats: Role of MRP4 (ABCC4). *Mol. Med.* **2015**, *21*, 58–67. [CrossRef]
22. Pandey, K.N. Molecular and genetic aspects of guanylyl cyclase natriuretic peptide receptor-A in regulation of blood pressure and renal function. *Physiol. Genom.* **2018**, *50*, 913–928. [CrossRef] [PubMed]
23. Moyes, A.J.; Hobbs, A.J. C-type Natriuretic Peptide: A Multifaceted Paracrine Regulator in the Heart and Vasculature. *Int. J. Mol. Sci.* **2019**, *20*, 2281. [CrossRef] [PubMed]
24. Rose, R.A.; Lomax, A.E.; Kondo, C.S.; Anand-Srivastava, M.B.; Giles, W.R. Effects of C-type natriuretic peptide on ionic currents in mouse sinoatrial node: a role for the NPR-C receptor. *Am. J. Physiol. Heart Circ. Physiol.* **2004**, *286*, H1970–H1977. [CrossRef]
25. Mouawad, R.; Li, Y.; Anand-Srivastava, M.B. Atrial natriuretic peptide-C receptor-induced attenuation of adenylyl cyclase signaling activates phosphatidylinositol turnover in A10 vascular smooth muscle cells. *Mol. Pharmacol.* **2004**, *65*, 917–924. [CrossRef] [PubMed]
26. Pandey, K.N. Guanylyl cyclase / atrial natriuretic peptide receptor-A: role in the pathophysiology of cardiovascular regulation. *Can. J. Physiol. Pharmacol.* **2011**, *89*, 557–573. [CrossRef] [PubMed]
27. Rubattu, S.; Sciarretta, S.; Morriello, A.; Calvieri, C.; Battistoni, A.; Volpe, M. NPR-C: a component of the natriuretic peptide family with implications in human diseases. *J. Mol. Med.* **2010**, *88*, 889–897. [CrossRef] [PubMed]
28. Rose, R.A.; Giles, W.R. Natriuretic peptide C receptor signalling in the heart and vasculature. *J. Physiol.* **2008**, *586*, 353–366. [CrossRef]
29. Dannewitz Prosseda, S.; Tian, X.; Kuramoto, K.; Boehm, M.; Sudheendra, D.; Miyagawa, K.; Zhang, F.; Solow-Cordero, D.; Saldivar, J.C.; Austin, E.D.; et al. FHIT, a Novel Modifier Gene in Pulmonary Arterial Hypertension. *Am. J. Respir Crit. Care Med.* **2019**, *199*, 83–98. [CrossRef]
30. Chen, Y.F. Atrial natriuretic peptide in hypoxia. *Peptides* **2005**, *26*, 1068–1077. [CrossRef]
31. Sun, J.-Z.; Oparil, S.; Lucchesi, P.; Thompson, J.A.; Chen, Y.-F. Tyrosine kinase receptor activation inhibits NPR-C in lung arterial smooth muscle cells. *Am. J. Physiol.-Lung Cell. Mol. Physiol.* **2001**, *281*, L155–L163. [CrossRef]
32. Adeoye, O.O.; Silpanisong, J.; Williams, J.M.; Pearce, W.J. Role of the sympathetic autonomic nervous system in hypoxic remodeling of the fetal cerebral vasculature. *J. Cardiovasc. Pharmacol.* **2015**, *65*, 308–316. [CrossRef] [PubMed]
33. Aguirre, J.; Morrell, N.; Long, L.; Clift, P.; Upton, P.; Polak, J.; Wilkins, M. Vascular remodeling and ET-1 expression in rat strains with different responses to chronic hypoxia. *Am. J. Physiol.-Lung Cell. Mol. Physiol.* **2000**, *278*, L981–L987. [CrossRef] [PubMed]
34. Chassagne, C.; Eddahibi, S.; Adamy, C.; Rideau, D.; Marotte, F.; Dubois-Rande, J.L.; Adnot, S.; Samuel, J.L.; Teiger, E. Modulation of angiotensin II receptor expression during development and regression of hypoxic pulmonary hypertension. *Am. J. Respir. Cell Mol. Biol.* **2000**, *22*, 323–332. [CrossRef] [PubMed]
35. Ouyang, N.; Ran, P.; Du, Z. Expression of FGF-b and c-myc in rats lung tissue affected by hypoxia. *Zhonghua Jie He He Hu Xi Za Zhi* **1997**, *20*, 22–24. [PubMed]
36. Myllyharju, J.; Schipani, E. Extracellular matrix genes as hypoxia-inducible targets. *Cell Tissue Res.* **2010**, *339*, 19–29. [CrossRef]
37. Baron, W.; Metz, B.; Bansal, R.; Hoekstra, D.; de Vries, H. PDGF and FGF-2 signaling in oligodendrocyte progenitor cells: regulation of proliferation and differentiation by multiple intracellular signaling pathways. *Mol. Cell. Neurosci.* **2000**, *15*, 314–329. [CrossRef]
38. Bonacina, F.; Baragetti, A.; Catapano, A.L.; Norata, G.D. Long pentraxin 3: experimental and clinical relevance in cardiovascular diseases. *Mediat. Inflamm.* **2013**, *2013*, 725102. [CrossRef]
39. Ehret, G.B.; Munroe, P.B.; Rice, K.M.; Bochud, M.; Johnson, A.D.; Chasman, D.I.; Smith, A.V.; Tobin, M.D.; Verwoert, G.C.; Hwang, S.-J. Genetic variants in novel pathways influence blood pressure and cardiovascular disease risk. *Nature* **2011**, *478*, 103.

40. Hu, Q.; Liu, Q.; Shasha Wang, X.Z.; Zhang, Z.; Lv, R.; Jiang, G.; Ma, Z.; He, H.; Li, D.; Liu, X. NPR-C gene polymorphism is associated with increased susceptibility to coronary artery disease in Chinese Han population: a multicenter study. *Oncotarget* **2016**, *7*, 33662. [CrossRef]
41. Fox, A.A.; Collard, C.D.; Shernan, S.K.; Seidman, C.E.; Seidman, J.G.; Liu, K.-Y.; Muehlschlegel, J.D.; Perry, T.E.; Aranki, S.F.; Lange, C. Natriuretic peptide system gene variants are associated with ventricular dysfunction after coronary artery bypass grafting. *Anesthesiol.* **2009**, *110*, 738–747. [CrossRef]
42. El Andalousi, J.; Li, Y.; Anand-Srivastava, M.B. Natriuretic peptide receptor-C agonist attenuates the expression of cell cycle proteins and proliferation of vascular smooth muscle cells from spontaneously hypertensive rats: role of Gi proteins and MAPkinase/PI3kinase signaling. *PLoS ONE* **2013**, *8*, e76183. [CrossRef] [PubMed]
43. Touyz, R.M. Intracellular mechanisms involved in vascular remodelling of resistance arteries in hypertension: role of angiotensin II. *Exp. Physiol.* **2005**, *90*, 449–455. [CrossRef] [PubMed]
44. Hashim, S.; Li, Y.; Anand-Srivastava, M.B. Small cytoplasmic domain peptides of natriuretic peptide receptor-C attenuate cell proliferation through Giα protein/MAP kinase/PI3-kinase/AKT pathways. *Am. J. Physiol. -Heart Circ. Physiol.* **2006**, *291*, H3144–H3153. [CrossRef] [PubMed]
45. Huntley, B.K.; Sandberg, S.M.; Noser, J.A.; Cataliotti, A.; Redfield, M.M.; Matsuda, Y.; Burnett, J.C., Jr. BNP-induced activation of cGMP in human cardiac fibroblasts: Interactions with fibronectin and natriuretic peptide receptors. *J. Cell. Physiol.* **2006**, *209*, 943–949. [CrossRef] [PubMed]
46. Sangaralingham, S.J.; Huntley, B.K.; Martin, F.L.; McKie, P.M.; Bellavia, D.; Ichiki, T.; Harders, G.E.; Chen, H.H.; Burnett, J.C., Jr. The aging heart, myocardial fibrosis, and its relationship to circulating C-type natriuretic peptide. *Hypertension* **2011**, *57*, 201–207. [CrossRef] [PubMed]
47. Egom, E.E.; Vella, K.; Hua, R.; Jansen, H.J.; Moghtadaei, M.; Polina, I.; Bogachev, O.; Hurnik, R.; Mackasey, M.; Rafferty, S.; et al. Impaired sinoatrial node function and increased susceptibility to atrial fibrillation in mice lacking natriuretic peptide receptor C. *J. Physiol.* **2015**, *593*, 1127–1146. [CrossRef] [PubMed]
48. Jansen, H.J.; Mackasey, M.; Moghtadaei, M.; Liu, Y.; Kaur, J.; Egom, E.E.; Tuomi, J.M.; Rafferty, S.A.; Kirkby, A.W.; Rose, R.A. NPR-C (Natriuretic Peptide Receptor-C) Modulates the Progression of Angiotensin II-Mediated Atrial Fibrillation and Atrial Remodeling in Mice. *Circ. Arrhythmia Electrophysiol.* **2019**, *12*, e006863. [CrossRef] [PubMed]
49. Mackasey, M.; Egom, E.E.; Jansen, H.J.; Hua, R.; Moghtadaei, M.; Liu, Y.; Kaur, J.; McRae, M.D.; Bogachev, O.; Rafferty, S.A.; et al. Natriuretic Peptide Receptor-C Protects Against Angiotensin II-Mediated Sinoatrial Node Disease in Mice. *Jacc. Basic Transl. Sci.* **2018**, *3*, 824–843. [CrossRef]
50. Rahali, S.; Li, Y.; Anand-Srivastava, M.B. Contribution of oxidative stress and growth factor receptor transactivation in natriuretic peptide receptor C-mediated attenuation of hyperproliferation of vascular smooth muscle cells from SHR. *PLoS ONE* **2018**, *13*, e0191743. [CrossRef]
51. Bubb, K.J.; Aubdool, A.A.; Moyes, A.J.; Lewis, S.; Drayton, J.P.; Tang, O.; Mehta, V.; Zachary, I.C.; Abraham, D.J.; Tsui, J.; et al. Endothelial C-Type Natriuretic Peptide Is a Critical Regulator of Angiogenesis and Vascular Remodeling. *Circulation* **2019**, *139*, 1612–1628. [CrossRef]

© 2019 by the author. Licensee MDPI, Basel, Switzerland. This article is an open access article distributed under the terms and conditions of the Creative Commons Attribution (CC BY) license (http://creativecommons.org/licenses/by/4.0/).

Review

Atrial Natriuretic Peptide: A Molecular Target of Novel Therapeutic Approaches to Cardio-Metabolic Disease

Valentina Cannone [1,2,*], Aderville Cabassi [2], Riccardo Volpi [2] and John C. Burnett Jr. [1]

1. Cardiorenal Research Laboratory, Circulatory Failure Division, Department of Cardiovascular Medicine, Mayo Clinic, Rochester, MN 55905, USA
2. Division of Clinical Medicine, Department of Medicine and Surgery, University of Parma, 43126 Parma, Italy
* Correspondence: cannone.valentina@mayo.edu

Received: 16 June 2019; Accepted: 29 June 2019; Published: 2 July 2019

Abstract: Atrial natriuretic peptide (ANP) is a cardiac hormone with pleiotropic cardiovascular and metabolic properties including vasodilation, natriuresis and suppression of the renin-angiotensin-aldosterone system. Moreover, ANP induces lipolysis, lipid oxidation, adipocyte browning and ameliorates insulin sensitivity. Studies on ANP genetic variants revealed that subjects with higher ANP plasma levels have lower cardio-metabolic risk. In vivo and in humans, augmenting the ANP pathway has been shown to exert cardiovascular therapeutic actions while ameliorating the metabolic profile. MANP is a novel designer ANP-based peptide with greater and more sustained biological actions than ANP in animal models. Recent studies also demonstrated that MANP lowers blood pressure and inhibits aldosterone in hypertensive subjects whereas cardiometabolic properties of MANP are currently tested in an on-going clinical study in hypertension and metabolic syndrome. Evidence from in vitro, in vivo and in human studies support the concept that ANP and related pathway represent an optimal target for a comprehensive approach to cardiometabolic disease.

Keywords: atrial natriuretic peptide; hypertension; heart failure; cardiometabolic disease; obesity; metabolic syndrome; cGMP; guanylyl cyclase receptor A; natriuretic peptides

---

1. Introduction

The pioneering work of De Bold established the heart as an endocrine organ with the discovery of atrial natriuretic peptide (ANP), which is a 28 amino-acid cardiac hormone synthesized and secreted by cardiomyocytes in response to myocardial stretch [1]. Atrial natriuretic peptide is an endogenous ligand that activates the particulate guanylyl cyclase A receptor (GC–A) determining the production of the second messenger cyclic guanosine monophosphate (cGMP). Through protein kinase G and ion channels cGMP mediates ANP biological actions. Atrial natriuretic peptide also binds to the natriuretic peptide clearance receptor (NPR–C) that degrades ANP whereas the enzymes neprilysin and insulin degrading enzyme inactivate ANP via rapid peptide degradation. Since its discovery it was clear that ANP is a key regulator of cardiovascular volume and pressure homeostasis. Indeed, ANP induces vasodilation, natriuresis, diuresis and it counteracts the renin-angiotensin-aldosterone system (RAAS). In the kidney, ANP increases glomerular filtration rate by increasing afferent arteriolar dilation in addition to efferent arteriolar constriction [2]. At different levels of the nephron ANP inhibits water and sodium reabsorption. Atrial natriuretic peptide antagonizes the RAAS by inhibiting renin secretion and aldosterone production. Moreover, ANP has antihypertrophic and antifibrotic properties and genetic deletion of GC–A results in cardiac hypertrophy, and fibrosis [3–8]. Atrial natriuretic peptide also inhibits the sympathetic nervous system while increasing vagal activity [2]. In pathologic conditions such as heart failure excessive RAAS activation, volume overload and consequently augmented myocardial stretch, ANP production is increased. Through its unloading properties, ANP functions as a compensatory response to the altered cardiovascular homeostasis. As discussed later, ANP serves as a key target for novel therapies such as sacubitril/valsartan for the treatment of heart failure.

Hypertension, one of the main risk factors for the development of heart failure, is also considered to be the result of inappropriately high activity of RAAS and sympathetic nervous system along with sodium retention [9]. Importantly, in vivo and recent epidemiological studies reveal a key role exerted by ANP deficiency in the pathophysiology of this disease. In 1995, Smithies and coworkers reported the seminal observation that mice heterozygotes for the disruption of the ANP gene developed hypertension when fed a high sodium diet [10]. Several years later Macheret et al. reported that subjects with pre-hypertension have significantly lower ANP values compared to normotensive individuals [11]. Further, patients with hypertension do not display any increase in ANP levels that might exert a compensatory response to their cardiovascular pathological status. Probably the lack of adequate circulating ANP contributes to the onset of hypertension and increases the risk for cardiovascular diseases. In addition, recent epidemiological studies reveal an inverse relationship between aldosterone and ANP circulating levels in the general community and hypertensive subjects, with aldosterone being higher in the presence of lower ANP levels [12,13]. Interestingly, heart failure also appears to be a state of ANP deficiency based on a study conducted by Reginauld and coworkers [14]. In a cohort of 112 subjects with acute decompensated heart failure, 26% of patients did not show any compensatory increase in ANP levels. In this subgroup, circulating values of the second messenger, cGMP were also not increased.

An emerging concept is that the heart not only regulates blood pressure homeostasis but is also a regulator of whole body metabolism. Indeed, several studies revealed that ANP is a modulator of metabolism. More specifically, ANP induces lipid mobilization and oxidation and enhances insulin sensitivity [15]. Indeed, infusion of ANP in humans determines a lipolytic effect with an increase in plasma levels of glycerol and non-esterified fatty-acids regardless of their body mass index [16,17], while it also enhances energy expenditure [18]. In addition, intravenous administration of ANP results in an increase in plasma levels of adiponectin [19,20]. This cytokine, which is secreted by adipocytes and cardiomyocytes, possesses cardioprotective properties and regulatory effects on glucose and lipid metabolism ameliorating insulin-sensitivity [21–24]. In vitro studies in subcutaneous adipocytes also showed that ANP inhibits the production of inflammatory cytokines involved in obesity-related inflammatory state and insulin resistance [25]. In rodents, ANP induces browning of adipocytes along with mitochondrial biogenesis [26], whereas in human skeletal muscle, the cardiac hormone increases mitochondrial oxidative metabolism and fat oxidation [27]. These important studies on ANP metabolic action are underscored by epidemiological studies reporting a relationship between ANP and metabolic diseases. Circulating levels of ANP are lower in obese compared to lean individuals [28] and are inversely related to each metabolic criterion of metabolic syndrome [29]. Low levels of ANP are also predictive of future development of diabetes [30]. The mechanisms underlying these associations might be related to an insulin/glucose-mediated regulation of the GC–A and NPR–C. Recent in vitro studies showed that insulin upregulates NPR–C expression in adipocytes in a glucose-dependent manner with an increased expression observed in the presence of higher glucose concentrations [31]. In humans, higher fasting levels of insulin and insulin resistance measured by homeostatic model assessment for insulin resistance (HOMA) positively correlate with adipose tissue NPRC expression. Indeed, expression of NPR–C is augmented, whereas GC–A is significantly decreased in adipose tissue of obese compared to normal weight subjects [32,33]. The ratio GC–A/NPR–C increases by losing weight and ameliorating insulin-sensitivity [34]. A similar pattern is observed for skeletal muscle with GC–A expression, which is inversely related to fat content, body mass index, fasting plasma insulin levels and insulin resistance, whereas NPR–C is upregulated in obese individuals with impaired glucose tolerance and type 2 diabetes [35]. Coue and coworkers provided new insights into the metabolic actions of ANP and reported that a potential mechanism through which ANP favors insulin sensitivity involves increasing glucose uptake in adipocytes [36]. Most interesting was the observation that the effect is attenuated in obesity, providing a further possible cross-talk between obesity and insulin resistance.

When these studies in vivo, in vitro and in humans are taken together, they show the important role played by ANP in cardiovascular and metabolic homeostasis, highlighting how this cardiac hormone could be an important therapeutic target in cardio-metabolic disease.

2. Atrial Natriuretic Peptide Genetic Variants

The natriuretic peptide precursor A gene (*NPPA*) is located on chromosome 1 and encodes the pre-pro-hormone from which ANP is obtained after cleavage. Over the last two decades key studies in humans have investigated the phenotype associated with single nucleotide polymorphisms of *NPPA* and revealed clinical findings, which further support the important role of ANP in defining the cardiovascular and metabolic phenotype.

The single nucleotide polymorphism rs5068, located in the 3'-UTR of *NPPA*, is associated with higher circulating levels of ANP [37–40]. In 2009, Newton-Cheh and coworkers showed that the minor G allele of rs5068 is related to higher plasma levels of ANP in general community-cohorts of whites from the United States and Northern Europe [37]. In line with ANP biological properties, the minor G allele is also associated with lower blood pressure and risk of hypertension. Cannone et al. investigated not only the cardiovascular but also the metabolic phenotype associated with this genetic variant in a general population of whites from the United States [38]. The carriers of rs5068 G minor allele, who have higher circulating levels of ANP and lower systolic blood pressure values, also have lower body mass index and waist circumference. A key finding of this genetic study was the lower prevalence of obesity and metabolic syndrome among the carriers of the minor allele. In addition, protective plasma high-density lipoprotein cholesterol was higher whereas C-reactive protein levels were lower in the AG/GG genotypes. Importantly, the association between rs5068 minor G allele and a clinical phenotype characterized by lower cardio-metabolic risk was replicated in a general community from the Mediterranean island of Sicily [41]. In non-diabetic Northern Europeans, the rs5068 minor G allele is associated with lower prevalence of left ventricular hypertrophy and decreased risk of developing diabetes in a 14-year follow up analysis [42,43]. The phenotype associated with rs5068 genotypes was also analyzed in African Americans and of note, subjects who are carriers of the G allele have lower triglycerides and insulin levels as well as higher high-density lipoprotein cholesterol [44]. Diabetes and metabolic syndrome are less prevalent among the AG/GG genotypes. The mechanism underlying the associations between rs5068 minor G allele and higher circulating levels of ANP was investigated by Arora et al. in an interesting study showing that this single nucleotide polymorphism does not allow micro-RNA 425 to attach to the complementary sequence and exert its inhibitory effect, resulting in a higher production of ANP [45].

While higher levels of ANP are protective, the emerging concept is that subjects who are exposed to lower circulating levels of ANP also have higher cardio-metabolic risk. Indeed, in a study aimed to identify genetic determinants of ANP plasma levels, Pereira et al. revealed that the ANP genetic variant rs5063 is associated with lower ANP levels, and the carriers of this single nucleotide polymorphism have higher diastolic blood pressure and risk of stroke [46].

Single nucleotide polymorphisms as rs5068 and rs5063, which are associated with variations of ANP circulating levels, provide the opportunity to investigate the phenotype related to a life-long exposure to higher or lower ANP plasma levels. The clinical characteristics observed in the carriers of rs5068 and rs5063 are consistent with the blood pressure lowering, lipolytic and insulin sensitizing effect of ANP and further support the concept of ANP as a therapeutic strategy in the treatment of cardio-metabolic disease.

3. Atrial Natriuretic Peptide as a Therapeutic for Cardio-Metabolic Disease

Metabolic Syndrome consists of several cardiovascular risk factors including elevated blood pressure, abdominal obesity, dyslipidemia and impaired fasting glucose [47]. Each factor is independently associated with the development of atherosclerotic cardiovascular disease and type 2 diabetes. Metabolic syndrome does not appear to be determined by a single cause but precipitated by two main underlying pathological conditions, which are abdominal obesity and insulin resistance. In the United States general adult population, the prevalence of the metabolic syndrome (including those with diabetes mellitus) is approximately 34% whereas diabetes mellitus and obesity, which also represent major risk factors for the development of cardiovascular disease, have a prevalence of 13% and 40%, respectively [48]. Hypertension is widely prevalent in the United States affecting around 32% of the adult population and represents one of the main features of metabolic syndrome [49,50]. If the most recent 2017 American College of Cardiology/American Heart Association guidelines for hypertension (defined as taking antihypertensive medication, or a systolic pressure ≥130 mmHg and/or a diastolic pressure ≥80 mmHg) are applied the prevalence raises to 46% based on the National Health and Nutrition Examination Survey data from 2011 to 2014 [51,52]. Hypertension and metabolic disease are closely interrelated, indeed, hypertension is present in 77% of the patients affected by metabolic syndrome and abdominal obesity is considered one of the most important risk factors for the development of hypertension [53,54]. One of the main pathophysiological pathways that link obesity and hypertension is the excessive activation of the sympathetic nervous system and the RAAS, which leads to an increase in sodium and water retention [55]. Both hypertension and metabolic syndrome represent a significant risk factor for the development of cardiovascular disease when considered individually [56–58]. If they coexist, the risk for cardiovascular events is further amplified, being almost double [59]. Heart failure affects 6.2 million (prevalence, 2.2%) of United States adults and hypertension, metabolic syndrome, obesity along with diabetes represent all well-known risk factors for its onset [48].

Previous epidemiological and physiological studies illustrated above support the concept that hypertension and heart failure as well as obesity and metabolic syndrome are conditions characterized by ANP deficiency. Moreover, with its favorable cardiovascular and metabolic properties ANP represents an appealing target for a combined approach to cardio-metabolic disease. In Japan, ANP in its recombinant form of carperitide has been successfully used for many years for the treatment of heart failure [60]. Infusion of carperitide has been shown to improve clinical conditions and degree of dyspnea in subjects with acute heart failure [61,62]. An 18-month follow-up analysis also revealed that low-dose carperitide infused for 72 h as the initial treatment in addition to standard therapy is associated with lower risk of re-hospitalization and mortality [63]. In 2014, the PARADIGM Trial demonstrated the beneficial therapeutic effect of sacubitril/valsartan (LCZ696), a combined angiotensin II receptor-neprilysin inhibitor, in subjects with heart failure with reduced ejection fraction [64]. Chronic heart failure patients who received sacubitril/valsartan as opposed to enalapril in addition to recommended standard therapy had lower risk of mortality and re-hospitalization. The benefit of LCZ696 included also an improvement in heart failure symptoms and physical limitations. When tested in the setting of acutely decompensated heart failure, sacubitril/valsartan resulted in a significant decrease in N-terminal pro-B-type natriuretic peptide as a marker of neurohormonal activation and hemodynamic stress [65]. A key mechanism of action of sacubitril/valsartan is the inhibition of neprilysin. Being neprilysin involved in the degradation of several peptides and related pathways including ANP, two recent interesting studies investigated the effect of sacubitril on the enzyme substrates [66,67]. Importantly, both studies report a significant and sustained increase of ANP in subjects receiving sacubitril/valsartan, which further support the concept that augmenting ANP circulating levels has a beneficial effect in the long-term treatment of cardiovascular diseases. Sacubitril/valsartan appears also to be more effective than valsartan alone in the treatment of hypertension [68]. In a multicenter, randomized, double-blind, cross-over study in Asian subjects with salt-sensitive hypertension, the use of valsartan/sacubitril lead to a greater reduction in both office and 24 h-ambulatory blood pressure over a 28 day period. The blood pressure lowering effect was accompanied by an acute significant increase of natriuresis and diuresis on the first day of therapy with angiotensin II receptor-neprilysin inhibitor. The metabolic effect of sacubitril/valsartan was tested in patients with cardio-metabolic disease. In a post-hoc analysis of the PARADIGM trial, subjects with heart failure and diabetes or HbA1c ≥ 6.5% who were randomized to sacubitril/valsartan had a better glycemic control than subjects receiving enalapril [69]. Over the three-year follow-up HbA1c concentrations and new use of insulin or oral antihyperglycaemic agents were all significantly lower when compared to the control group. In a cohort of subjects with hypertension and obesity, eight-week treatment with sacubitril/valsartan increased insulin-sensitivity and abdominal subcutaneous adipose tissue lipolysis [70].

These studies discussed above clearly support the therapeutic potential of ANP in cardio-metabolic disease. However, ANP has a half-life of 2–5 min, which renders this cardiac peptide unsuitable for a single or twice daily administration. Conversely, MANP is a 40 amino acid peptide with a 12 amino acid extension to the carboxyl-terminus of ANP and has a half-life of 45 min. MANP was engineered at Mayo Clinic to represent a novel GC-A activator with biological actions that are superior to native ANP. Indeed, the novel 12 amino acid extension on the carboxyl-terminus results in MANP being highly resistant to enzymatic degradation by both neprilysin and insulin-degrading enzyme, but does not alter the high affinity for GC-A [71,72]. When compared to native ANP intravenous infusion of MANP in normal canines resulted in a greater activation of the second messenger cGMP and consequent greater and more sustained blood pressure lowering effect, increase in renal blood flow and glomerular filtration rate [73]. Diuresis and natriuresis were augmented in addition to a greater and more sustained inhibition of angiotensin II and aldosterone. MANP was also tested in a canine model of hypertension obtained by continuous infusion of angiotensin II [74]. Intravenous administration of MANP determined a significant decrease in mean arterial pressure, which was sustained up to two hours after MANP infusion. Systemic vascular resistance decreased along with pulmonary capillary wedge pressure, pulmonary artery pressure and right atrial pressure. In spite of the systemic blood pressure lowering effect, renal blood flow and glomerular filtration rate increased in addition to a marked rise in water and sodium excretion. Proximal and distal sodium reabsorption of sodium decreased during MANP infusion and for the following hour. The biological properties of MANP, which emerged from these two in vivo studies, made MANP an unprecedented therapeutic candidate for being tested in a large animal model of heart failure with hypertension [75]. Indeed, MANP was infused for 45 min and its effects were compared with nitroglycerin, which is a therapeutic agent commonly used in the clinical setting of heart failure. MANP administration resulted in a greater and sustained increase in the plasma levels of cGMP whereas nitroglycerin did not change circulating levels of the second messenger. While both MANP and nitroglycerin lowered mean arterial pressure, MANP effect was more sustained in time. Both compounds reduced systemic and renal vascular resistance in addition to pulmonary capillary wedge pressure, pulmonary and right atrial pressures. Importantly, only MANP was found to have a renal hemodynamic effect inducing a significant increase in renal blood flow, which continued for 120 min after the administration. Glomerular filtration rate, diuresis and natriuresis were also markedly increased during the infusion of MANP and for the following 60 min. In contrast to nitroglycerin, which did not modify aldosterone levels, during MANP administration circulating aldosterone values were significantly lower than baseline.

Recent studies further tested the cardiovascular and metabolic actions of MANP in vitro and in vivo. When human subcutaneous and visceral pre-adipocytes were incubated with MANP the cGMP pathway was activated inducing a significant production of the second messenger [76]. In normal Sprague–Dawley rats, MANP acute infusion significantly decreased mean arterial pressure while increasing non-esterified fatty acids, which is a marker of lipolysis. On-going studies are also testing MANP properties in a rodent model of hypertension and metabolic syndrome. Acute intravenous administration of MANP reduced blood pressure and induced an increase in circulating levels of adiponectin. Importantly, these in vitro and in vivo findings highlight how MANP not only retains cardiovascular and metabolic properties of native ANP but also might display greater and more sustained biological actions.

A first in human study of MANP was recently completed [77]. MANP was administered to humans with stable hypertension withdrawn from anti-hypertensive medications for two weeks. The study was designed as a single ascending dose administered subcutaneously as a single injection. The goal was to determine the maximal tolerated dose as well as to establish safety and tolerability. Three different doses (1, 2.5 and 5 µg/kg) of MANP were tested in three different groups. Each group included four subjects receiving MANP and one subject receiving a placebo in addition to their usual antihypertensive therapy. Blood pressure was assessed over a 24 h period. Preliminary data demonstrated that MANP was well tolerated with no serious adverse effects. Mild adverse events included mild headache; transient light-headedness and transient orthostatic vasovagal syncope lasting for 7 s in one subject. No significant changes in laboratory values or electrocradiogram from baseline were observed. Blood pressure was reduced with all doses. Specifically, the maximal systolic blood pressure reduction was 20 mmHg (with 5.0 µg/kg); maximal duration of systolic blood pressure reduction was 24 hr (with 2.5 and 5.0 µg/kg); maximal diastolic blood pressure reduction was 12 mmHg (with 2.5 and 5.0 µg/kg) and maximal duration of diastolic blood pressure reduction was 24 hr (with 5.0 µg/kg). Cyclic GMP plasma levels were increased whereas aldosterone circulating levels were descreased with all MANP doses administered. Importantly, an on-going clinical study is currently evaluating the cardiovascular and metabolic actions of MANP in subjects with hypertension and metabolic syndrome with the primary goal of assessing the potential therapeutic effect of MANP in cardio-metabolic disease. To date, no studies have compared the recombinant form of human ANP, Carperitide, with MANP. Future assessments might evaluate the cardio-metabolic effects of the two compounds in the clinical setting of heart failure.

In conclusion, ANP is a cardiac hormone with pleiotropic biological actions (Figure 1). In vivo, in vitro, epidemiological and in human studies support the concept that augmenting ANP and related pathways might be a successful strategy in the treatment of cardio-metabolic disease. MANP, a novel designer ANP-based peptide, which has been tested and is currently being tested in clinical trials, represents a potentially safe, effective and comprehensive therapeutic approach to cardio-metabolic dysfunction.

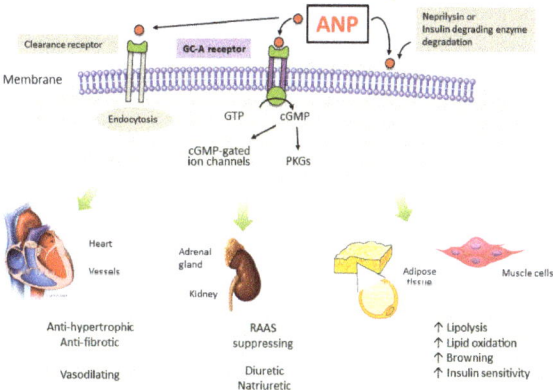

Figure 1. Atrial natriuretic peptide binds to the guanylyl cyclase A receptor (GC–A) resulting in the production of the second messenger cyclic guanosine monophosphate (cGMP). Protein kinase G (PKG) and cGMP-gated ion channels are then activated and mediate ANP biological actions, which are summarized in the figure.

Funding: This research was funded by the National Heart Lung and Blood Institute (RO1 HL36634 and RO1 HL136340), American Heart Association (16SDG29930003) and Marie Ingalls Cardiovascular Research Career Development Fund in honor of Dr Alexander Schirger.

Conflicts of Interest: Mayo Clinic has licensed MANP to Zumbro Discovery; John C. Burnett Jr. is Co-Founder of Zumbro Discovery and holds equity.

Abbreviations

ANP	Atrial natriuretic peptide
cGMP	Cyclic guanosine monophosphate
GC–A	Guanylyl cyclase A receptor
HOMA	Homeostatic model assessment for insulin resistance
NPPA	Natriuretic peptide precursor A gene
NPR–C	Natriuretic peptide clearance receptor
RAAS	Renin-angiotensin-aldosterone system

References

1. Garbers, D.L.; Chrisman, T.D.; Wiegn, P.; Katafuchi, T.; Albanesi, J.P.; Bielinski, V.; Barylko, B.; Redfield, M.M.; Burnett, J.C., Jr. Membrane guanylyl cyclase receptors: An update. *Trends Endocrinol. Metab.* **2006**, *17*, 251–258. [CrossRef] [PubMed]
2. Volpe, M.; Carnovali, M.; Mastromarino, V. The natriuretic peptides system in the pathophysiology of heart failure: From molecular basis to treatment. *Clin. Sci. (Lond.)* **2016**, *130*, 57–77. [CrossRef] [PubMed]
3. Patel, J.B.; Valencik, M.L.; Pritchett, A.M.; Burnett, J.C., Jr.; McDonald, J.A.; Redfield, M.M. Cardiac-specific attenuation of natriuretic peptide A receptor activity accentuates adverse cardiac remodeling and mortality in response to pressure overload. *Am. J. Physiol. Heart Circ. Physiol.* **2005**, *289*, H777–H784. [CrossRef] [PubMed]
4. Holtwick, R.; van Eickels, M.; Skryabin, B.V.; Baba, H.A.; Bubikat, A.; Begrow, F.; Schneider, M.D.; Garbers, D.L.; Kuhn, M. Pressure-independent cardiac hypertrophy in mice with cardiomyocyte-restricted inactivation of the atrial natriuretic peptide receptor guanylyl cyclase-A. *J. Clin. Investig.* **2003**, *111*, 1399–1407. [CrossRef] [PubMed]
5. Calderone, A.; Thaik, C.M.; Takahashi, N.; Chang, D.L.; Colucci, W.S. Nitric oxide, atrial natriuretic peptide, and cyclic GMP inhibit the growth-promoting effects of norepinephrine in cardiac myocytes and fibroblasts. *J. Clin. Investig.* **1998**, *101*, 812–818. [CrossRef] [PubMed]
6. Vellaichamy, E.; Kaur, K.; Pandey, K.N. Enhanced activation of pro-inflammatory cytokines in mice lacking natriuretic peptide receptor-A. *Peptides* **2007**, *28*, 893–899. [CrossRef] [PubMed]
7. Wang, D.; Oparil, S.; Feng, J.A.; Li, P.; Perry, G.; Chen, L.B.; Dai, M.; John, S.W.; Chen, Y.F. Effects of pressure overload on extracellular matrix expression in the heart of the atrial natriuretic peptide-null mouse. *Hypertension* **2003**, *42*, 88–95. [CrossRef] [PubMed]
8. Oliver, P.M.; Fox, J.E.; Kim, R.; Rockman, H.A.; Kim, H.S.; Reddick, R.L.; Pandey, K.N.; Milgram, S.L.; Smithies, O.; Maeda, N. Hypertension, cardiac hypertrophy, and sudden death in mice lacking natriuretic peptide receptor A. *Proc. Natl. Acad. Sci. USA* **1997**, *94*, 14730–14735. [CrossRef] [PubMed]
9. Goodfriend, T.L.; Calhoun, D.A. Resistant hypertension, obesity, sleep apnea, and aldosterone: Theory and therapy. *Hypertension* **2004**, *43*, 518–524. [CrossRef] [PubMed]
10. John, S.W.; Krege, J.H.; Oliver, P.M.; Hagaman, J.R.; Hodgin, J.B.; Pang, S.C.; Flynn, T.G.; Smithies, O. Genetic decreases in atrial natriuretic peptide and salt-sensitive hypertension. *Science* **1995**, *267*, 679–681. [CrossRef]
11. Macheret, F.; Heublein, D.; Costello-Boerrigter, L.C.; Boerrigter, G.; McKie, P.; Bellavia, D.; Mangiafico, S.; Ikeda, Y.; Bailey, K.; Scott, C.G.; et al. Human hypertension is characterized by a lack of activation of the antihypertensive cardiac hormones ANP and BNP. *J. Am. Coll. Cardiol.* **2012**, *60*, 1558–1565. [CrossRef] [PubMed]
12. Buglioni, A.; Cannone, V.; Cataliotti, A.; Sangaralingham, S.J.; Heublein, D.M.; Scott, C.G.; Bailey, K.R.; Rodeheffer, R.J.; Dessi-Fulgheri, P.; Sarzani, R.; et al. Circulating aldosterone and natriuretic peptides in the general community: Relationship to cardiorenal and metabolic disease. *Hypertension* **2015**, *65*, 45–53. [CrossRef] [PubMed]
13. Cannone, V.; Buglioni, A.; Sangaralingham, S.J.; Scott, C.; Bailey, K.R.; Rodeheffer, R.; Redfield, M.M.; Sarzani, R.; Burnett, J.C., Jr. Aldosterone, hypertension, and antihypertensive therapy: Ifrom a general population. *Mayo. Clin. Proc.* **2018**, *93*, 980–990. [CrossRef] [PubMed]

14. Reginauld, S.H.; Cannone, V.; Iyer, S.; Scott, C.; Baily, K.; Schaefer, J.; Chen, Y.; Sangaralingham, S.J.; Burnett, J.C., Jr. Differential regulation of ANP and BNP in acute decompensated heart failure - deficiency of ANP. *J. Am. Coll. Cardiol.* **2019**, in press.
15. Coue, M.; Moro, C. Natriuretic peptide control of energy balance and glucose homeostasis. *Biochimie* **2016**, *124*, 84–91. [CrossRef] [PubMed]
16. Birkenfeld, A.L.; Boschmann, M.; Moro, C.; Adams, F.; Heusser, K.; Franke, G.; Berlan, M.; Luft, F.C.; Lafontan, M.; Jordan, J. Lipid mobilization with physiological atrial natriuretic peptide concentrations in humans. *J. Clin. Endocrinol. Metab.* **2005**, *90*, 3622–3628. [CrossRef] [PubMed]
17. Galitzky, J.; Sengenes, C.; Thalamas, C.; Marques, M.A.; Senard, J.M.; Lafontan, M.; Berlan, M. The lipid-mobilizing effect of atrial natriuretic peptide is unrelated to sympathetic nervous system activation or obesity in young men. *J. Lipid Res.* **2001**, *42*, 536–544. [PubMed]
18. Birkenfeld, A.L.; Budziarek, P.; Boschmann, M.; Moro, C.; Adams, F.; Franke, G.; Berlan, M.; Marques, M.A.; Sweep, F.C.; Luft, F.C.; et al. Atrial natriuretic peptide induces postprandial lipid oxidation in humans. *Diabetes* **2008**, *57*, 3199–3204. [CrossRef] [PubMed]
19. Tsukamoto, O.; Fujita, M.; Kato, M.; Yamazaki, S.; Asano, Y.; Ogai, A.; Okazaki, H.; Asai, M.; Nagamachi, Y.; Maeda, N.; et al. Natriuretic peptides enhance the production of adiponectin in human adipocytes and in patients with chronic heart failure. *J. Am. Coll. Cardiol.* **2009**, *53*, 2070–2077. [CrossRef] [PubMed]
20. Birkenfeld, A.L.; Boschmann, M.; Engeli, S.; Moro, C.; Arafat, A.M.; Luft, F.C.; Jordan, J. Atrial natriuretic peptide and adiponectin interactions in man. *PLoS ONE* **2012**, *7*, e43238. [CrossRef]
21. Maeda, N.; Shimomura, I.; Kishida, K.; Nishizawa, H.; Matsuda, M.; Nagaretani, H.; Furuyama, N.; Kondo, H.; Takahashi, M.; Arita, Y.; et al. Diet-induced insulin resistance in mice lacking adiponectin/ACRP30. *Nat. Med.* **2002**, *8*, 731–737. [CrossRef] [PubMed]
22. Shibata, R.; Ouchi, N.; Ito, M.; Kihara, S.; Shiojima, I.; Pimentel, D.R.; Kumada, M.; Sato, K.; Schiekofer, S.; Ohashi, K.; et al. Adiponectin-mediated modulation of hypertrophic signals in the heart. *Nat. Med.* **2004**, *10*, 1384–1389. [CrossRef] [PubMed]
23. Shibata, R.; Sato, K.; Pimentel, D.R.; Takemura, Y.; Kihara, S.; Ohashi, K.; Funahashi, T.; Ouchi, N.; Walsh, K. Adiponectin protects against myocardial ischemia-reperfusion injury through AMPK- and COX-2-dependent mechanisms. *Nat. Med.* **2005**, *11*, 1096–1103. [CrossRef] [PubMed]
24. Costello-Boerrigter, L.C.; Burnett, J.C., Jr. A new role for the natriuretic peptides: Metabolic regulators of the adipocyte. *J. Am. Coll. Cardiol.* **2009**, *53*, 2078–2079. [CrossRef] [PubMed]
25. Moro, C.; Klimcakova, E.; Lolmede, K.; Berlan, M.; Lafontan, M.; Stich, V.; Bouloumie, A.; Galitzky, J.; Arner, P.; Langin, D. Atrial natriuretic peptide inhibits the production of adipokines and cytokines linked to inflammation and insulin resistance in human subcutaneous adipose tissue. *Diabetologia* **2007**, *50*, 1038–1047. [CrossRef] [PubMed]
26. Bordicchia, M.; Liu, D.; Amri, E.Z.; Ailhaud, G.; Dessi-Fulgheri, P.; Zhang, C.; Takahashi, N.; Sarzani, R.; Collins, S. Cardiac natriuretic peptides act via p38 MAPK to induce the brown fat thermogenic program in mouse and human adipocytes. *J. Clin. Investig.* **2012**, *122*, 1022–1036. [CrossRef] [PubMed]
27. Engeli, S.; Birkenfeld, A.L.; Badin, P.M.; Bourlier, V.; Louche, K.; Viguerie, N.; Thalamas, C.; Montastier, E.; Larrouy, D.; Harant, I.; et al. Natriuretic peptides enhance the oxidative capacity of human skeletal muscle. *J. Clin. Investig.* **2012**, *122*, 4675–4679. [CrossRef]
28. Wang, T.J.; Larson, M.G.; Levy, D.; Benjamin, E.J.; Leip, E.P.; Wilson, P.W.; Vasan, R.S. Impact of obesity on plasma natriuretic peptide levels. *Circulation* **2004**, *109*, 594–600. [CrossRef]
29. Wang, T.J.; Larson, M.G.; Keyes, M.J.; Levy, D.; Benjamin, E.J.; Vasan, R.S. Association of plasma natriuretic peptide levels with metabolic risk factors in ambulatory individuals. *Circulation* **2007**, *115*, 1345–1353. [CrossRef]
30. Magnusson, M.; Jujic, A.; Hedblad, B.; Engstrom, G.; Persson, M.; Struck, J.; Morgenthaler, N.G.; Nilsson, P.; Newton-Cheh, C.; Wang, T.J.; et al. Low plasma level of atrial natriuretic peptide predicts development of diabetes: The prospective Malmo Diet and Cancer study. *J. Clin. Endocrinol. Metab.* **2012**, *97*, 638–645. [CrossRef]
31. Bordicchia, M.; Ceresiani, M.; Pavani, M.; Minardi, D.; Polito, M.; Wabitsch, M.; Cannone, V.; Burnett, J.C., Jr.; Dessi-Fulgheri, P.; Sarzani, R. Insulin/glucose induces natriuretic peptide clearance receptor in human adipocytes: A metabolic link with the cardiac natriuretic pathway. *Am. J. Physiol. Regul. Integr. Comp. Physiol.* **2016**, *311*, R104–R114. [CrossRef] [PubMed]

32. Ryden, M.; Backdahl, J.; Petrus, P.; Thorell, A.; Gao, H.; Coue, M.; Langin, D.; Moro, C.; Arner, P. Impaired atrial natriuretic peptide-mediated lipolysis in obesity. *Int. J. Obes. (Lond.)* **2016**, *40*, 714–720. [CrossRef] [PubMed]
33. Verboven, K.; Hansen, D.; Moro, C.; Eijnde, B.O.; Hoebers, N.; Knol, J.; Bouckaert, W.; Dams, A.; Blaak, E.E.; Jocken, J.W. Attenuated atrial natriuretic peptide-mediated lipolysis in subcutaneous adipocytes of obese type 2 diabetic men. *Clin. Sci. (Lond.)* **2016**, *130*, 1105–1114. [CrossRef] [PubMed]
34. Kovacova, Z.; Tharp, W.G.; Liu, D.; Wei, W.; Xie, H.; Collins, S.; Pratley, R.E. Adipose tissue natriuretic peptide receptor expression is related to insulin sensitivity in obesity and diabetes. *Obesity (Silver Spring)* **2016**, *24*, 820–828. [CrossRef] [PubMed]
35. Coue, M.; Badin, P.M.; Vila, I.K.; Laurens, C.; Louche, K.; Marques, M.A.; Bourlier, V.; Mouisel, E.; Tavernier, G.; Rustan, A.C.; et al. Defective natriuretic peptide receptor signaling in skeletal muscle links obesity to type 2 diabetes. *Diabetes* **2015**, *64*, 4033–4045. [CrossRef] [PubMed]
36. Coue, M.; Barquissau, V.; Morigny, P.; Louche, K.; Lefort, C.; Mairal, A.; Carpene, C.; Viguerie, N.; Arner, P.; Langin, D.; et al. Natriuretic peptides promote glucose uptake in a cGMP-dependent manner in human adipocytes. *Sci. Rep.* **2018**, *8*, 1097. [CrossRef]
37. Newton-Cheh, C.; Larson, M.G.; Vasan, R.S.; Levy, D.; Bloch, K.D.; Surti, A.; Guiducci, C.; Kathiresan, S.; Benjamin, E.J.; Struck, J.; et al. Association of common variants in NPPA and NPPB with circulating natriuretic peptides and blood pressure. *Nat. Genet.* **2009**, *41*, 348–353. [CrossRef] [PubMed]
38. Cannone, V.; Boerrigter, G.; Cataliotti, A.; Costello-Boerrigter, L.C.; Olson, T.M.; McKie, P.M.; Heublein, D.M.; Lahr, B.D.; Bailey, K.R.; Averna, M.; et al. A genetic variant of the atrial natriuretic peptide gene is associated with cardiometabolic protection in the general community. *J. Am. Coll. Cardiol.* **2011**, *58*, 629–636. [CrossRef]
39. Cannone, V.; Barlera, S.; Pileggi, S.; Masson, S.; Franzosi, M.G.; Latini, R.; Scardulla, C.; Clemenza, F.; Maggioni, A.P.; Nicolosi, G.L.; et al. The Anp genetic variant Rs5068 and circulating levels of natriuretic peptides in patients with chronic heart failure. *Int. J. Cardiol.* **2014**, *176*, 1249–1251. [CrossRef]
40. Ellis, K.L.; Newton-Cheh, C.; Wang, T.J.; Frampton, C.M.; Doughty, R.N.; Whalley, G.A.; Ellis, C.J.; Skelton, L.; Davis, N.; Yandle, T.G.; et al. Association of genetic variation in the natriuretic peptide system with cardiovascular outcomes. *J. Mol. Cell. Cardiol.* **2011**, *50*, 695–701. [CrossRef]
41. Cannone, V.; Cefalu, A.B.; Noto, D.; Scott, C.G.; Bailey, K.R.; Cavera, G.; Pagano, M.; Sapienza, M.; Averna, M.R.; Burnett, J.C., Jr. The atrial natriuretic peptide genetic variant rs5068 is associated with a favorable cardiometabolic phenotype in a Mediterranean population. *Diabetes Care.* **2013**, *36*, 2850–2856. [CrossRef] [PubMed]
42. Jujic, A.; Leosdottir, M.; Ostling, G.; Gudmundsson, P.; Nilsson, P.M.; Melander, O.; Magnusson, M. A genetic variant of the atrial natriuretic peptide gene is associated with left ventricular hypertrophy in a non-diabetic population–the Malmo preventive project study. *BMC Med. Genet.* **2013**, *14*, 64. [CrossRef] [PubMed]
43. Jujic, A.; Nilsson, P.M.; Engstrom, G.; Hedblad, B.; Melander, O.; Magnusson, M. Atrial natriuretic peptide and type 2 diabetes development–biomarker and genotype association study. *PLoS ONE* **2014**, *9*, e89201. [CrossRef] [PubMed]
44. Cannone, V.; Scott, C.G.; Decker, P.A.; Larson, N.B.; Palmas, W.; Taylor, K.D.; Wang, T.J.; Gupta, D.K.; Bielinski, S.J.; Burnett, J.C., Jr. A favorable cardiometabolic profile is associated with the G allele of the genetic variant rs5068 in African Americans: The Multi-Ethnic Study of Atherosclerosis (MESA). *PLoS ONE* **2017**, *12*, e0189858. [CrossRef] [PubMed]
45. Arora, P.; Wu, C.; Khan, A.M.; Bloch, D.B.; Davis-Dusenbery, B.N.; Ghorbani, A.; Spagnolli, E.; Martinez, A.; Ryan, A.; Tainsh, L.T.; et al. Atrial natriuretic peptide is negatively regulated by microRNA-425. *J. Clin. Investig.* **2013**, *123*, 3378–3382. [CrossRef] [PubMed]
46. Pereira, N.L.; Tosakulwong, N.; Scott, C.G.; Jenkins, G.D.; Prodduturi, N.; Chai, Y.; Olson, T.M.; Rodeheffer, R.J.; Redfield, M.M.; Weinshilboum, R.M.; et al. Circulating atrial natriuretic peptide genetic association study identifies a novel gene cluster associated with stroke in whites. *Circ. Cardiovasc. Genet.* **2015**, *8*, 141–149. [CrossRef] [PubMed]
47. Eckel, R.H.; Alberti, K.G.; Grundy, S.M.; Zimmet, P.Z. The metabolic syndrome. *Lancet* **2010**, *375*, 181–183. [CrossRef]
48. Benjamin, E.J.; Muntner, P.; Alonso, A.; Bittencourt, M.S.; Callaway, C.W.; Carson, A.P.; Chamberlain, A.M.; Chang, A.R.; Cheng, S.; Das, S.R.; et al. Heart disease and stroke statistics-2019 update: A report from the American Heart Association. *Circulation* **2019**, *139*, e56–e528. [CrossRef]

49. Yoon, S.S.; Carroll, M.D.; Fryar, C.D. Hypertension prevalence and control among adults: United States, 2011-2014. *NCHS Data Brief.* **2015**, *220*, 1–8.
50. Egan, B.M.; Li, J.; Hutchison, F.N.; Ferdinand, K.C. Hypertension in the United States, 1999 to 2012: Progress toward Healthy People 2020 goals. *Circulation* **2014**, *130*, 1692–1699. [CrossRef]
51. Whelton, P.K.; Carey, R.M.; Aronow, W.S.; Casey, D.E., Jr.; Collins, K.J.; Dennison Himmelfarb, C.; DePalma, S.M.; Gidding, S.; Jamerson, K.A.; Jones, D.W.; et al. 2017 ACC/AHA/AAPA/ABC/ACPM/AGS/APhA/ASH/ASPC/NMA/PCNA Guideline for the prevention, detection, evaluation, and management of high blood pressure in adults: A report of the American College of Cardiology/American Heart Association task force on clinical practice guidelines. *Hypertension* **2018**, *71*, e13–e115. [PubMed]
52. Muntner, P.; Carey, R.M.; Gidding, S.; Jones, D.W.; Taler, S.J.; Wright, J.T., Jr.; Whelton, P.K. Potential US population impact of the 2017 ACC/AHA high blood pressure guideline. *Circulation* **2018**, *137*, 109–118. [CrossRef] [PubMed]
53. Mozaffarian, D.; Benjamin, E.J.; Go, A.S.; Arnett, D.K.; Blaha, M.J.; Cushman, M.; de Ferranti, S.; Despres, J.P.; Fullerton, H.J.; Howard, V.J.; et al. Heart disease and stroke statistics–2015 update: A report from the American Heart Association. *Circulation* **2015**, *131*, e29–e322. [CrossRef] [PubMed]
54. Sironi, A.M.; Gastaldelli, A.; Mari, A.; Ciociaro, D.; Positano, V.; Buzzigoli, E.; Ghione, S.; Turchi, S.; Lombardi, M.; Ferrannini, E. Visceral fat in hypertension: Influence on insulin resistance and beta-cell function. *Hypertension* **2004**, *44*, 127–133. [CrossRef] [PubMed]
55. Rahmouni, K.; Correia, M.L.; Haynes, W.G.; Mark, A.L. Obesity-associated hypertension: New insights into mechanisms. *Hypertension* **2005**, *45*, 9–14. [CrossRef]
56. Chobanian, A.V.; Bakris, G.L.; Black, H.R.; Cushman, W.C.; Green, L.A.; Izzo, J.L., Jr.; Jones, D.W.; Materson, B.J.; Oparil, S.; Wright, J.T., Jr.; et al. Seventh report of the Joint National Committee on Prevention, Detection, Evaluation, and Treatment of High Blood Pressure. *Hypertension* **2003**, *42*, 1206–1252. [CrossRef] [PubMed]
57. Ford, E.S. Risks for all-cause mortality, cardiovascular disease, and diabetes associated with the metabolic syndrome: A summary of the evidence. *Diabetes Care* **2005**, *28*, 1769–1778. [CrossRef]
58. Galassi, A.; Reynolds, K.; He, J. Metabolic syndrome and risk of cardiovascular disease: A meta-analysis. *Am. J. Med.* **2006**, *119*, 812–819. [CrossRef]
59. Schillaci, G.; Pirro, M.; Vaudo, G.; Gemelli, F.; Marchesi, S.; Porcellati, C.; Mannarino, E. Prognostic value of the metabolic syndrome in essential hypertension. *J. Am. Coll. Cardiol.* **2004**, *43*, 1817–1822. [CrossRef]
60. Saito, Y. Roles of atrial natriuretic peptide and its therapeutic use. *J. Cardiol.* **2010**, *56*, 262–270. [CrossRef]
61. Suwa, M.; Seino, Y.; Nomachi, Y.; Matsuki, S.; Funahashi, K. Multicenter prospective investigation on efficacy and safety of carperitide for acute heart failure in the 'real world' of therapy. *Circ. J.* **2005**, *69*, 283–290. [CrossRef] [PubMed]
62. Nomura, F.; Kurobe, N.; Mori, Y.; Hikita, A.; Kawai, M.; Suwa, M.; Okutani, Y. Multicenter prospective investigation on efficacy and safety of carperitide as a first-line drug for acute heart failure syndrome with preserved blood pressure: COMPASS: Carperitide Effects Observed Through Monitoring Dyspnea in Acute Decompensated Heart Failure Study. *Circ. J.* **2008**, *72*, 1777–1786. [PubMed]
63. Hata, N.; Seino, Y.; Tsutamoto, T.; Hiramitsu, S.; Kaneko, N.; Yoshikawa, T.; Yokoyama, H.; Tanaka, K.; Mizuno, K.; Nejima, J.; et al. Effects of carperitide on the long-term prognosis of patients with acute decompensated chronic heart failure: The PROTECT multicenter randomized controlled study. *Circ. J.* **2008**, *72*, 1787–1793. [CrossRef] [PubMed]
64. McMurray, J.J.; Packer, M.; Desai, A.S.; Gong, J.; Lefkowitz, M.P.; Rizkala, A.R.; Rouleau, J.L.; Shi, V.C.; Solomon, S.D.; Swedberg, K.; et al. Angiotensin-neprilysin inhibition versus enalapril in heart failure. *N. Engl. J. Med.* **2014**, *371*, 993–1004. [CrossRef] [PubMed]
65. Velazquez, E.J.; Morrow, D.A.; DeVore, A.D.; Duffy, C.I.; Ambrosy, A.P.; McCague, K.; Rocha, R.; Braunwald, E.; Investigators, P.-H. Angiotensin-neprilysin inhibition in acute decompensated heart failure. *N. Engl. J. Med.* **2019**, *380*, 539–548. [CrossRef]
66. Nougue, H.; Pezel, T.; Picard, F.; Sadoune, M.; Arrigo, M.; Beauvais, F.; Launay, J.M.; Cohen-Solal, A.; Vodovar, N.; Logeart, D. Effects of sacubitril/valsartan on neprilysin targets and the metabolism of natriuretic peptides in chronic heart failure: A mechanistic clinical study. *Eur. J. Heart Fail.* **2019**, *21*, 598–605. [CrossRef] [PubMed]

67. Ibrahim, N.E.; McCarthy, C.P.; Shrestha, S.; Gaggin, H.K.; Mukai, R.; Szymonifka, J.; Apple, F.S.; Burnett, J.C., Jr.; Iyer, S.; Januzzi, J.L., Jr. Effect of neprilysin inhibition on various natriuretic peptide assays. *J. Am. Coll. Cardiol.* **2019**, *73*, 1273–1284. [CrossRef]
68. Wang, T.D.; Tan, R.S.; Lee, H.Y.; Ihm, S.H.; Rhee, M.Y.; Tomlinson, B.; Pal, P.; Yang, F.; Hirschhorn, E.; Prescott, M.F.; et al. Effects of sacubitril/valsartan (LCZ696) on natriuresis, diuresis, blood pressures, and NT-proBNP in salt-sensitive hypertension. *Hypertension* **2017**, *69*, 32–41. [CrossRef]
69. Seferovic, J.P.; Claggett, B.; Seidelmann, S.B.; Seely, E.W.; Packer, M.; Zile, M.R.; Rouleau, J.L.; Swedberg, K.; Lefkowitz, M.; Shi, V.C.; et al. Effect of sacubitril/valsartan versus enalapril on glycaemic control in patients with heart failure and diabetes: A post-hoc analysis from the PARADIGM-HF trial. *Lancet. Diabetes Endocrinol.* **2017**, *5*, 333–340. [CrossRef]
70. Jordan, J.; Stinkens, R.; Jax, T.; Engeli, S.; Blaak, E.E.; May, M.; Havekes, B.; Schindler, C.; Albrecht, D.; Pal, P.; et al. Improved insulin sensitivity with angiotensin receptor neprilysin inhibition in individuals with obesity and hypertension. *Clin. Pharmacol. Ther.* **2017**, *101*, 254–263. [CrossRef]
71. Dickey, D.M.; Yoder, A.R.; Potter, L.R. A familial mutation renders atrial natriuretic Peptide resistant to proteolytic degradation. *J. Biol. Chem.* **2009**, *284*, 19196–19202. [CrossRef] [PubMed]
72. Ralat, L.A.; Guo, Q.; Ren, M.; Funke, T.; Dickey, D.M.; Potter, L.R.; Tang, W.J. Insulin-degrading enzyme modulates the natriuretic peptide-mediated signaling response. *J. Biol. Chem.* **2011**, *286*, 4670–4679. [CrossRef] [PubMed]
73. McKie, P.M.; Cataliotti, A.; Huntley, B.K.; Martin, F.L.; Olson, T.M.; Burnett, J.C., Jr. A human atrial natriuretic peptide gene mutation reveals a novel peptide with enhanced blood pressure-lowering, renal-enhancing, and aldosterone-suppressing actions. *J. Am. Coll. Cardiol.* **2009**, *54*, 1024–1032. [CrossRef] [PubMed]
74. McKie, P.M.; Cataliotti, A.; Boerrigter, G.; Chen, H.H.; Sangaralingham, S.J.; Martin, F.L.; Ichiki, T.; Burnett, J.C., Jr. A novel atrial natriuretic peptide based therapeutic in experimental angiotensin II mediated acute hypertension. *Hypertension* **2010**, *56*, 1152–1159. [CrossRef] [PubMed]
75. McKie, P.M.; Cataliotti, A.; Ichiki, T.; Sangaralingham, S.J.; Chen, H.H.; Burnett, J.C., Jr. M-atrial natriuretic peptide and nitroglycerin in a canine model of experimental acute hypertensive heart failure: Differential actions of 2 cGMP activating therapeutics. *J. Am. Heart Assoc.* **2014**, *3*, e000206. [CrossRef] [PubMed]
76. Cannone, V.; Huntley, B.K.; Heublein, D.M.; Sandberg, S.M.; Harders, G.E.; Sangaralingham, J.S.; Martin, F.L.; Burnett, J.C., Jr. MANP: A novel designer natriuretic peptide for cardiometabolic disease. *J. Card Fail.* **2012**, *18*. [CrossRef]
77. Chen, H.H.; Neutel, J.; Smith, D.; Heublein, D.; Burnett, J. A first-in-human trial of a novel designer natriuretic peptide ZD100 in human hypertension. *J. Am. Soc. Hypertension* **2016**, *10*, e23. [CrossRef]

© 2019 by the authors. Licensee MDPI, Basel, Switzerland. This article is an open access article distributed under the terms and conditions of the Creative Commons Attribution (CC BY) license (http://creativecommons.org/licenses/by/4.0/).

Review

ARNi: A Novel Approach to Counteract Cardiovascular Diseases

Massimo Volpe [1,2,*], Speranza Rubattu [1,2] and Allegra Battistoni [1]

[1] Department of Clinical and Molecular Medicine; School of Medicine and Psychology, Sapienza University of Rome, 00189 Rome, Italy; rubattu.speranza@neuromed.it (S.R.); alle.battistoni@gmail.com (A.B.)
[2] IRCCS Neuromed, 86077 Pozzilli, Italy
* Correspondence: massimo.volpe@uniroma1.it; Tel.: +39-0633775979

Received: 22 March 2019; Accepted: 25 April 2019; Published: 28 April 2019

Abstract: Cardiovascular diseases (CVDs) still represent the greatest burden on healthcare systems worldwide. Despite the enormous efforts over the last twenty years to limit the spread of cardiovascular risk factors, their prevalence is growing and control is still suboptimal. Therefore, the availability of new therapeutic tools that may interfere with different pathophysiological pathways to slow the establishment of clinical CVDs is important. Previously, the inhibition of neurohormonal systems, namely the renin–angiotensin–aldosterone system (RAAS) and the sympathetic nervous system, has proven to be useful in the treatment of many CVDs. Attempts have recently been made to target an additional hormonal system, that of the natriuretic peptides (NPs), which, when dysregulated, can also play a role in the development CVDs. Indeed, a new class of drug, the angiotensin receptor–neprilysin inhibitors (ARNi), has the ability to counteract the effects of angiotensin II as well as to increase the activity of NPs. ARNi have already been proven to be effective in the treatment of heart failure with reduced ejection fraction. New evidence has suggested that, in the next years, the field of ARNi application will widen to include other CVDs, such as heart failure, with preserved ejection fraction and hypertension.

Keywords: angiotensin receptor–neprilysin inhibitor; natriuretic peptides; renin–angiotensin system; heart failure; arterial hypertension

---

1. Inhibition of Neurohormonal Systems in Cardiovascular Diseases

The worldwide prevalence of cardiovascular diseases (CVDs) is still alarmingly high. Although the CV mortality has declined between 1990 and 2015, mostly thanks to advantages conferred by interventional cardiology and anti-ischemic therapy, rates have plateaued in recent years, accounting for almost 18 million deaths/year in 2015 [1,2]. Accordingly, there were more than 422 million cases of CVDs/year globally, representing a major issue for healthcare systems [1,2]. Indeed, the prevalence of CVDs—and especially of congestive heart failure (HF), which is the final stage of different CVDs—has not yet started to decrease. The prognosis of patients affected by HF, both with preserved (HFpEF) and reduced ejection fraction (HFrEF), is still poor, highlighting the need for more effective preventive and treatment strategies [1].

Since the recognition of neurohormonal systems as being responsible for the development and progression of HF, neuroendocrine modulation with beta blockers targeting the sympathetic nervous system (SNS); angiotensin-converting enzyme inhibitors (ACEis), angiotensin receptor blockers (ARBs), and mineralocorticoid receptor antagonists (MRAs) for the renin–angiotensin–aldosterone system (RAAS) have been pivotal in the treatment of chronic HF and reducing associated morbidity and mortality. No additional effective pharmacotherapies have since been discovered [3]. A notable exception is the most recently discovered single molecule with a dual component, sacubitril/valsartan, which combines an ARB with the neutral endopeptidase inhibitor (NEPi) neprilysin, increasing the availability of natriuretic peptides (NPs) [4,5]. The development of this drug comes after decades of attempts to use NPs as therapeutic weapons in HF.

2. Natriuretic Peptides—Biological Properties

The NP family includes three different molecules—atrial natriuretic peptide (ANP), brain natriuretic peptide (BNP), and C-type natriuretic peptide (CNP)—that play a key role in cardiorenal homeostasis [6]. They are mostly synthesized within the heart in response to volume overload and myocyte stress, although synthesis in response to neuroendocrine regulation has also been shown [7,8]. Although they are synthesized as pre-prohormones their biologically active domain is the α-carboxy-terminal peptide [7,9]. NP effects are mediated by guanylyl cyclase (GC) receptors, with the NPR-A receptor being the main effector of both ANP and BNP, whereas the NPR-B receptor mediates CNP actions [7,9]. Both ANP and BNP contribute to regulating vascular tone, mainly due to their vasodilating properties. Indeed, ANP, through the activation of the cGMP-dependent protein kinase G (PKG), leads to increased production of nitric oxide (NO) with subsequent relaxation of the vascular smooth muscle cells and a decrease in blood pressure (BP) [10]. ANP may also induce Ca^{2+}/calmodulin-dependent endothelial NO synthase in the aorta, ventricle, and kidney [11,12]. NPs also enhance diuresis and natriuresis, leading to lower BP levels. Indeed, ANP may enhance the glomerular filtration rate (GFR) through a direct vasodilating effect on the afferent arterioles with an increase in blood ultrafiltration. Moreover, ANP boosts natriuresis by acting directly on Na^+ channels in the nephron, decreasing renin release from juxtaglomerular cells, and by inhibiting the synthesis and release of aldosterone [13,14]. NPs may also influence left ventricle afterload, not only acting on vascular tone and body fluid homeostasis, but also by decreasing adverse vascular remodeling over time. Indeed, high BNP levels may reduce the expression of transforming growth factor β (TGF-β), which is an important profibrotic molecule, and facilitate the degradation of extracellular fibrotic component [15–17]. In the heart, ANP counteracts the sympathetic nerve activity and increases vagal activity, leading to a decrease of heart rate and of cardiac output [18]. Moreover, NPs, by interacting with their G-coupled receptor, might antagonize pathological signaling leading to hypertrophy, such as cGMP-PKG deterioration, with titin hypophosphorylation [19], NO reduction with endothelial dysfunction [20], inflammation, and fibrosis [21].

Due to their multiple functions, the biological signature of NPs is to reduce body fluid and maintain BP and CV homeostasis. On this basis, NPs are physiological antagonists of both SNS and RAAS, and they are fundamental actors in HF.

3. Mechanisms of Degradation of NPs

The type C natriuretic peptide receptor (NPR-C) is mostly expressed in the kidney, adrenals, lungs, brain, heart, as well as vascular wall [22]. Unlike NPR-A and NPR-B, NPR-C exerts its biological action through receptor–ligand internalization, followed by lysosomal delivery of its ligand (NP) for degradation [23]. Therefore, NPR-C has been designated as the "clearance receptor". Circulating NPs may also be cleared via proteolytic cleavage by neutral endopeptidase (NEP). Neprilysin is a membrane-bound zinc-dependent metallopeptidase acting on the amino side of hydrophobic residues. It is expressed in different tissues, including the myocardium, kidneys, brain, and vessels [24].

It has been shown that whenever levels of NPs are high, such as in HF, NEP becomes the main source for their metabolism [25]. The affinity of NEP for NPs depends, in part, on their ability to match their structure with the active site of NEP [26]. In this regard, ANP represents the major target of NEP action. Furthermore, amino-terminal NPs are not cleared by NEP [27].

NEP is responsible for the metabolism of more than 50 putative substrates. Indeed, NEP also accounts for the degradation of bradykinin, substance P, adrenomedullin, angiotensin II, and endothelin-1. As a result, its activity or, on the contrary, its blockade, may lead to complex effects [24]. Therefore, the benefits derived from NEP inhibition could be more that those related just to the increase in NP availability. In particular, the decrease in bradykinin metabolism may be important since high bradykinin levels enhance NO-mediated vasodilation and may modulate ischemic preconditioning [28–30]. On the other hand, it should also be noted that bradykinin, as well as substance P, may increase vascular permeability and, therefore, they may also be implicated in the development of angioedema [31], which is a potential side effect of NEP inhibition. NEP inhibition also increases adrenomedullin level. Proadrenomedullin may have a hypotensive effect by decreasing peripheral catecholamine release, increasing natriuresis and vasodilation through the cyclic adenosine monophosphate (cAMP), NO, and renal prostaglandin systems [32,33].

On the other hand, the inhibition of NEP increases angiotensin II and endothelin-1 levels which are both involved in vessel contraction and fibrosis, resulting in a consequent reduction of the beneficial effects derived from NP increase. Moreover, NEP converts angiotensin-I to angiotensin 1–7 [31] which conveys vasodilating, antiproliferative, and natriuretic effects through the activation of the Mas receptor [34]. Therefore, the inhibition of NEP will enhance substrate conversion to angiotensin-I, thereby potentiating the RAAS and neutralizing the advantages of NP augmentation [5,35,36]. This is the rationale for the need for a concomitant RAAS blockade. In the case of endothelin-1, NEP also hydrolyzes the big endothelin-1 precursor peptide. Thus, the effect of a NEP inhibitor on endothelin-1 levels will depend on the net effect of hydrolysis of both big-endothelin-1 and endothelin-1 [37]. The latter may lead to vasodilation through NO and prostaglandin by binding to the endothelin B receptor on endothelia. Conversely, binding to the endothelin A and endothelin B receptors on vascular smooth muscle cells induces vasoconstriction. In the myocardium, endothelin stimulates fibroblast synthesis of collagen and promotes cardiac hypertrophy [38,39].

Given its multiple substrates, the net effect deriving from the inhibition of NEP may be difficult to foresee as it catalyzes molecules with opposite effects on CV homeostasis.

4. Therapeutic Strategies Involving NP Metabolism in Cardiovascular Diseases

For therapeutic purposes, synthetic NPs that attempt to reproduce the beneficial effects of NPs were first developed. Among the synthetic peptides, anaritide and carperitide are synthetic forms of ANP, whereas nesiritide is a synthetic form of BNP [40]. Ularitide and cenderitide are the synthetic forms of urodilatin and CNP, respectively. These drugs have shown some benefits in the treatment of HF [9], but their effects are usually scant and their tolerability inadequate, so that their clinical use is not supported. On the other hand, inhibition of endogenous NP degradation has been attempted. Inhibition of NPR-C seemed unreliable because of its multiple functions other than NP clearance. The NEPi was first proposed for use in monotherapy. In the early 1990s, candoxatril was initially studied as an antihypertensive agent, but it was not found to have a sustained antihypertensive effect. It later became evident that the lack of an antihypertensive effect with NEPi alone was secondary to the increased levels of the vasoconstrictors such as angiotensin II and endothelin-1 [30,37,41]. Therefore, drugs combining a NEPi and an ACEi were developed [42]. However, these drugs were discharged because of the higher occurrence of angioedema that was dependent on the dual mechanism of action. Indeed, both ACE and NEP are enzymes responsible for the metabolism of bradykinin, which may cause vasodilation, angioedema, and airway obstruction.

Eventually, it was the time for a new class of medications, known as angiotensin receptor–neprilysin inhibitors (ARNi). Indeed, ARNi combine the NEP inhibition, due to sacubitril, with the ATII receptor I inhibition by valsartan (an ARB), offering the benefits of this two-step approach whilst avoiding the side effects of an increase in bradykinin due to dual inhibition of its metabolism. Indeed, the first-in-class ARNi, sacubitril/valsartan or LCZ696, contains valsartan and a NEPi prodrug, sacubitril (AHU377), in a 1:1 molar ratio. Upon ingestion, sacubitril is metabolized into an active NEPi, sacubitrilat (LBQ657). The target dose of 97/103 mg BID of sacubitril/valsartan resulted in equivalent plasma concentrations as valsartan 160 mg BID and a rise in cGMP, representing an increase in NPR activation, secondary to effective NEP inhibition [43]. Recent data suggest that the reduction in NEP activity results in a favorable impact of sacubitril/valsartan on HF progression, due especially to an increase in ANP and possibly CNP, rather than BNP [44].

5. Clinical Applications of ARNi

5.1. Heart Failure with Reduced Ejection Fraction

The prospective comparison of angiotensin receptor–neprilysin inhibitor with ACEi to determine impact on global mortality and morbidity in heart failure (PARADIGM-HF) trial first tested sacubitril/valsartan in HFrEF [5]. In PARADIGM-HF, sacubitril/valsartan was compared to enalapril in a cohort of patients affected by symptomatic HFrEF with left ventricular ejection fraction (LVEF) ≤ 35% and elevated B-type NP levels or hospitalization for HF within the previous year. Sacubitril/valsartan proved to be more effective than enalapril in reducing the primary outcome, a composite of death from CV causes or first hospitalization for HF (Table 1). Although there were more events of symptomatic hypotension in the case of using sacubitril/valsartan, more participants assigned to enalapril discontinued their study medication due to adverse effects. In clinical practice, this would mean a higher proportion of patients achieving optimal RAAS inhibition (as well as the additional benefits associated with concomitant neprilysin inhibition) with an ARNi rather than with enalapril. [5] Therefore, the latest American and European guidelines for the management of HF added sacubitril/valsartan as a first-line therapy for outpatients affected by chronic HFrEF [45,46].

Table 1. Main clinical trials about the effects of LCZ696 on cardiovascular outcomes.

Study; Aim	Study Population	Design	Outcomes
PARADIGM-HF (and post hoc analysis) Comparison of efficacy of LCZ696 versus enalapril in patients with HFrEF (LVEF ≤35%) [5,47]	Patients with symptomatic HFrEF (NYHA functional classes II to IV) and elevated B-type natriuretic peptide levels or hospitalization for HF within the previous 12 months; $n = 8442$	Multicenter, randomized, double-blind study	LCZ696 reduced the composite primary of CV death or HF hospitalization more than enalapril; LCZ696 reduced secondary endpoint more than enalapril: • any CV death; • first worsening HF hospitalization; • all-cause mortality Moreover, LCZ696 group had fewer hospitalizations for worsening HF, less necessity to receive intensive care, intravenous positive inotropic agents, and to have implantation of a HF device or cardiac transplantation.
TRANSITION To assess the safety and tolerability of starting a therapy with LCZ696 while still in the hospital or after discharge [48]	HFrEF patients hospitalized for ADHF, after stabilization $n = 1002$	Multicenter, randomized, open-label, parallel-group study	The percentage of patients taking target dose of sacubitril/valsartan 200 mg BID at 10 weeks post randomization was the same among patients who started taking LCZ696 during hospitalization or after discharge
PIONEER-HF To assess the percentage change from baseline in NTproBNP levels with LCZ696 [49]	HFrEF patients hospitalized for ADHF after stabilization $n = 736$	Multicenter, randomized, double-blind study	LCZ696 led to a reduction in the NTproBNP concentration than a therapy with enalapril at 4 and 8 weeks; LCZ696 led to a reduction in the level of high-sensitivity cardiac troponin T; LCZ696 led to lower rate of rehospitalization for HF
TITRATION To assess the tolerability of initiating/uptitrating LCZ696 from 50 to 200 mg BID over 3 and 6 weeks [50]	Patients with symptomatic HFrEF (NYHA functional classes II to IV) + one or more of the following additional eligibility requirements: for outpatients currently treated with ACEi/ARB, the dose must have been stable for at least 2 weeks; to be classified as ACEi/ARB-naïve, the patient must not have taken ACEi/ARB for at least 4 weeks; hospitalized patients had to be either ACEi/ARB-naïve, or on a tolerated dose of an ACEi/ARB at screening $n = 429$	Multicenter, randomized, double bind, parallel study	Initiation/uptitration of LCZ696 from 50 to 200 mg BID had a tolerability profile in line with other HF treatments.

Table 1. Cont.

Study; Aim	Study Population	Design	Outcomes
PARAMOUNT To assess the efficacy of LCZ96 versus valsartan to change NTproBNP levels from baseline [51]	Patients with signs and symptoms of HF, ≥40 years, with NTproBNP ≥400 pg/mL and a LVEF ≥45%, while on active diuretic therapy n = 301	Multicenter, randomized, double-blind study	The decline in NTproBNP at 12 weeks after initiation of the treatment was greater in the LCZ696 group. LCZ969 was also able to ameliorate LA size and NHYA class (secondary endpoints)
PARAMETER To assess the efficacy of LCZ696 versus olmesartan in reducing arterial stiffness [52]	Elderly patients (aged ≥60 years) with systolic hypertension and pulse pressure >60 mmHg n = 454	Multicenter, randomized, double-blind study	LCZ696 reduced central aortic SBP more than olmesartan and reduced mean 24-hour ambulatory brachial and central aortic SBP

ACEi: angiotensin converting enzyme inhibitors; ARB: angiotensin II receptor I blockers; CV: cardiovascular; ADHF: acute decompensated heart failure; BID: bis in die; LVEF: left ventricular ejection fraction; HFrEF: heart failure with reduced ejection fraction; HFrpEF: heart failure with preserved ejection fraction; NTproBNP: amino-terminal pro-brain natriuretic peptide; NYHA: New York Heart Association; SBP: systolic blood pressure.

Improvement in the prognosis of patients assigned to sacubitril/valsartan also remained consistent in the subgroup of prediabetic, undiagnosed diabetic, and diagnosed diabetic patients, who are at a higher risk of adverse CV outcomes [53]. This evidence agrees with previous preclinical data demonstrating the cardio- and nephroprotective effects of ARNi [54–57].

A subsequent analysis of the PARADIGM trial reported that sacubitril/valsartan use was associated with further evidence of clinical benefit in comparison with enalapril, including fewer visits to an emergency department for HF, a reduced need for intensification of the treatment for HF, and a lower requirement for intensive care, HF devices, or cardiac transplantation [47]. Moreover, another subsequent analysis of PARADIGM trial, which has enrolled almost half of the patients with a high CV risk, showed fewer coronary events in those treated with sacubitril/valsartan [58]. A recent experimental study in rats provided insight into the differential effects of sacubitril and valsartan in a model of HF. In particular, it has been shown that sacubitril in association with valsartan significantly improves load-dependent left ventricle contractility and relaxation with a reduction of myocardial collagen content, while the improvement in load-independent left ventricular contractility is due to valsartan [59].

Following the evidence for chronic HF, the PIONEER-HF study, a multicenter trial, has been designed to investigate the role of sacubitril/valsartan in patients affected by HFrEF hospitalized for an episode of acute HF (AHF), after hemodynamic stabilization, regardless of the duration of diagnosis or background HF therapy, and without a preceding run-in period. Thus, this trial has been performed in treatment-naïve hospitalized patients. The primary endpoint of PIONEER-HF was the proportional change in amino-terminal pro-brain natriuretic peptide (NTproBNP) level from baseline through one month and then two months. The main result was that sacubitril/valsartan led to a greater reduction in the NTproBNP concentration than enalapril from the first week of treatment, as well as to a decrease of markers of myocardial injury. Furthermore, in-hospital initiation of sacubitril/valsartan therapy was associated with a subsequent lower rate of rehospitalizations for HF. The rates of experienced side effects did not differ significantly between the sacubitril/valsartan group and the enalapril group [49].

More insights about the management of patients hospitalized for HF have been retrieved by the TRANSITION trial. This is a randomized, phase IV, multicenter, open-label study which assessed the safety and tolerability of introducing a therapy with sacubitril/valsartan in 1002 patients hospitalized for decompensated acute HFrEF still in the hospital or once discharged. Almost one-third of patients were newly diagnosed with HFrEF, and one-quarter were naïve to ACEi or ARB. The primary endpoint of achieving the target dose of sacubitril/valsartan 200 mg BID at 10 weeks after randomization has been achieved in 45% of patients that started taking sacubitril/valsartan in hospital, and in 50.4% of the post-discharge group, without any significant difference in adverse effects between the two groups [48].

Recently, subsequent analyses of previous trials have given more insightful data about a specific subset of patients. Indeed, a post hoc analysis of the PARADIGM trial investigated the effects of sacubitril/valsartan in diabetic patients, showing that this treatment leads to a better glycemic profile (reduction of Hb1Ac and less need to undertake insulin therapy or oral hypoglycemic agents) in the long term, independent of the reduction in body weight [60]. Similar beneficial effects of sacubitril/valsartan on lipid and glucose metabolism have also been reported in hypertensive obese patients as [61]. Preclinical models of diabetes seem to indicate that this beneficial effect of sacubitril/valsartan depends on the rise in NP levels, bradykinin, glucagon-like peptide 1, and on the reduction in angiotensin II levels that would result in subsequently improved insulin sensitivity [62,63]. These data may have great clinical relevance since diabetes is not only a comorbidity widely present among HF patients, but its evolution can substantially modify the patient prognosis. Moreover, recent interesting preclinical data seem to confirm the beneficial effects of sacubitril/valsartan on vascular and neural complications in type 2 diabetes, giving room for hypotheses about possible wider future applications of ARNi in diabetic patients [64].

5.2. Heart Failure with Preserved Ejection Fraction

Up to now, sacubitril/valsartan does not have an official indication in patients with HFpEF. However, it has also been supposed to play a beneficial effect in HFpEF by blocking a profibrotic/prohypertrophic mechanism (valsartan) while stimulating an antifibrotic/antihypertrophic mechanism (sacubitril) [65]. Indeed, the RAAS may play a pivotal role in enhancing the inflammation, endothelial dysfunction, and remodeling implicated in the progression of HFpEF. Despite this, RAAS inhibitors have failed to demonstrate mortality benefits in this setting. Beyond the single RAAS inhibition, the sacubitril/valsartan might ameliorate several key pathways in the development of HFpEF, namely cardiac remodeling, such as left ventricular hypertrophy and stiffness, microvascular dysfunction, and oxidative stress by increasing NPs levels [65]. Sacubitril/valsartan has been tested in a phase II trial in patients affected by HFpEF, the PARAMOUNT (prospective comparison of ARNi with ARB on management of heart failure with preserved ejection fraction) trial [66]. In this trial, sacubitril/valsartan 200 mg BID was compared with valsartan 160 mg BID in patients symptomatic for HFpEF, aged ≥40 years, with NTproBNP ≥400 pg/mL, while on diuretic therapy. The primary endpoint, the decline in NTproBNP at 12 weeks, was greater in the sacubitril/valsartan group. Furthermore, after 9 months, the left atrial dimension, which may indicate diastolic function [66], declined more in the sacubitril/valsartan arm as well as markers of fibrosis [51]. Subsequent analyses have shown that these sacubitril/valsartan effects were independent of the BP-lowering effect [67]. In addition, patients in the sacubitril/valsartan arm had greater improvements in New York Heart Association (NYHA) class and preserved better renal function compared to the valsartan group [68].

Given these favorable results, the current phase III PARAGON (prospective comparison of sacubitril/valsartan with ARB global outcome in HF with preserved ejection fraction) trial has been designed to determine whether sacubitril/valsartan can reduce CV death or total HF hospitalizations in patients with HFpEF. This trial has enrolled symptomatic patients with LVEF ≥45% and elevated NP, or history of HF hospitalization within 9 months and evidence of structural heart disease. The results of this trial are expected in 2019 [69].

Finally, the randomized, 24-week, double-blind multicenter controlled study comparing sacubitril/valsartan with medical therapy for comorbidities in HFpEF patients (PARALLAX) is currently recruiting participants to test the superiority of LCZ696 in reducing NTproBNP levels and improving HF symptoms and exercise function in HFpEF patients [70].

5.3. Hypertension

Preclinical studies have given insights on how may ARNi favorably exert antihypertensive and cardioprotective effects in animal models of hypertension [71]. In fact, a significant reduction of BP and proteinuria levels and a full prevention from stroke was observed over long-term treatment with sacubitril/valsartan, as compared to valsartan, in the high-salt-fed, stroke-prone, spontaneously hypertensive rat [71]. Furthermore, in a model of spontaneous hypertensive rat, sacubitril/valsartan proved to be as effective as valsartan in improving endothelium-dependent and -independent vasorelaxation [72]. Moreover, sacubitril/valsartan has shown an improved ability to reduce BP levels compared to valsartan, regardless of the amount of salt intake. This effect was associated with a significant increase of urinary sodium excretion and suppression of sympathetic activity. In addition, it reduced myocardial inflammation, remodeling, and endothelial dysfunction, also ameliorating coronary circulation [73].

In hypertensive patients, a proof-of-concept trial enrolling mostly white, mild-to-moderate hypertensive patients demonstrated that compared with valsartan or AHU377 alone, sacubitril/valsartan treatment for 2 months provided additional reduction of BP, systolic, diastolic, and pulse pressures, both sitting and ambulatory, without any excess in serious adverse effects [74]. In the PARAMETER (prospective comparison of angiotensin receptor–neprilysin inhibitor with angiotensin receptor blocker measuring arterial stiffness in the elderly) study, sacubitril/valsartan demonstrated efficacy in reducing arterial stiffness in the elderly with systolic hypertension and pulse pressure >60 mmHg [52]. At 3 months, sacubitril/valsartan reduced central aortic systolic BP more than olmesartan and reduced mean 24-hour ambulatory brachial and central aortic systolic BP, therefore, fewer patients in the sacubitril/valsartan group required add-on antihypertensives [52]. Similarly, a recent study in a cohort of elderly Asiatic patients affected by isolated systolic hypertension showed that sacubitril/valsartan was more effective than olmesartan in reducing mean systolic BP and pulse pressure [75]. Tolerability of sacubitril/valsartan and olmesartan was the same.

Furthermore, Ruilope and colleagues demonstrated that LCZ696 monotherapy was dose-dependently superior to valsartan monotherapy by clinical and ambulatory BP measurements for all tested doses [74]. Therefore, available evidence seems to support an application of ARNi as an antihypertensive compound with adequate tolerability and effectiveness throughout 24 hours.

6. Future Perspective of NP-Based Therapies

There are several ongoing studies to help understand ARNi doses and tolerability in different clinical settings, as well as to increase its possible fields of application.

Concerning HF, there are six ongoing trials investigating the possible benefit of ARNi on different endpoints: biomarker changes and ventricular remodeling among patients with HFrEF (PROVE-HF, NCT02887183) [76], changes in aortic impedance among patients affected by HF and hypertension (EVALUATE-HF, NCT02874794) [77], changes in functional mitral regurgitation (PRIME, NCT02687932) in patients with LVEF between 25% and 50% [78], and changes in mean pulmonary artery pressure in patients with reduced LVEF (PARENT, NCT02788656) [79]. The HFN-LIFE study (ClinicalTrials.gov Identifier: NCT02816736) will help to assess the safety and tolerability of the lowest dose of sacubitril/valsartan in patients with HFrEF symptomatic at rest [80]. Finally, the PARADOR (comparing ARNi with ACE inhibitor on endothelial function) trial is a multicenter, randomized, double-blind trial designed to compare the effects of sacubitril/valsartan vs. enalapril on endothelial function in patients with HFrEF [81].

Preliminary evidence in a mouse model of acute myocardial infarction (AMI) showed that LCZ696 significantly suppressed the production of proinflammatory cytokines, matrix metalloproteinase-9 activity, and aldosterone [82]. In a rat model, the association of sacubitril and valsartan has been found to protect myocardial ischemic damage by possibly ameliorating oxidative stress [83]. In addition, in two animal models, AMI sacubitril/valsartan has shown to determine the short- and long-term benefits in preventing MI-induced ventricular dysfunction compared to valsartan alone, with a reduction of fibrosis and myocardial scar and increased perfusion to the infarcted areas [84,85]. These findings provide a promising experimental basis to investigate the cardioprotective effects of sacubitril/valsartan in AMI patients. Therefore, a phase III, randomized, controlled PARADISE-AMI (prospective ARNi versus ACE inhibitor trial to determine superiority in reducing heart failure events after MI) study is currently recruiting post-AMI patients without prior HF, and with reduced LVEF or pulmonary congestion [86]. This trial will evaluate the benefit of sacubitril/valsartan versus ramipril in reducing the occurrence of the primary composite endpoint of CV death, HF hospitalization, and outpatient HF.

Lastly, preclinical data seem to support the hypothesis of a more beneficial effect of sacubitril/valsartan over valsartan alone in CV abnormalities associated with chronic kidney disease [87].

7. Conclusions

The most recent discovery in the field of CV therapy concerns the chance of interfering with NP metabolism. The development of the first drug used in clinical practice in this sense, thanks to its second active domain active as ARB, resulted in an improvement in mortality and morbidity in patients affected by HFrEF, for which the drug was primarily developed. It is also likely that deeper knowledge of NPs and NEP activity could soon make us reconsider the principles of therapy of many CVDs.

Author Contributions: A.B., M.V. and S.R. have equally contributed in writing this review and have approved the submitted version and agree to be personally accountable for the contributions and for ensuring that questions related to the accuracy or integrity of any part of the work.

Conflicts of Interest: The authors declare no conflict of interest.

References

1. Roth, G.A.; Johnson, C.; Abajobir, A.; Abd-Allah, F.; Abera, S.F.; Abyu, G.; Ahmed, M.; Aksut, B.; Alam, T.; Alam, K.; et al. Global, regional, and national burden of cardiovascular diseases for 10 causes, 1990 to 2015. *J. Am. Coll. Cardiol.* **2017**, *70*, 1–25. [CrossRef] [PubMed]
2. Mozaffarian, D.; Benjamin, E.J.; Go, A.S.; Arnett, D.K.; Blaha, M.J.; Cushman, M.; Das, S.R.; de Ferranti, S.; Després, J.P.; Fullerton, H.; et al. Heart disease and stroke statistics-2016 update: A report from the American Heart Association. *Circulation* **2016**, *133*, e38–e360. [CrossRef]
3. Yandrapalli, S.; Aronow, W.S.; Mondal, P.; Chabbott, D.R. The evolution of natriuretic peptide augmentation in management of heart failure and the role of sacubitril/valsartan. *Arch. Med. Sci.* **2017**, *13*, 1207–1216. [CrossRef]
4. Gu, J.; Noe, A.; Chandra, P.; Al-Fayoumi, S.; Ligueros-Saylan, M.; Sarangapani, R.; Maahs, S.; Ksander, G.; Rigel, D.F.; Jeng, A.Y.; et al. Pharmacokinetics and pharmacodynamics of LCZ696, a novel dual-acting angiotensin receptor-neprilysin inhibitor (ARNi). *J. Clin. Pharmacol.* **2010**, *50*, 401–414. [CrossRef]
5. McMurray, J.J.; Packer, M.; Desai, A.S.; Gong, J.; Lefkowitz, M.P.; Rizkala, A.R.; Rouleau, J.L.; Shi, V.C.; Solomon, S.D.; Swedberg, K.; et al. PARADIGM-HF Investigators and Committees. Angiotensin-neprilysin inhibition versus enalapril in heart failure. *N. Engl. J. Med.* **2014**, *371*, 993–1004. [CrossRef]
6. Levin, E.R.; Gardner, D.G.; Samson, W.K. Natriuretic peptides. *N. Engl. J. Med.* **1998**, *339*, 321–328. [PubMed]
7. Potter, L.R.; Yoder, A.R.; Flora, D.R.; Antos, L.K.; Dickey, D.M.; et al. Natriuretic peptides: Their structures, receptors, physiologic functions and therapeutic applications. *Handb. Exp. Pharmacol.* **2009**, *191*, 341–366.
8. Melo, L.G.; Steinhelper, M.E.; Pang, S.C.; Tse, Y.; Ackermann, U. ANP in regulation of arterial pressure and fluid-electrolyte balance: Lessons from genetic mouse models. *Physiol. Genomics* **2000**, *3*, 4–58. [CrossRef]
9. Volpe, M.; Battistoni, A.; Rubattu, S. Natriuretic peptides in heart failure: Current achievements and future perspectives. *Int. J. Cardiol.* **2019**, *281*, 186–189. [CrossRef]
10. Elesgaray, R.; Caniffi, C.; Ierace, D.R.; Jaime, M.F.; Fellet, A.; Arranz, C.; Costa, M.A. Signaling cascade that mediates endothelial nitric oxide synthase activation induced by atrial natriuretic peptide. *Regul. Pept.* **2008**, *151*, 130–134. [CrossRef] [PubMed]
11. Costa, M.A.; Elesgaray, R.; Balaszczuk, A.M.; Arranz, C. Role of NPR-C natriuretic receptor in nitric oxide system activation induced by atrial natriuretic peptide. *Regul. Pept.* **2006**, *135*, 63–68. [CrossRef] [PubMed]
12. Theilig, F.; Wu, Q. ANP-induced signaling cascade and its implications in renal pathophysiology. *Am. J. Physiol. Renal Physiol.* **2015**, *308*, F1047–F1055. [CrossRef]
13. Kurtz, A.; Della Bruna, R.; Pfeilschifter, J.; Taugner, R.; Bauer, C. Atrial natriuretic peptide inhibits renin release from juxtaglomerular cells by a cGMP-mediated process. *Proc. Natl. Acad. Sci. USA* **1986**, *83*, 4769–4773. [CrossRef] [PubMed]
14. Brenner, B.M.; Ballermann, B.J.; Gunning, M.E.; Zeidel, M.L. Diverse biological actions of atrial natriuretic peptide. *Physiol. Rev.* **1990**, *70*, 665–699. [CrossRef]
15. Woods, R.L. Cardioprotective functions of atrial natriuretic peptide and B-type natriuretic peptide: A brief review. *Clin. Exp. Pharmacol. Physiol.* **2004**, *31*, 791–794. [CrossRef] [PubMed]

16. Holditch, S.J.; Schreiber, C.A.; Nini, R.; Tonne, J.M.; Peng, K.W.; Geurts, A.; Jacob, H.J.; Burnett, J.C.; Cataliotti, A.; Ikeda, Y. B-type natriuretic peptide deletion leads to progressive hypertension, associated organ damage, and reduced survival: Novel model for human hypertension. *Hypertension.* **2015**, *66*, 199–210. [CrossRef]
17. Tamura, N.; Ogawa, Y.; Chusho, H.; Nakamura, K.; Nakao, K.; Suda, M.; Kasahara, M.; Hashimoto, R.; Katsuura, G.; Mukoyama, M.; et al. Cardiac fibrosis in mice lacking brain natriuretic peptide. *Proc. Natl. Acad. Sci. USA* **2000**, *97*, 4239–4244. [CrossRef]
18. Volpe, M.; Cuocolo, A.; Vecchione, F.; Mele, A.F.; Condorelli, M.; Trimarco, B. Vagal mediation of the effects of atrial natriuretic factor on blood pressure and arterial baroreflexes in the rabbit. *Circ. Res.* **1987**, *60*, 747–755. [CrossRef]
19. Van Heerebeek, L.; Hamdani, N.; Falcão-Pires, I.; Leite-Moreira, A.F.; Begieneman, M.P.; Bronzwaer, J.G.; van der Velden, J.; Stienen, G.J.; Laarman, G.J.; Somsen, A. Low myocardial protein kinase G activity in heart failure with preserved ejection fraction. *Circulation* **2012**, *126*, 830–839. [CrossRef]
20. Franssen, C.; Chen, S.; Unger, A.; Korkmaz, H.I.; De Keulenaer, G.W.; Tschöpe, C.; Leite-Moreira, A.F.; Musters, R.; Niessen, H.W.; Linke, W.A. Myocardial microvascular inflammatory endothelial activation in heart failure with preserved ejection fraction. *JACC Heart Fail.* **2016**, *4*, 312–324. [CrossRef]
21. Ahluwalia, A.; MacAllister, R.J.; Hobbs, A.J. Vascular actions of natriuretic peptides. Cyclic GMP-dependent and -independent mechanisms. *Basic Res. Cardiol.* **2004**, *99*, 83–89. [CrossRef]
22. Rubattu, S.; Sciarretta, S.; Morriello, A.; Calvieri, C.; Battistoni, A.; Volpe, M. NPR-C: A component of the natriuretic peptide family with implications in human diseases. *J. Mol. Med. (Berl.)* **2010**, *88*, 889–897. [CrossRef] [PubMed]
23. Nussenzveig, D.; Lewicki, J.; Maack, T. Cellular mechanisms of the clearance function of Type C receptors of atrial natriuretic factor. *J. Biol. Chem.* **1990**, *265*, 20952–20958. [PubMed]
24. Mangiafico, S.; Costello-Boerrigter, L.C.; Andersen, I.A.; Cataliotti, A.; Burnett, J.C., Jr. Neutral endopeptidase inhibition and the natriuretic peptide system: An evolving strategy in cardiovascular therapeutics. *Eur. Heart J.* **2013**, *34*, 886c–893c. [CrossRef] [PubMed]
25. Potter, L.R. Natriuretic peptide metabolism, clearance and degradation. *FEBS J.* **2011**, *278*, 1808–1817. [CrossRef] [PubMed]
26. Pankow, K.; Schwiebs, A.; Becker, M.; Siems, W.E.; Krause, G.; Walther, T. Structural substrate conditions required for neutral endopeptidase-mediated natriuretic peptide degradation. *J. Mol. Biol.* **2009**, *393*, 496–503. [CrossRef]
27. Palmer, S.C.; Yandle, T.G.; Nicholls, M.G.; Frampton, C.M.; Richards, A.M. Regional clearance of amino-terminal pro-brain natriuretic peptide from human plasma. *Eur. J. Heart Fail.* **2009**, *11*, 832–839. [CrossRef] [PubMed]
28. Palmer, S.C.; Yandle, T.G.; Nicholls, M.G.; Frampton, C.M.; Richards, A.M. Neprilysin inhibitors potentiate effects of bradykinin on B2 receptor. *Hypertension* **2002**, *39*, 619–623.
29. Kozlovski, V.I.; Lomnicka, M.; Jakubowski, A.; Chlopicki, S. Inhibition of neutral endopeptidase by thiorphan does not modify coronary vascular responses to angiotensin I, angiotensin II and bradykinin in the isolated guinea pig heart. *Pharmacol. Rep.* **2007**, *59*, 421–427. [PubMed]
30. Dalzell, J.R.; Seed, A.; Berry, C.; Whelan, C.J.; Petrie, M.C.; Padmanabhan, N.; Clarke, A.; Biggerstaff, F.; Hillier, C.; McMurray, J.J. Effects of neutral endopeptidase (neprilysin) inhibition on the response to other vasoactive peptides in small human resistance arteries: Studies with thiorphan and omapatrilat. *Cardiovasc. Ther.* **2014**, *32*, 13–18. [CrossRef] [PubMed]
31. Byrd, J.B.; Touzin, K.; Sile, S.; Gainer, J.V.; Yu, C.; Nadeau, J.; Adam, A.; Brown, N.J. Dipeptidyl peptidase IV in angiotensin-converting enzyme inhibitor-associated angioedema. *Hypertension* **2008**, *51*, 141–147. [CrossRef]
32. Nakamura, M.; Yoshida, H.; Hiramori, K. Comparison of vasodilator potency of adrenomedulling and proadrenomedullin N-terminal 20 peptide in human. *Life Sci.* **1999**, *65*, 2151–2156. [CrossRef]
33. Wilkinson, I.B.; McEniery, C.M.; Bongaerts, K.H.; MacCallum, H.; Webb, D.J.; Cockcroft, J.R. Adrenomedullin (ADM) in the human forearm vascular bed: Effect of neutral endopeptidase inhibition and comparison with proadrenomedullin NH2-terminal 20 peptide (PAMP). *Br. J. Clin. Pharmacol.* **2001**, *52*, 159–164. [CrossRef]
34. Pinheiro, S.V.; Simones, E.; Silva, A.C. Angiotensin converting enzyme 2, angiotensin-(1–7), and receptor MAS axis in the kidney. *Int. J. Hypertens.* **2012**, *22*, 224–233. [CrossRef]

35. Newby, D.E.; McDonagh, T.; Currie, P.F.; Northridge, D.B.; Boon, N.A.; Dargie, H.J. Candoxatril improves exercise capacity in patients with chronic heart failure receiving angiotensin converting enzyme inhibition. *Eur. Heart J.* **1998**, *19*, 1808–1813. [CrossRef]
36. Stephenson, S.L.; Kenny, A.J. Metabolism of neuropeptides. Hydrolysis of the angiotensins, bradykinin, substance P and oxytocin by pig kidney microvillar membranes. *Biochem. J.* **1987**, *241*, 237–247. [CrossRef]
37. Ferro, C.J.; Spratt, J.C.; Haynes, W.G.; Webb, D.J. Inhibition of neutral endopeptidase causes vasoconstriction of human resistance vessels in vivo. *Circulation* **1998**, *97*, 2323–2330. [CrossRef]
38. Kawanabe, Y.; Nauli, S.M. Endothelin. *Cell. Mol. Life Sci.* **2011**, *68*, 195–203. [CrossRef]
39. Shubeita, H.E.; McDonough, P.M.; Harris, A.N.; Knowlton, K.U.; Glembotski, C.C.; Brown, J.H.; Chien, K.R. Endothelin induction of inositol phospholipid hydrolysis, sarcomere assembly, and cardiac gene expression in ventricularmyocytes: A paracrine mechanism for myocardial cell hypertrophy. *J. Biol. Chem.* **1990**, *265*, 20555–20562. [PubMed]
40. Rubattu, S.; Calvieri, C.; Pagliaro, B.; Volpe, M. Atrial natriuretic peptide and regulation of vascular function in hypertension and heart failure: Implications for novel therapeutic strategies. *J. Hypertens.* **2013**, *31*, 1061–1072. [CrossRef]
41. McDowell, G.; Coutie, W.; Shaw, C.; Buchanan, K.D.; Struthers, A.D.; Nicholls, D.P. The Effect of the Neutral Endopeptidase Inhibitor Drug, Candoxatril, on Circulating Levels of Two of the Most Potent Vasoactive Peptides. *Br. J. Clin. Pharmacol.* **1997**, *43*, 329–332. [CrossRef] [PubMed]
42. Packer, M.; Califf, R.M.; Konstam, M.A.; Krum, H.; McMurray, J.J.; Rouleau, J.L.; Swedberg, K. Comparison of omapatrilat and enalapril in patients with chronic heart failure: The Omapatrilat Versus Enalapril Randomized Trial of Utility in Reducing Events (OVERTURE). *Circulation* **2002**, *106*, 920–926. [CrossRef] [PubMed]
43. Cohn, J.N.; Tognoni, G. A randomized trial of the angiotensin-receptor blocker valsartan in chronic heart failure. *N. Engl. J. Med.* **2001**, *345*, 1667–1675. [CrossRef] [PubMed]
44. Nougué, H.; Pezel, T.; Picard, F.; Sadoune, M.; Arrigo, M.; Beauvais, F.; Launay, J.M.; Cohen-Solal, A.; Vodovar, N.; Logeart, D. Effects of sacubitril/valsartan on neprilysin targets and the metabolism of natriuretic peptides in chronic heart failure: A mechanistic clinical study. *Eur. J. Heart Fail.* **2018**. [CrossRef]
45. Ponikowski, P.; Voors, A.A.; Anker, S.D.; Bueno, H.; Cleland, J.G.; Coats, A.J.; Falk, V.; González-Juanatey, J.R.; Harjola, V.P.; Jankowska, E.A.; et al. 2016 ESC Guidelines for the diagnosis and treatment of acute and chronic heart failure: The Task Force for the diagnosis and treatment of acute and chronic heart failure of the European Society of Cardiology (ESC). *Eur. J. Heart Fail.* **2016**, *18*, 891–975. [CrossRef] [PubMed]
46. Yancy, C.W.; Jessup, M.; Bozkurt, B.; Butler, J.; Casey, D.E., Jr.; Colvin, M.M.; Drazner, M.H.; Filippatos, G.S.; Fonarow, G.C.; Givertz, M.M.; et al. 2017 ACC/AHA/HFSA Focused Update of the 2013 ACCF/AHA Guideline for the Management of Heart Failure: A Report of the American College of Cardiology/American Heart Association Task Force on Clinical Practice Guidelines and the Heart Failure Society of America. *Circulation* **2017**, *136*, e137–e161.
47. Packer, M.; McMurray, J.J.; Desai, A.S.; Gong, J.; Lefkowitz, M.P.; Rizkala, A.R.; Rouleau, J.L.; Shi, V.C.; Solomon, S.D.; Swedberg, K.; et al. For the PARADIGM-HF investigators and coordinators. Angiotensin receptor neprilysin inhibition compared with enalapril on the risk of clinical progression in surviving patients with heart failure. *Circulation* **2015**, *131*, 54–61. [CrossRef]
48. Vicent, L.; Cinca, J.; Vazquez-García, R.; Gonzalez-Juanatey, J.R.; Rivera, M.; Segovia, J.; Pascual-Figal, D.; Bover, R.; Worner, F.; Delgado-Jiménez, J.; et al. Discharge treatment with ACE inhibitor/ARB after a heart failure hospitalization is associated with a better prognosis irrespectively of left ventricular ejection fraction. *Intern. Med. J.* **2019**. [CrossRef]
49. Velazquez, E.J.; Morrow, D.A.; DeVore, A.D.; Duffy, C.I.; Ambrosy, A.P.; McCague, K.; Rocha, R.; Braunwald, E.; PIONEER-HF Investigators. Angiotensin-Neprilysin Inhibition in Acute Decompensated Heart Failure. *N. Engl. J. Med.* **2019**, *380*, 539–548. [CrossRef] [PubMed]
50. Senni, M.; McMurray, J.J.V.; Wachter, R.; McIntyre, H.F.; Anand, I.S.; Duino, V.; Sarkar, A.; Shi, V.; Charney, A. Impact of systolic blood pressure on the safety and tolerability of initiating and up-titrating sacubitril/valsartan in patients with heart failure and reduced ejection fraction: Insights from the TITRATION study. *Eur. J. Heart Fail.* **2018**, *20*, 491–500. [CrossRef] [PubMed]

51. Zile, M.R.; Jhund, P.S.; Baicu, C.F.; Claggett, B.L.; Pieske, B.; Voors, A.A.; Prescott, M.F.; Shi, V.; Lefkowitz, M.; McMurray, J.J.; et al. Prospective comparison of ARNi with ARB on management of heart failure with preserved ejection fraction (PARAMOUNT) investigators. Plasma biomarkers reflecting profibrotic processes in heart failure with a preserved ejection fraction: Data from the prospective comparison of ARNi with ARB on management of heart failure with preserved ejection fraction study. *Circ. Heart Fail.* **2016**, *9*, e002551. [PubMed]
52. Williams, B.; Cockcroft, J.R.; Kario, K.; Zappe, D.H.; Brunel, P.C.; Wang, Q.; Guo, W. Effects of sacubitril/valsartan versus olmesartan on central hemodynamics in the elderly with systolic hypertension: The PARAMETER study. *Hypertension* **2017**, *69*, 411–420. [CrossRef]
53. Kristensen, S.L.; Preiss, D.; Jhund, P.S.; Squire, I.; Cardoso, J.S.; Merkely, B.; Martinez, F.; Starling, R.C.; Desai, A.S.; Lefkowitz, M. Risk Related to Pre-Diabetes Mellitus and Diabetes Mellitus in Heart Failure With Reduced Ejection Fraction: Insights From Prospective Comparison of ARNI With ACEI to Determine Impact on Global Mortality and Morbidity in Heart Failure Trial. *Circ. Heart Fail.* **2016**, *9*, e002560. [CrossRef]
54. Suematsu, Y.; Miura, S.; Goto, M.; Matsuo, Y.; Arimura, T.; Kuwano, T.; Imaizumi, S.; Iwata, A.; Yahiro, E.; Saku, K. LCZ696, an angiotensin receptor-neprilysin inhibitor, improves cardiac function with the attenuation of fibrosis in heart failure with reduced ejection fraction in streptozotocin-induced diabetic mice. *Eur. J. Heart Fail.* **2016**, *18*, 386–393. [CrossRef] [PubMed]
55. Roksnoer, L.C.; van Veghel, R.; de Vries, R.; Garrelds, I.M.; Bhaggoe, U.M.; Friesema, E.C.; Leijten, F.P.; Poglitsch, M.; Domenig, O.; Clahsen-van Groningen, M.C.; et al. Optimum AT1 receptor-neprilysin inhibition has superior cardioprotective effects compared with AT1 receptor blockade alone in hypertensive rats. *Kidney Int.* **2015**, *88*, 109–120. [CrossRef] [PubMed]
56. Roksnoer, L.C.; van Veghel, R.; van Groningen, M.C.; de Vries, R.; Garrelds, I.M.; Bhaggoe, U.M.; van Gool, J.M.; Friesema, E.C.; Leijten, F.P.; Hoorn, E.; et al. Blood pressure-independent renoprotection in diabetic rats treated with AT1 receptor-neprilysin inhibition versus AT1 receptor blockade alone. *Clin. Sci.* **2016**, *30*, 1209–1220. [CrossRef]
57. Habibi, J.; Aroor, A.R.; Das, N.A.; Manrique-Acevedo, C.M.; Johnson, M.S.; Hayden, M.R.; Nistala, R.; Wiedmeyer, C.; Chandrasekar, B.; DeMarco, V.G.; et al. The combination of a neprilysin inhibitor (sacubitril) and angiotensin-II receptor blocker (valsartan) attenuates glomerular and tubular injury in the Zucker Obese rat. *Cardiovasc. Diabetol.* **2019**, *18*, 40. [CrossRef]
58. Mogensen, U.M.; Køber, L.; Kristensen, S.L.; Jhund, P.S.; Gong, J.; Lefkowitz, M.P.; Rizkala, A.R.; Rouleau, J.L.; Shi, V.C.; Swedberg, K.; et al. The effects of sacubitril/valsartan on coronary outcomes in PARADIGM-HF. *Am. Heart J.* **2017**, *188*, 35–41. [CrossRef]
59. Maslov, M.Y.; Foianini, S.; Mayer, D.; Orlov, M.V.; Lovich, M.A. Synergy Between Sacubitril and Valsartan 1 Leads to Hemodynamic, Antifibrotic, and Exercise Tolerance Benefits in Rats with Preexisting Heart Failure. *Am. J. Physiol. Heart Circ. Physiol.* **2019**, *316*, H289–H297. [CrossRef] [PubMed]
60. Seferovic, J.P.; Claggett, B.; Seidelmann, S.B.; Seely, E.W.; Packer, M.; Zile, M.R.; Rouleau, J.L.; Swedberg, K.; Lefkowitz, M.; Shi, V.; et al. Effect of sacubitril/valsartan versus enalapril on glycaemic control in patients with heart failure and diabetes: A post-hoc analysis from the PARADIGM-HF trial. *Lancet Diabetes Endocrinol.* **2017**, *5*, 333–340. [CrossRef]
61. Jordan, J.; Stinkens, R.; Jax, T.; Engeli, S.; Blaak, E.E.; May, M.; Havekes, B.; Schindler, C.; Albrecht, D.; Pal, P.; et al. Improved Insulin Sensitivity with Angiotensin Receptor Neprilysin Inhibition in Individuals with Obesity and Hypertension. *Clin. Pharmacol. Ther.* **2017**, *101*, 254–263. [CrossRef] [PubMed]
62. Wang, C.H.; Leung, N.; Lapointe, N.; Szeto, L.; Uffelman, K.D.; Giacca, A.; Rouleau, J.L.; Lewis, G.F. Vasopeptidase inhibitor omapatrilat induces profound insulin sensitization and increases myocardial glucose uptake in Zucker fatty rats: Studies comparing a vasopeptidase inhibitor, angiotensin-converting enzyme inhibitor, and angiotensin II type I receptor blocker. *Circulation* **2003**, *107*, 1923–1929. [PubMed]
63. Plamboeck, A.; Holst, J.J.; Carr, R.D.; Deacon, C.F. Neutral endopeptidase 24.11 and dipeptidyl peptidase IV are both mediators of the degradation of glucagon-like peptide 1 in the anaesthetised pig. *Diabetologia* **2005**, *48*, 1882–1890. [CrossRef]
64. Davidson, E.P.; Coppey, L.J.; Shevalye, H.; Obrosov, A.; Yorek, M.A. Vascular and Neural Complications in Type 2 Diabetic Rats: Improvement by Sacubitril/Valsartan Greater Than Valsartan Alone. *Diabetes* **2018**, *67*, 1616–1626. [CrossRef] [PubMed]

65. Gori, M.; D'Elia, E.; Senni, M. Sacubitril/valsartan therapeutic strategy in HFpEF: Clinical insights and perspectives. *Int. J. Cardiol.* **2019**, *281*, 158–165. [CrossRef]
66. Solomon, S.D.; Zile, M.; Pieske, B.; Voors, A.; Shah, A.; Kraigher-Krainer, E.; Shi, V.; Bransford, T.; Takeuchi, M.; Gong, J.; et al. The angiotensin receptor neprilysin inhibitor LCZ696 in heart failure with preserved ejection fraction: A phase 2 double-blind randomised controlled trial. *Lancet* **2012**, *380*, 1387–1395. [CrossRef]
67. Jhund, P.S.; Claggett, B.; Packer, M.; Zile, M.R.; Voors, A.A.; Pieske, B.; Lefkowitz, M.; Shi, V.; Bransford, T.; McMurray, J.J.; et al. Independence of the blood pressure lowering effect and efficacy of the angiotensin receptor neprilysin inhibitor, LCZ696, in patients with heart failure with preserved ejection fraction: An analysis of the PARAMOUNT trial. *Eur. J. Heart Fail.* **2014**, *16*, 671–677. [CrossRef]
68. Voors, A.A.; Gori, M.; Liu, L.C.; Claggett, B.; Zile, M.R.; Pieske, B.; McMurray, J.J.; Packer, M.; Shi, V.; Lefkowitz, M. For the PARAMOUNT investigators. Renal effects of the angiotensin receptor neprilysin inhibitor LCZ696 in patients with heart failure and preserved ejection fraction. *Eur. J. Heart Fail.* **2015**, *17*, 510–517. [CrossRef]
69. Solomon, S.D.; Rizkala, A.R.; Lefkowitz, M.P.; Shi, V.C.; Gong, J.; Anavekar, N.; Anker, S.D.; Arango, J.L.; Arenas, J.L.; Atar, D.; et al. Baseline Characteristics of Patients with Heart Failure and Preserved Ejection Fraction in the PARAGON-HF Trial. *Circ. Heart Fail.* **2018**, *11*, e004962. [CrossRef] [PubMed]
70. A Randomized, Double-Blind Controlled Study Comparing LCZ696 to Medical Therapy for Comorbidities in HFpEF Patients (PARALLAX). Available online: http://clinicaltrials.gov/ct2/show/NCT03066804 (accessed on 28 December 2018).
71. Rubattu, S.; Cotugno, M.; Forte, M. Effects of dual angiotensin type 1 receptor/neprilysin inhibition vs. angiotensin type 1 receptor inhibition on target organ injury in the stroke-prone spontaneously hypertensive rat. *J. Hypertens.* **2018**, *36*, 1902–1914. [CrossRef]
72. Seki, T.; Goto, K.; Kansui, Y.; Ohtsubo, T.; Matsumura, K.; Kitazono, T. Angiotensin II Receptor–Neprilysin Inhibitor Sacubitril/Valsartan Improves Endothelial Dysfunction in Spontaneously Hypertensive Rats. *J. Am. Heart Assoc.* **2017**, *6*, e006617. [CrossRef]
73. Kusaka, H.; Sueta, D. LCZ696, Angiotensin II Receptor-Neprilysin Inhibitor, Ameliorates High-Salt-Induced Hypertension and Cardiovascular Injury More Than Valsartan Alone. *Am. J. Hypertens.* **2015**, *28*, 1409–1417. [CrossRef]
74. Ruilope, L.M.; Dukat, A.; Böhm, M.; Lacourcière, Y.; Gong, J.; Lefkowitz, M.P. Blood pressure reduction with LCZ696, a novel dualacting inhibitor of the angiotensin II receptor and neprilysin: A randomised, double-blind, placebocontrolled, active comparator study. *Lancet* **2010**, *375*, 1255–1266. [CrossRef]
75. Supasyndh, O.; Wang, J.; Hafeez, K.; Zhang, Y.; Zhang, J.; Rakugi, H. Efficacy and Safety of Sacubitril/Valsartan (LCZ696) Compared with Olmesartan in Elderly Asian Patients (≥65 Years) With Systolic Hypertension. *Am. J. Hypertens.* **2017**, *30*, 1163–1169. [CrossRef]
76. Novartis. Effects of Sacubitril/Valsartan Therapy on Biomarkers, Myocardial Remodeling and Outcomes. (PROVE-HF). Available online: https://clinicaltrials.gov/ct2/show/NCT02887183 (accessed on 28 December 2018).
77. Novartis. Study of Effects of Sacubitril/Valsartan vs. Enalapril on Aortic Stiffness in Patients with Mild to Moderate HF with Reduced Ejection Fraction (EVALUATE-HF). Available online: https://clinicaltrials.gov/ct2/show/NCT02874794 (accessed on 28 December 2018).
78. Novartis. Pharmacological Reduction of Functional, Ischemic Mitral Regurgitation (PRIME). Available online: https://clinicaltrials.gov/ct2/show/NCT02687932 (accessed on 28 December 2018).
79. Novartis. Pulmonary artery pressure reduction with Entresto (sacubitril/valsartan) (PARENT). Available online: https://clinicaltrials.gov/ct2/show/NCT02788656 (accessed on 28 December 2018).
80. Novartis. Entresto TM (LCZ696) In Advanced Heart Failure (LIFE Study) (HFN-LIFE). Available online: https://clinicaltrials.gov/ct2/show/NCT02816736 (accessed on 28 December 2018).
81. Novartis. Comparing ARNi with ACE inhibitor on endothelial function (PARADOR). Available online: https://clinicaltrials.gov/ct2/show/NCT03119623 (accessed on 28 December 2018).
82. Ishii, M.; Kaikita, K.; Sato, K.; Sueta, D.; Fujisue, K.; Arima, Y.; Oimatsu, Y.; Mitsuse, T.; Onoue, Y.; Araki, S.; et al. Cardioprotective effects of LCZ696 (sacubitril/valsartan) after experimental acute myocardial infarction. *JACC Basic Transl. Sci.* **2017**, *2*, 655–668. [CrossRef]

83. Imran, M.; Hassan, M.Q.; Akhtar, M.S.; Rahman, O.; Akhtar, M.; Najmi, A.K. Sacubitril and valsartan protect from experimental myocardial infarction by ameliorating oxidative damage in Wistar rats. *Clin. Exp. Hypertens.* **2019**, *41*, 62–69. [CrossRef] [PubMed]
84. Torrado, J.; Cain, C.; Mauro, A.G.; Romeo, F.; Ockaili, R.; Chau, V.Q.; Nestler, J.A.; Devarakonda, T.; Ghosh, S.; Das, A.; et al. Sacubitril/Valsartan Averts Adverse Post-Infarction Ventricular Remodeling and Preserves Systolic Function in Rabbits. *J. Am. Coll. Cardiol.* **2018**, *72*, 2342–2356. [CrossRef]
85. Pfau, D.; Thorn, S.L.; Zhang, J.; Mikush, N.; Renaud, J.M.; Klein, R.; deKemp, R.A.; Wu, X.; Hu, X.; Sinusas, A.J.; et al. Angiotensin Receptor Neprilysin Inhibitor Attenuates Myocardial Remodeling and Improves Infarct Perfusion in Experimental Heart Failure. *Sci. Rep.* **2019**, *9*, 5791. [CrossRef]
86. Novartis. Prospective ARNi vs. ACE Inhibitor Trial to Determine Superiority in Reducing Heart Failure Events After MI (PARADISE-MI). Available online: http://clinicaltrials.gov/ct2/show/NCT02924727 (accessed on 28 December 2018).
87. Suematsu, Y.; Jing, W.; Nunes, A.; Kashyap, M.L.; Khazaeli, M.; Vaziri, N.D.; Moradi, H. LCZ696 (Sacubitril/valsartan), an Angiotensin-Receptor Neprilysin Inhibitor, Attenuates Cardiac Hypertrophy, Fibrosis and Vasculopathy in a Rat Model of Chronic Kidney Disease. *J. Card. Fail.* **2018**, *24*, 266–275. [CrossRef] [PubMed]

© 2019 by the authors. Licensee MDPI, Basel, Switzerland. This article is an open access article distributed under the terms and conditions of the Creative Commons Attribution (CC BY) license (http://creativecommons.org/licenses/by/4.0/).

MDPI
St. Alban-Anlage 66
4052 Basel
Switzerland
Tel. +41 61 683 77 34
Fax +41 61 302 89 18
www.mdpi.com

International Journal of Molecular Sciences Editorial Office
E-mail: ijms@mdpi.com
www.mdpi.com/journal/ijms